*Points, Lines
and Walls*

# Points, Lines and Walls

## In liquid crystals, magnetic systems and various ordered media

M. KLÉMAN
*Université Paris-Sud, Orsay, France*

*A Wiley–Interscience Publication*

# JOHN WILEY & SONS

Chichester · New York · Brisbane · Toronto · Singapore

PHYSICS

6806-5167

*Library of Congress Cataloging in Publication Data:*
Kléman, Maurice.
    Points, lines, and walls.

    'A Wiley–Interscience publication.'
    Bibliography: p.
    Includes index.
    1. Crystals — Defects.   I. Title.
QD921.K644     548'.842     81-21976
ISBN 0 471 10194 X          AACR2


*British Library Cataloguing in Publication Data:*
Kléman, M.
    Points, lines and walls: in liquid crystals, magnetic
    systems and various ordered media.
    1. Crystals — Defects
    I. Title
    548'.81     QD931

    ISBN 0 471 10194 X

Typeset by Speedlith Photo Litho Ltd., Manchester and printed in the United States of America.

*Le père polyèdre engendre réguliers*
*les cinq fils de l'espace où voltigeait Euclide*
*mais la coupe sévère et les tranches obliques*
*multiplient la figure à l'hydre minéral*
*ils se mussent secrets sous la couche d'argile*
*ils se tronquent la face en la guangue ductile*
*ils se clivent le tronc en des poses fragiles*
*et s'entassent obscurs cependant parallèles*

. . . . . . . . .

Raymond Queneau

*Petite cosmogonie portative*

*Deuxième chant, vers* 116–123

(*Paris, Ed. Gallimard,* 1969. *Reproduced by permission* © *Editions Gallimard,* 1950)

# Contents

# Preface

This monograph originates from a series of lectures given at the Ecole Polytechnique (for undergraduates) and at the Laboratoire de Physique des Solides, at the University of Paris-Orsay (for the advanced students and research workers).

For the former, elementary concepts in the *physics of defects* were discussed in depth, starting from first principles and applying them to non-classical cases, essentially liquid crystals and systems of ordered spins. For the latter, new developments in the theory of the classification of defects were presented with illustrations chosen in the same materials, and also in the various phases of $^3$He and $^4$He. All these concepts are gathered here, in a progression which follows the progression of these sets of lectures and also, indeed, the progression of my own research in this field, which started in the late sixties at the Laboratoire de Physique des Solides. I had there the great chance to work in close collaboration with experimentalists who were, as I was and as I still am, passionate for the methods and concepts which gave such an impetus to physical metallurgy, in the years following World War II, I mean the concepts of dislocations and the methods of observation. I hope that the spirit of these advances will be perceptible in this text, where I have tried to present together, in the way they stimulated us, experiment and theory.

The matters treated in this book are introduced in the first chapter. In this preface I want to limit myself to a brief summary of its contents.

The classical viewpoint in the theory of defects (the Volterra process in media of various symmetries) is developed in the first chapters, with detailed applications to singular lines (disclinations, dislocations) and singular points in liquid crystals (nematic, cholesteric and smectic phases), as well as to singular surfaces in magnetic media (magnetic walls). This last application is only one example in a whole class of objects (like grain boundaries) whose elastic properties can be described as due to disclination lines bounding surfaces of dislocations. Magnetic systems are in fact preferred here since they mix properties related to singularities in spin ordering and in elastic behaviour, but the general theory is anyway presented.

With regard to elastic properties (curvature elasticity for liquid crystals, involving torques, and usual stress elasticity) we have developed at length the energy calculations in the linear approximation, insisting on the difference between internal and applied torques (resp. stresses). Our presentation of these matters, which do not present difficulties for mechanicians aware of the theory of mathematical elasticity but are seldom well known by physicists, is as pedagogical as possible.

An important feature of the physics of defects in liquid crystals is the role played by geometrical concepts. Nematics are, in a way, bundles of lines (the lines of force of the optical axis) and smectics are sets of surfaces (the molecular layers). This approach dates back to G. Friedel and F. Grandjean, who discovered focal domains in smectic phases. A large part of this book reflects this point of view (which yields results which have to be compared to those obtained through applying the laws of elasticity) for nematics, various types of smectic phases, but also gauge 'fields', which are met by the physicist interested for example in helium phases, and are akin to fields of trihedra, a special case of fields already considered in the mechanics of continua at the beginning of this century by the Cosserat brothers. The importance of geometry clearly sets itself forward because all these media have continuous symmetry groups, and this characteristic also imposes peculiar properties to defects, particularly in their classification which reflects the topological properties of the symmetry groups.

The last part (essentially the last chapter) of this book deals therefore with the topology of the ordered media. I developed these concepts in collaboration first with G. Toulouse, and then L. Michel. I have chosen here to present them as the conclusion of a reflection which begins with the nature of the Volterra process, its geometrical meaning and experimental illustration. I hope that this presentation will appear complementary to those already published in various review papers; at the same time, in my mind, it has the advantage of not letting the reader ignore some physical content of the Volterra process which is partly obliterated in the topological classification of defects.

This book owes much to all those with whom I have had the advantage of collaborating or discussing these subjects in depth. First of all, I want to mention J. Friedel, who has been my teacher. I had numerous discussions and I learnt much from Sir Ch. Frank and L. Michel. I was at the time a careful listener to P. G. de Gennes' brilliant lectures; he introduced me to the physics of liquid crystals. I benefited greatly from the friendly and always attentive criticisms of J. P. Poirier. Last, but not least, I found inspiration for many of the novel points developed here in discussing with experimentalists and theoreticians in physics of defects working in Orsay, especially Laurette Bourdon, Patricia Cladis, Michel Labrune, Christiane Nourtier, Jacques Miltat, Youri Ryschenkow and Claudine Williams. Finally, I want to thank all those who gave me their permission to reproduce the photographs which illustrate this book.

This book is a new version with many additions (in particular in Chaps. 5, 7 and 10) of an original text published in French by Les Editions de Physique, in two small volumes (1977, 1978), entitled "Points, lignes, parois". The author wants to thank the anonymous translators for a first version of the text which enabled him to complete it without too much effort.

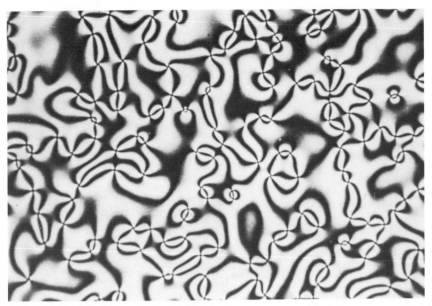

I   Thin film of nematic between two glass slides observed using optical microscopy; with crossed nicols. Numerous nuclei are observed, seen along the axis of the wedge disclination of integral order (see sections 3.3 and 3.6). (Courtesy: G. Ryschenkow)

III   Nematic droplets on a gelatine support; the different anchoring conditions on the support and the free surface bring about the presence of one singular point per droplet (3.4). Photographed using optical microscopy. (Courtesy: P. Piéranski)

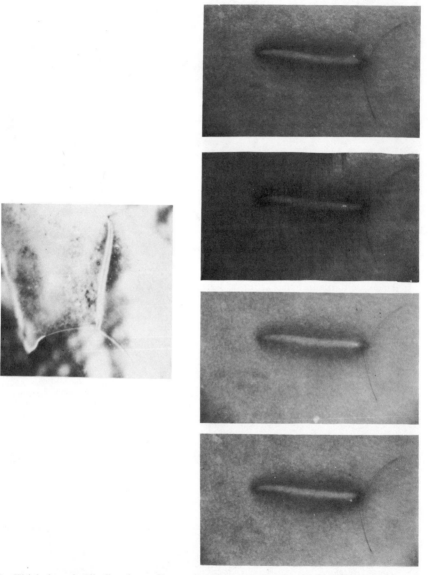

II Thick threads (disclinations of integral order) and thin threads (half-integral order) in a nematic. This optical micrograph illustrates the conservation laws of the order and the optical phenomena due to birefringence, from which the molecular distribution can be calculated. (The splitting of a point source of 5 $\mu$m diameter as it crosses the thick filament is observed). (Courtesy: P. E. Cladis)

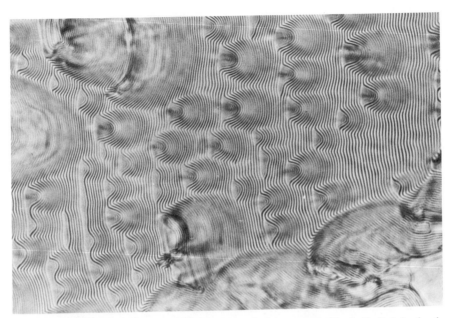

IV  Elliptical features in a cholesteric phase (see Figure 4.20). The half-pitch is clearly visible. Photographed using optical microscopy. (Courtesy: Y. Bouligand)

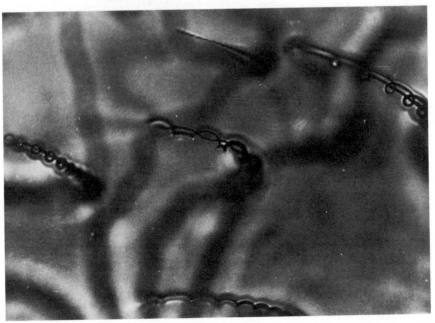

V  Helicoidal lines in oblique position in relation to the cholesteric axis (see section 4.4). Photographed using optical microscopy. (Courtesy: J. Rault)

VI    Focal domains in a thermotropic smectic (section 5.5). Photographed using optical microscopy. (Courtesy: C. E. Williams)

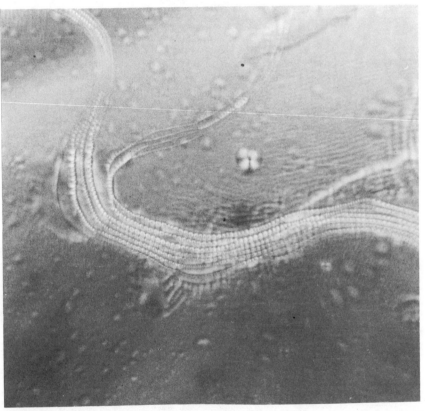

VII  Decoration of Grandjean terraces by focal domains. Thermotropic smectic (section 5.7). Photographed using optical microscopy. (Courtesy: C. E. Williams)

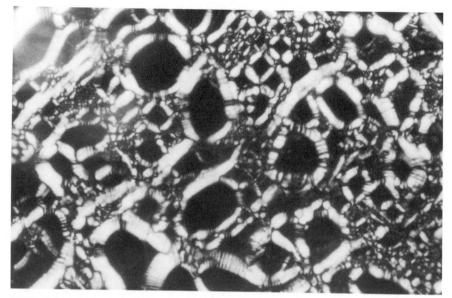

VIII Oily streaks in a homeotropic sample of a lyotropic smectic (lecithin–water). (section 5.7). Photographed using optical microscopy. (Courtesy: M. Veyssié)

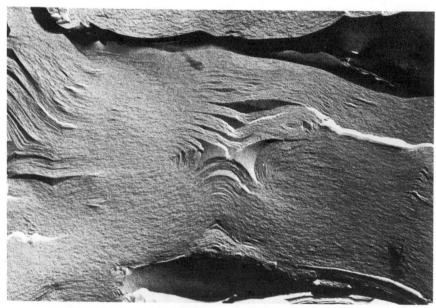

IX Fracture (freeze-etching) in a lecithin–water phase. We observe the folding of the layers into tori (section 5.5) (Courtesy: M. J. Costello and T. Gulik-Krzywicki)

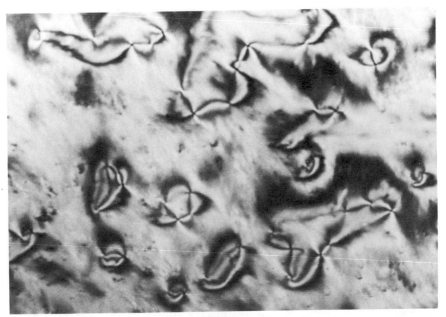

X    Smectic C (DOBCP). Horizontal layers, crossed nicols: texture with nuclei (Schlieren texture)

XI    Smectic C (DOBCP). Horizontal unidirectional anchoring on a glass slide with vertical layers along two directions symmetrical with respect to anchoring. Crossed nicols. Two regions of different orientation can be seen, separated by vertical boundaries. In DOBCP the optical axis is virtually at 45° to the layers

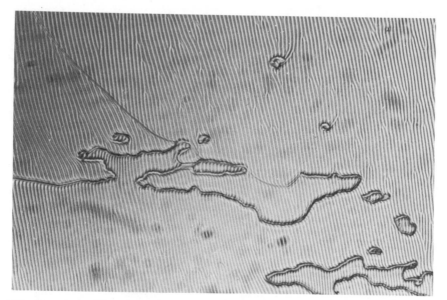

XII   Smectic C (DOBCP). Free droplet on glass with unidirectional anchoring formation of ribbons (Courtesy: Allet, Kléman and Vidal, *Journal de Physique* **39**, 181 (1978))

XIII   Smectic E (p-phenyl benzylidene aminocinnamate). (Courtesy: L. Strzelecki, L. Liebert and P. Keller)

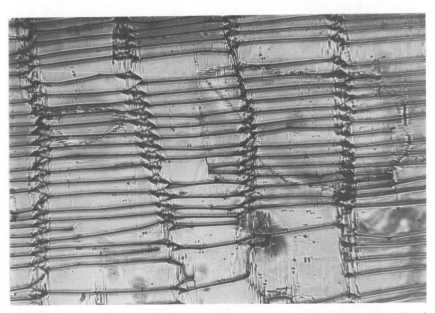

XIV  Smectic C* (DOBCP + active molecules). Vertical layers parallel to the singularities. Sets of parallel $2\pi$ singularities, situated at distances equal to the pitch of the smectic C*. (Courtesy: M. Brunet)

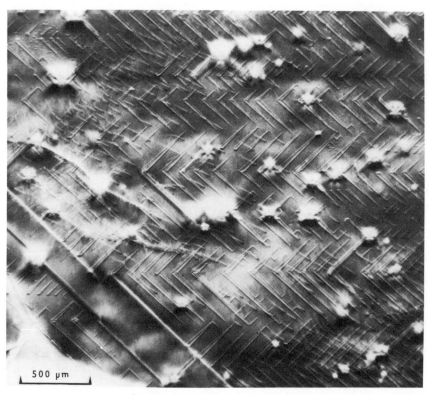

500 μm

XV    Monocrystalline Fe 3% Si of orientation close to [100] observed by Lang's method. Sets of surface domains in firtree branches, aligned along the wall at 180° and vertical in the photo. The oblique walls to the lower left are 90° Bloch walls. (Courtesy: J. Miltat)

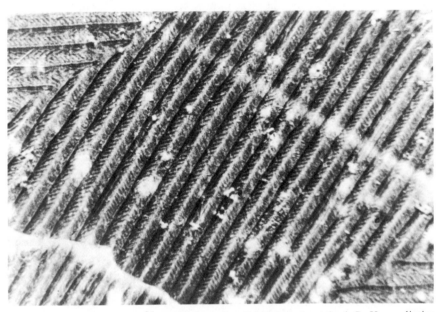

XVI   Monocrystalline Fe 3% Si, of orientation [111]. Lang's method. Co $K\alpha_1$ radiation, reflection (011), horizontal diffraction vector. Closure domains on the surface. (Courtesy: M. Labrune)

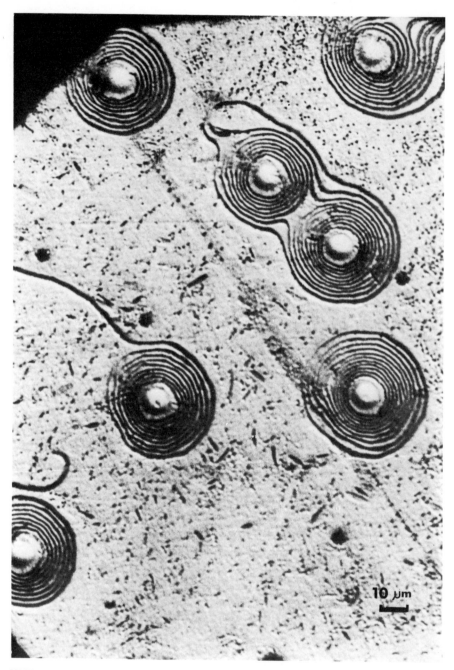

XVII   Magnetic bubbles surrounded by 36° walls in garnets covered with a layer of 100 Å of permalloy. Bitter's method. (Courtesy: I: B. Puchalska and H. Jouve, *I.E.E.E. Trans. Mag.* **13** (Sept. 1977)

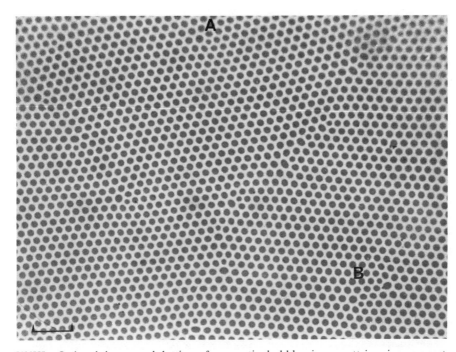

XVIII  Ordered hexagonal lattice of magnetic bubbles in an yttrium-iron garnet. Faraday effect. A granular structure is apparent, with dislocations of the hexagonal lattice: the interface dislocations have a core formed by a bubble with five nearest neighbours and a bubble with seven. Isolated dislocation at A. Dipole of dislocations at B. Scale 50 μ.
(Courtesy: J. Miltat)

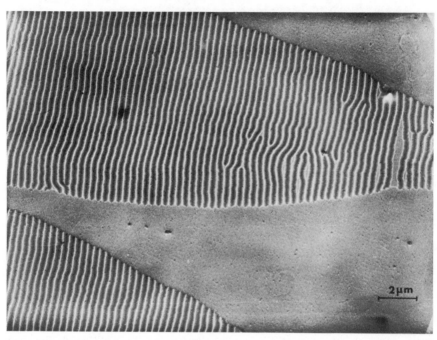

XIX   Striped domains in a layer of permalloy 83 Ni–17 Fe of thickness 2500 Å;
observation by high voltage microscopy. (Courtesy: I. B. Puchalska, *Acta Physica Polonica* **36**, 4, 10 (1969)

# Chapter 1
# Introduction

## 1.1 GENERAL HISTORY

The concept of a dislocation first arose in the work of Italian scientists (Volterra, Somigliana) on the singularities of solutions to equations of linear elasticity. These singularities have two characteristics.

First, they are surrounded by a field of *distortions* and a field of *internal stresses* is created, which must be distinguished from the effects of forces due to external fields applied either on the surface or within the bulk. A solid, therefore, even when free from all exterior interactions, possesses an *energy* dependent on the distribution of these internal singularities, and is in a metastable state. Taking this concept further, a *density* of singularities (Nye, Kröner) can be defined. The solid described with the inclusion of these densities is said to be in a state known as the *natural state*, as opposed to the so-called perfect state.

Another aspect of these singularities is the existence of a *topological characteristic* which can be defined either by an operational schema (*the Volterra process*) or by the *Burgers circuit* concept. The Volterra process can be used to classify the singularities as functions of the *symmetries* of the solid under consideration (crystallographic group).

By virtue of their dual characteristics of topological stability and energy properties, these singularities of crystalline solids (translation dislocations) play a part in many microscopic and macroscopic physical properties. However, the role of these singularities is most important in the plastic properties of solids (J. Friedel). The understanding of plasticity in metals, especially, has been greatly advanced in the last twenty years. There has also been an expansion of knowledge of the mechanisms governing the interactions between translation dislocations, their nucleation and their multiplication. Currently, the accent is on the problems of the *core* (the region in the immediate vicinity of the line, where linear elasticity no longer adequately describes the phenomena) and on materials new to this research (ionic solids, covalent solids and molecular crystals). The physics of translation dislocations therefore still attracts a great deal of interest (J. Friedel, 1980).

Experimental evidence of *translation dislocations* in crystalline solids is relatively recent (electron microscope observations by, for example, Whelan *et al.*, 1957). Evidence of *rotation dislocations* came much earlier, from work with

1

*liquid crystals* (Lehmann, 1904, 1909, 1910). Liquid crystal states were not formalized until recently, although a considerable amount of experimental work on mesomorphic phases, by G. Friedel and F. Grandjean as well as by O. Lehmann, did produce a precise description of the principal configurations (G. Friedel, 1922). From observations of these defects, G. Friedel was able to establish his famous classification of the mesomorphic phases into the nematic, cholesteric and smectic phases, thus illustrating the close relationship between these types of defects and the symmetry of the ordered medium. These predictions were soon confirmed by structural studies using X-rays by E. Friedel (1925), among others. Rotation dislocations (or, as they are usually called, disclinations) are linked to elements of rotation symmetry; translation dislocations are linked to translation symmetries. Harris (1970a) coined the term *dispiration* for defects which are linked to translation–rotation symmetries.

Frank's (1958) article revived the interest in liquid crystals after several decades of neglect. While work on macroscopic properties continued (see the general text by de Gennes, 1974) much research was carried out on the singularities of molecular distribution and the grouping together of these singularities (textures). As in the case of distortions in elastic solids, whose sources are translation dislocations, these singularities, as classified by the Volterra process, can be considered as being the actual origin of the molecular distributtins which accompany them.

It was immediately obvious that liquid crystals were an enormous new area for research into dislocation theory. Particular characteristics in this field, which could be omitted in the study of crystalline solids, are as follows.

(a) *From the energy point of view:* the necessity to consider molecular distribution taking into account large angular variations, therefore leading to a non-linear elasticity of couples and gradients of rotation. The energy aspect of defects therefore loses much of the simplicity which is evident when dealing with solids.

(b) *From the topological point of view:* the existence not only of defect lines (one-dimensional) but of singular points (zero-dimensional) and also of surface defects (two-dimensional) which are all linked to crystallographic symmetries. The need to use advanced mathematical concepts (theory of surfaces, theory of continuous groups and algebraic topology) to describe large angular distortions and the laws of conservation linked to non-commutable rotation symmetries.

At the same time as the first results on defects in mesmorphic phases were made known, studies appeared on disclinations in bidimensional crystals (Harris, 1970b; Nabarro, 1970) and on the concept of disclination density, thus bringing into being the generalization of a dislocation density (Anthony, 1970a). The typical bidimensional crystals which bear out these theories are of a biological nature (capsids of bacterial viruses, etc.). They are also involved in the symmetries of stacking of vortex lines in superconductors (Abrikosov, 1957; Traüble and Essman, 1968). The development of biological research is not confined to bidimensional crystals. In fact, the concept of rotation dislocation and models supplied by liquid crystals are widely utilized by Bouligand, and some of the analogies between liquid crystals and biological materials are still attracting the attention of researchers. Many biological tissues have liquid crystal structures or show pseudomorphoses of mesomorphic phases. Models of phenomena related to the origin of life have even been looked for in the astonishing growth properties of liquid crystals observed using optical microscopy (Lehmann, 1907)!

Let us also stress the new importance of defects in the theory of phase transitions. It was proposed long ago that the solid → liquid transition of a three dimensional crystal is driven by the nucleation of a large number of dislocation loops (Mizushima, 1960). This theory has been revived recently (Edwards and Warner). But the most numerous achievements are now in the field of two dimensional transitions, starting with the work of Kosterlitz and Thouless (1973), and continuing with the recent proposal by Halperin and Nelson (1978) of a new phase (the so-called hexatic phase) which is stabilized by defects. Experiments are active in this field in order to verify this prediction.

The concept of dislocation is also used in order to describe the nucleation and propagation of earthquakes (Madariaga, 1980; Savage, 1980). In this approach the stress is clearly upon elastic rather than topological properties.

There are important analogies between rotation dislocations in liquid crystals, defects of ordered spin phases (ferromagnets, helimagnets, etc.), vortex lines of classical hydrodynamics, superconductors and superfluids (Feynman, Onsager). All these media can be characterized by a unit vector (or director) defined as being along the length of the molecule in liquid crystals, or along the direction of magnetization in magnetic materials, or along the phase gradient in super-conductors and superfluid phases. In order to extend the analogy, and attempt a general classification of defects in various media, a useful concept is the *order parameter*, which is characterized by its covariance (scalar for the superfluid phase, vectorial for ferromagnets and tensorial if the local state is defined by a deformable or indeformable trihedron ($^3$He, biaxial nematic or elementary lattice structure of crystalline solids)). It is also characterized by certain topological properties of dimensionality and connectivity. Recent research has therefore been oriented towards a generalization of the concept of a defect, an aim which has been aided by experimental research on phases having unusual symmetries and by the introduction of mathematical methods not common in physics.

## 1.2  LAYOUT OF THIS TEXT

This book is planned around various current aspects of dislocation theory as they have been revealed by research on *rotation* dislocations. The contents of the chapters are ordered in such a way as to relate the concepts which have been successively introduced into the physics of materials. Some chapters have an experimental bias, while others are rather more theoretical. The structure of the book as a whole is determined by the conceptual development of this field of research.

Chapter 2 opens with an account of the classification of defects by the Volterra process. Chapters 3–5 deal more specifically with the rotation dislocations in

4

classical liquid crystals (i.e. those identified by G. Friedel). Chapters 6–10 extend the physics of rotation dislocations to anisotropic fluids and also to crystalline and ferromagnetic solids. The final chapter is devoted to an account of the classification of defects by homotopy group theory.

In Chapter 2, after a discussion of the concept of the ordered medium, an account is given of Volterra's classification, with numerous applications to rotation dislocations. Certain topological aspects of disclinations in bidimensional crystals are dealt with here in order to indicate immediately the particular aspects which cause the profound difference between disclinations and translation dislocations.

Since the theory originated in the experimental study of liquid crystals, it was natural to devote several chapters (i.e. 3, 4, 5 and 7) to mesomorphic phases. The essentials of what is now known about defects in the nematic, cholesteric and smectic phases will be found in Chapters 3, 4 and 5 combined with 7, respectively. It is both easy and fascinating to observe defects in liquid crystals; because of the fluidity of the phase, the crystal anneals well and the defect density is low. Furthermore, because of the large double refraction in these phases, liquid crystals can be studied directly using a polarizing microscope. These observations will be shown as often as possible in their experimental context. The peculiar elasticity of these phases will be developed with emphasis on the phenomena due to large curvatures. Knowledge of this elasticity is necessary for the study of defect stability. In some cases, the geometric description of defects may necessitate the use of concepts from the theory of curves and surfaces, which will be described. In the particular case of nematics, after a detailed description of G. Friedel's 'threads' and 'nuclei',† the phenomena of surface anchoring will be described in some detail. This experimental aspect, the theoretical importance of which was recognized by Grandjean (1916), leads on to the concept of surface disclinations. The core of these objects will be described by a Peierls-type theory. Cholesterics will be the subject of a shorter chapter. After the description of the principal singularities (for which few energy calculations have been made) a short but abundantly illustrated survey will be made of the principal textures. However, it is not intended to reproduce Bouligand's very comprehensive articles on this subject. Smectics A will be studied in detail. These lamellar phases possess the interesting property of having both rotation symmetries and non-trivial translation symmetries. This brings about very curious relations between rotation and translation dislocations, and the presence of conjugated lines in the form of *focal conics*. It is at this point that the topological concepts of the theory of some of the relevant surfaces are introduced.

Smectic A phases constitute a state of intermediate order between liquid crystals in the strictest sense and crystalline solids. However, their fluidity is the essential feature. Smectic C phases, while still liquid, are more ordered than the A phases, while the B and H phases, etc. constitute lamellar states which are clearly much closer to traditional crystalline solids because of the appearance of an order

† *Translators' note:* In French, 'fils' and 'noyaux'.

in the layers. Before the study of these new smectic phases could be carried out (Chapter 7) it was necessary to return to the definition of rotation dislocations in Chapter 6 and extend it to systems in which the local state is not defined by a vector or director, but by a *trihedron* (Cosserat medium). After this introduction, certain specific characteristics of the geometry of these media are stressed. This is not well known to physicists, although engineers have made some use of it.

The state of distortion of an incompressible Cosserat medium at rest can be defined by the gradient of rotation from one trihedron to the next. A $K_{ij}$ tensor with nine components (Kröner contortion tensor) plays the same role for a fluid medium as the deformation gradient $\beta_{ij}$ used in elastic solids. In particular, just as for a given $\beta_{ij}$ field there is a corresponding condition which says that the deformations due to this field do not give rise to densities of translation dislocations (or isolated dislocations), in the same way there is a correspondence between a given $K_{ij}$ field and compatibility conditions which says that the associated field of trihedrons does not give rise to densities of rotation dislocations (or isolated dislocations).

Note that a deformation of pure contortion can be defined in a crystalline solid; if $e_{ij} \equiv 0$, local rotations of trihedrons can exist which can be described by the tensor $K_{ij}$. This is the generalized glide first introduced by Nye. The latter author was restricted to small rotation gradients and had not considered the singularities of the $K_{ij}$ field. In solids, as in anisotropic fluids with trihedrons, these are rotation dislocations. They can be represented by the quantity

$$\theta_{ij} = \varepsilon_{ikl} K_{lj,k} - \tfrac{1}{2} \varepsilon_{ipq} \varepsilon_{jrs} K_{pr} K_{qs}. \tag{1.1}$$

The biaxial nematic is certainly the type of material which corresponds best to the definition of a fluid Cosserat medium. The A phase of $^3$He also has a trihedron for an order parameter, but the existence of gauge invariance allows a simpler expression to be chosen than use of the contortion tensor (Delrieu, 1977). Smectics C and B, etc. are also Cosserat media, in which because of the existence of layers,† an order parameter can be chosen, which is simpler than the local trihedron and associated contortion. However, in this work, all Cosserat media will be described by this tensor whose simple relationship to the order parameter trihedron makes it easier to imagine the geometric nature of these singularities.

An ordinary nematic is also a Cosserat medium; one of the axes of the local trihedron is along the length of the molecule, the two others can be chosen at random. It will be shown that a particular choice of these axes can be made by identifying the disclination lines with the singularities of a Nye field, i.e. ultimately with focal curves.

The very general geometric considerations of Chapter 6, which are largely uñpublished, are applied in Chapter 7 to the study of smectics C, chiral ferroelectric phases, smectics B, etc. It is shown first how the formalism of $K_{ij}$ makes it possible to write phenomenologically the free energy of the smectic C

---

† There is probably an analogy to be drawn between the gauge invariance condition of $^3$He and the condition for the existence of layers (i.e. the invariance of twist of a fixed vector of the trihedron) in smectics C.

phases in the case of large curvatures and leads to a local geometric interpretation of the different elasticity coefficients (this method can be extended to all Cosserat media). The same method will be used to study the role of the compatibility conditions ($\theta_{ij} = 0$) in Chapter 9, which ultimately play a role at least as important as the free energy minimization equations in the search for permitted configurations. The defects which affect the symmetry considerations in the different smectic phases will also be discussed. In connection with this, chiral phases of smectics C must, owing to the coexistence of two incommensurable periodicities along the same axis, show defects of the dispiration type.

Ordered layer phases (B, H, etc.) pose the interesting problem of seeking the molecular disposition of the least energy on a layer having any curvature. The concept of a superficial Nye distribution will be defined bringing in the theory of the conformal representation of surfaces.

A general theory of rotation dislocations in solids is developed in Chapter 8. The concept is easier to visualize here than in anisotropic fluids, and several types of rotation dislocation can be distinguished. The most general deformations of a crystalline solid are described with a contortion tensor $K_{ij}$ and a strain tensor $e_{ij}$. We are naturally limited here to small deformations and the singularities of the $K_{ij}$ field can be clearly separated from those of the $e_{ij}$ field.

The former are already known to us: they are the rotation dislocations of the lattice (or the trihedron field of the lattice to return to the language used above) given by equation (1.1). Disorientation walls (*diremption*) should also be included in the singularities of the $K_{ij}$ field. From the work of Frank, it is known that these walls can be described as surface arrays of translation dislocations. This result is presented here as a property of rotation dislocations: the boundary of a disorientation wall is a disclination loop which possesses *wedge* parts (parallel to the rotation vector), *twist* parts (perpendicular to the rotation vector) and *mixed* parts. It can be shown, using the Volterra process, that translation dislocations are necessarily attached to the twist parts; they therefore mark the surface section of the disclination and constitute the disorientation wall in question. This result is demonstrated in Chapter 2 and is also true for anisotropic fluids. However, in this case, translation dislocations can undergo a viscous relaxation towards dislocations of infinitesimal Burgers vector lines, because of the symmetries of the medium, and finally filling not a surface bound by the loop but a volume. Their existence therefore has lost is physical meaning.

The $e_{ij}$ field gives rise to a second type of rotation dislocation, of which the most simple example is shown by considering the contact surface between two crystalline solids of different parameters. It is known from Frank that this wall (*distranslation*) can be described as an assembly of epitaxial dislocations. By analogy with diremptions, we consider the edge of such a wall to be a rotation dislocation. More generally, Kroner's incompatibility tensor,

$$\eta_{ij} = \varepsilon_{ipq}\varepsilon_{jlm}e_{pl,qm},\tag{1.2}$$

represents densities of rotation dislocations of this type.

The densities of translation dislocations are not independent of densities (1.1) and (1.2). It may be interesting to define elementary defects other than the two types of rotation dislocation described here. Chapter 8 contains various illustrations of this case.

Chapter 9 deals with materials with spin (ferromagnets and helimagnets) which are important for the purpose of this book. They possess defects characteristic of an anisotropic fluid system (through the spin field) and the rotation dislocations characteristic of a solid. This chapter is constructed in the same way as those specifically on liquid crystals, that is to say it includes an energy study and a topological study of the various defects. The defects of the spin system have topological resemblances to defects in liquid crystals, but the energy aspects are very different. An inhomogeneous distribution of spin, i.e: the presence of walls, lines or points, causes *magnetostrictive internal stresses*. These internal stresses can be attributed to three types of source:

(a) dislocation walls (situated on the magnetic walls);
(b) disclinations of a weak rotation vector (situated on the junctions of the walls); and
(c) dislocations or dipoles (situated on Bloch or Néel lines).

Experimental illustrations of these theories can be obtained by X-ray topography. It is reasonable to think that these analyses constitute a necessary basis for all subsequent study of the interactions between lattice dislocations and magnetic walls, and of the plasticity of magnetic materials.[†]

Candidates for detailed research into these phenomena in years to come will be found in the field of rare earth metals, in which magnetostriction is considerable.

Chapter 10 is devoted to very recent and on-going research, concerning a deep topological analysis of singularities of the order parameter. This leads to a classification of topologically stable defects based on homotopy groups of the variety representative of the properties of connectivity and of the order parameter (manifold of internal states $V$) first introduced by Finkelstein. This approach has several advantages over classification by the Volterra process: defects of all dimensionalities can be included in the same analysis (singular points, lines and walls) as well as topologically stable but not singular defects (such as Bloch and Néel walls or solitons in one-dimensional media). This classification also naturally includes the lines which the Volterra process, somewhat artificially, attributes to infinitesimal dislocation densities (such as twist lines), or requires to be profoundly altered (non-orientable lines, focal conics). Numerous examples will be taken from preceding chapters and looked at in this new light. In this chapter the concept of the Burgers vector will also be generalized by attaching to a circuit surrounding a defect its image in the space of the euclidean group $G$. This image is an open circuit $\Gamma$, beginning with the identity element of the $G$ group and ending with an $h$ element of the symmetry group $H$ of the ordered medium, $H$ being a sub-group of $G$.

---

† We have ignored in this monograph all aspects of lattice dislocation–magnetization interactions (see, on this subject, our review article to be published in Nabarro (1982).

## 1.3 CONCLUSIONS; PERSPECTIVES

Despite the variety of viewpoints (pure theory and the physics of materials under both theoretical and experimental aspects) this work should give the opportunity to both students and researchers interested in the physics of defects, whether theoreticians or experimentalists, to catch up on the most recent research. Also, since the content of the book presupposes only an elementary knowledge of translation dislocations, this work should serve as a general introduction for those who wish to make a start in this area of physics.

The diversity of interest is perhaps one of the most interesting characteristics of this field of study. Researchers from many areas have taken an interest in it: metallurgists (who were for many years the only physicists to take an interest in dislocations, sometimes to the scorn of their colleagues who found the study of imperfections impure); materials scientists; biologists; and theoreticians now attracted by the use of methods of algebraic topology. These methods are similar to those which have had such great success in the fields of particle physics. Additionally, there are hopes that the theory of defects could be connected to both the reflections of d'Arcy Thompson on the forms and phenomena of growth and the more recent thoughts of René Thom on morphogenesis. Both these authors wish to raise qualitative phenomena to a more privileged state in order to bring them to a scientific description. For d'Arcy Thompson, this clearly means that the science of forms and growth must come closer to physics and mathematics, and in particular to geometry. He is aware, however, of the originality of qualitative phenomena, and a mere return to the quantitative description is not what he has in mind. This point of view is amplified by Thom; he sees in the theory of dynamic systems, which deals with the universal behaviour of differential equations and singularities of their solutions (instabilities, bifurcations and catastrophes) the characteristic mathematical framework of qualitative phenomena. Finally, there are two aspects of defect research which appear here: the naturalistic aspect — this research is often closer on the experimental side to the science of observation than to the exact sciences — and the topological aspect — defects are always defined as universally invariant phenomena.

The future development of the physics of dislocations must therefore follow different paths, and be related by cross fertilization to very diversified fields of physics. On the experimental side it is clear that much remains to be done in the *dynamics* of defects in anisotropic fluids, for example in the phases of liquid crystals and $^3$He. It should already have been noted that this work does not deal with dynamic phenomena. The future programme of the rheology of fluid anisotropic media is to explain the hydrodynamic properties of these media by the behaviour of their singularities, as translation defects explain the plasticity of crystalline solids. On the theoretical side it is to be hoped that the methods of classification of defects by algebraic topology will be complemented with methods of differential topology (dynamic systems) so that the energy aspects of the description of defects can be included. These mathematical methods have already been used to give a model of turbulence and the approach to it (Ruelle and Takens). Since turbulent chaos is a statistical distribution of singularities of the vortex type, an approach to this phenomenon via the theory of defects could be fruitful.

# Chapter 2
# Translation and rotation dislocations. The Volterra process

There is a topological relationship between the types of defects found in an ordered medium and the symmetry elements of that medium. This chapter deals with the Volterra process, which illustrates this relationship for linear defects (dislocations and disclinations). Another approach to the classification of defects will be proposed in a later chapter.

## 2.1 ORDER AND SYMMETRY

### 2.1.1 Symmetries of a solid crystal

If a solid crystal is observed at 0 K, it will consist of a pattern, which is a collection of similar or different atoms or molecules, repeated in three directions in space by the translations (Figure 2.1)

$$\mathbf{b} = n_1\mathbf{b}_1 + n_2\mathbf{b}_2 + n_3\mathbf{b}_3 \tag{2.1}$$

which form the *lattice* (Bravais lattice). $n_1$, $n_2$ and $n_3$ are any integers and $\mathbf{b}_1$, $\mathbf{b}_2$ and $\mathbf{b}_3$ are non-coplanar vectors which define the unit cell of the lattice. The

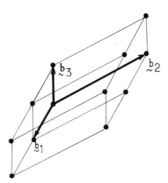

Figure 2.1. Bravais lattice. The nodes of the lattice, represented here by dots, constitute the equivalent sites produced by translation

lengths $b_1$, $b_2$ and $b_3$ are *parameters* of the lattice. The choice of $\mathbf{b}_1$, $\mathbf{b}_2$ or $\mathbf{b}_3$ is somewhat arbitrary, although in practice the shortest possible lengths are always chosen. Therefore for an atom on a lattice site, the atoms at $\mathbf{b}_1$, $\mathbf{b}_2$ and $\mathbf{b}_3$ are the nearest equivalent neighbours (i.e. with the same neighbourhood reproduced by translation).

The arrangements of atoms along the directions $\mathbf{b}_1$, $\mathbf{b}_2$ and $\mathbf{b}_3$ are then *dense arrangements* and the planes $b_i b_j$ are *dense* planes.

The Bravais lattice, i.e. the collection of lattice sites (which differs from the pattern), possesses a symmetry group by which it can be reproduced. This group naturally includes the sub-group of $\mathbf{b}_i$ translations, but also includes rotations and reflections. In particular, the Bravais lattice always includes centres of symmetry. There is a difference between symmetry operations of the first kind (translations and rotations) which consist of displacements analogous to those of a non-deformable body, and symmetry operations of the second kind, which always contain a reflection. If the pattern is now superimposed on the Bravais lattice, certain symmetries of the Bravais lattice are no longer necessarily respected, for example the centre of symmetry if the pattern is asymmetrical.

The purpose of crystallography is the classification of symmetry groups. Here we shall content ourselves with the following results.

(1) The classification of elements of symmetry leads to the property that the only possible rotation symmetries in the Bravais lattice are at the angles $\pm 2\pi/n$, where $n = 1, 2, 3, 4$ and $6$.

(2) The Bravais lattice does not only include nodes which have translational equivalents; other nodes can be equivalent to the former by the other operations of the Bravais group (which is also called the point group). This leads to the classification of 14 *crystalline systems*; in metals we mostly find:

   (a) body centred cubic (b.c.c.) systems;
   (b) face centred cubic (f.c.c.) systems; and
   (c) hexagonal lattice systems.

(3) If the symmetries of the pattern are taken into account, then there are 230 *space groups*! The study of these would require a detailed investigation.

The crystal has been defined at $0\,\mathrm{K}$ where thermal vibrations are nil. At temperatures higher than $0\,\mathrm{K}$, the effect of vibrations would be to vary the distances $b_1$, $b_2$ and $b_3$ in an inhomogeneous manner. The mean quadratic values of these deviations, $\sim b^2 (K_B T)/U$, where $U$ is an interaction energy between neighbouring atoms, are small as long as the temperature is far below the melting point of the solid. A solid crystal is therefore characterized by strong correlations of position between atoms.

## 2.1.2 Symmetries of directional media

The preponderance of correlations of position in solids means that they can be imagined as being formed on the model of the Bravais point lattice. A collection of particular directions can be associated with each point of the Bravais lattice, represented, for example, by a vector. If these directions are absolutely random, the new lattice possesses the same symmetry elements as the original. This is true for the paramagnetic crystal, where a spin vector can be associated with each atom. On the other hand, in a ferromagnetic or helimagnetic crystal, the directions are correlated. In a ferromagnet all the spins are parallel (Figure 2.2): this does not change the symmetry properties of translation, but can cause certain rotation symmetries to disappear (all the rotation symmetries of an angle equal to $\pi$ about every axis perpendicular to the direction of spin will disappear).

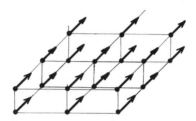

Figure 2.2. Cubic ferromagnet. The spins are shown here along one of the axes of the cube

In a typical helimagnet (e.g. rare earth metal) a helicoidal distribution of spins is associated with the **c** axis of a hexagonal crystal, the spins being perpendicular to the **c** axis. Suppose first of all that the pitch of the helix is commensurable with the parameter $c$ of the lattice (along **c**): the **c** axis is then no longer a rotation axis of order 6, but a rotation–translation axis (helicoidal rotation). At the same time, certain rotation axes perpendicular to **c** may disappear. But if the pitch is not commensurable with $c$, all the symmetry elements which are linked to the repetition of the motif along **c** and which existed before taking the spins into consideration will then disappear.

The modification of the symmetries of the Bravais lattice can be carried further by supposing that the position correlations disappear. Take the example of the above helimagnet: the crystalline characteristic disappears (it is now amorphous or liquid) but the *angular correlations* between spins remain. The symmetry elements therefore reappear linked to these correlations: the **c** axis is an axis of helicoidal rotation of a pitch equal to that of the angular correlations.

Now consider a ferromagnet where the position correlations between atoms have disappeared (amorphous ferromagnet) but where the angular correlations remain. The system then has new symmetry elements:

(a) all the translations (the ferromagnet reproduces every translation); and
(b) all those rotations (at whatever angle) along the axes parallel to the directions of spin; the disappearance of the quantization rule, where $n = 1$, 2, 3, 4 and 6, should be noted.

These different examples are a good indication of the rich and fluctuating nature of the order concept. We shall define directional media as media characterized by angular correlations between directions. A well-known case is that of Cosserat media (Cosserat and Cosserat, 1909) for which there is a trihedron of directions at each site.

The directional media, which have been studied most from the point of view of dislocations, are *liquid crystals*. Most of them are formed from organic rod-shaped molecules and they belong to one of the following phases (G. Friedel, 1922) (Figure 2.3).

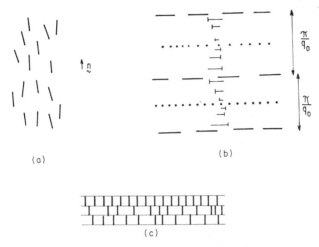

(a)

(b)

(c)

Figure 2.3. In this schema, a *nail* represents a molecule lying at an angle $\beta$ to the plane of the diagram; the length of the nail is proportional to cos $\beta$; the head of the nail is below the plane of the diagram and its point is directed towards the observer. Figure 2.3.(b) therefore represents a right-handed cholesteric

*Nematic state.* The molecules are arranged parallel to each other, but their centres of gravity are distributed at random. The unit direction vector, **n**, along the molecule, is equivalent to the direction $-\mathbf{n}$, for at random there are as many molecules in one direction as in the other. The symmetry elements are:

(a) all translations;
(b) all rotations about the **n** axis; and
(c) all rotations of $\pi$ (or multiples of $\pi$) about every axis perpendicular to the molecule (Figure 2.3a).

*Cholesteric state.* The molecules form a spiral with a pitch $p = 2\pi/q_0$ around a constant direction, called the cholesteric axis, denoted $\chi$. In every plane perpendicular to this axis (cholesteric plane) the molecules are parallel to each other with a disorder in position of the centre of gravity. As in the nematic state, $\mathbf{n} \sim -\mathbf{n}$. A cholesteric medium is said to be *right handed* if the helicoidal rotation of the molecules goes in the direction of the *Maxwell corkscrew*; to be *left handed* if the helix is in the opposite direction. We have $q_0 > 0$ for a right-hand screw, the frame of reference being assumed as direct ($n_x = \cos q_0 z$, $n_y = \sin q_0 z$, $n_z = 0$). The elements of crystalline symmetry are the following (Figure 2.3b):

(a) all translations in the cholesteric plane;
(b) all rotation–translation movement along the cholesteric axes (rotation $2\pi\alpha$ and translation $2\pi\alpha/q_0$, for any $\alpha$);
(c) rotations of $\pi$ (or multiples of $\pi$) along the axes perpendicular to $\chi$, parallel or perpendicular to a molecule;
(d) rotations of $\pi$ (or multiples of $\pi$) about the $\chi$ axes; and
(e) a translation $\pi/q_0$ along $\chi$.

*Smectic state.* The molecules are arranged in planar layers of thickness $d$. In smectics A, $\mathbf{n}(\sim -\mathbf{n})$ is perpendicular to the layer. The symmetry elements are (Figure 2.3c):

(a) all translations in the plane of the layers;
(b) all rotations about the axes perpendicular to the layers;
(c) a translation $\mathbf{d}$ along the normal to te layers; and
(d) a rotation of a multiple of $\pi$ about all axes perpendicular to the molecules situated in the middle or at the end of a layer.

Crystallographically, smectics A represent the transition phases between solid (point) crystals and directional media (angular). The other smectics (C, B, etc.) will be described in the chapter reserved for them.

New liquid crystal states have been discovered recently, whose building units are organic plate-like molecules (Chandrasekhar *et al.*, 1977; Billard, 1980). The states in which these molecules are arranged are far from being fully understood, but it is worth mentioning the existence of a *hexagonal state* in which the molecules are stacked in fluid parallel columns which form a hexagonal pattern (see also Levelut, 1979).

In this section long range order (or disorder) relations have been considered. But on the scale of a few atoms or molecules a perfectly disordered medium (amorphous, liquid, paramagnetic etc.) may possess a local order (positional or directional). This important concept will hardly ever be used in this monograph.

### 2.1.3 Various generalizations of the order concept

We have been restricted to the description of relatively simple spatially ordered media. For solid crystals, the existence of complex *patterns* can be considered as constituting a generalization of the Bravais lattice concept. For directional media, the obvious generalization is where order is not defined by a single vector (or director), but by several. This immediately brings to mind the trihedron formed by the axes $\mathbf{b}_1$, $\mathbf{b}_2$ and $\mathbf{b}_3$. A particularly interesting example is given by a recently discovered phase of helium, $^3\mathrm{He}$, where the local state is defined by three vectors (one spin vector $\gamma$ and two vectors orthogonal to $\gamma$, $\Delta'$ and $\Delta''$, describing the orbital state (cf. de Gennes, 1973a)).

A second type of less trivial generalization consists in the concept of the *order parameter* (cf. Landau and Lifshitz, 1967). In all systems which have been studied, the physical state is characterized in equilibrium by a thermodynamic potential (free energy, internal energy, etc.) or a hamiltonian, an essential property of which is that it is *invariant* for the translation and rotation symmetries envisaged. The necessity for invariance will be used when establishing, for example, the phenomenological form of the free energy of nematics (Chapter 3). However, invariances can be envisaged in relation to the more general operations which have no geometrical meaning in the crystal. A classic example is the gauge invariance of superfluids. The question of the order parameter can be presented in the usual way: consider a physical system which is followed through a phase transition from a *disordered* phase (generally at high temperature) where the potential is invariant under the operations of a certain continuous group $G_0$, to an *ordered* phase (generally at a low temperature) where the potential is only invariant under the operations of a sub-group $G_r$ of $G_0$. Among the operations of $G_r$, as already noted, there may be some which are not spatial rotations and translations. They operate on a physically observable entity with $n$ real components $S_1, S_2, \ldots, S_n$ (the order parameter). In the case of spin systems, the order parameter consists of the three components $M_\alpha$ of magnetization $\mathbf{M}$. $G_0$ is the complete group of rotations in the disordered state and $G_r$ is the group of rotations about $\mathbf{M}$ (taken along a determined, but arbitrary, direction) if the ordered state is ferromagnetic. This is the case shown above. In the case of a superfluid or a superconductor, the order parameter consists of two components of the complex wave function $\psi = \psi' + i\psi''$. $G_0$ is gauge invariant while $G_r$ corresponds to a determined (but arbitrary) choice of gauge.

The concept of director or constant modulus vector media lends itself to an important generalization, to which we will return in a later chapter (Chapter 10). For every vector field, $\gamma(\mathbf{r})$, which is arbitrary in length and position, but continuous, a correspondence can be made with a field of vectors of constant modulus $\mathbf{n}(r) = \gamma(r)/|\gamma(r)|$. In the same way, for every field of lines there is a corresponding field of directors (at every point on the line) which is tangential to the line. Certain properties of these fields are therefore analogous to those of deformed spin crystals (for a field of vectors) or nematic crystals (for a field of lines) and can be analysed, with the necessary changes, by similar topological arguments. The study of vector fields (for example) has now been greatly extended as an adequate geometric representation of the solutions of ordinary differential equations (Arnold, 1974). Therefore one can expect some cross-fertilization between the theory of defects and the theory of differential equations.

### 2.1.4 The Volterra process (J. Friedel, 1964; Nabarro, 1967).

From now on a *perfect crystal* will be understood as being every perfectly ordered medium in the sense which has just been described. The Volterra process describes the creation of a *line* singularity (dislocation) in such a medium, where the only invariances are spatial translations and rotations (isometries).

16

Consider, in a perfect crystal, a line $L$, and cut the crystal along a surface $\Sigma$ bound by $L$, but otherwise unspecified. Two lips, $\Sigma_1$ and $\Sigma_2$ (Figure 2.4), on $\Sigma$ can be made from this operation.

(a) Cut along $L$ a cylinder of material of small radius, which is taken out of the medium. The medium is thus no longer simply connected.

(b) Displace rigidly the two lips $\Sigma_1$ and $\Sigma_2$ from each other by a quantity $\mathbf{d}(\mathbf{r})$ which consists of a translation $\mathbf{b}$ and a rotation $(\Omega, \mathbf{v})$, symbolized by†

$$\mathbf{d}(\mathbf{r}) = \mathbf{b} + 2\sin\frac{\Omega}{2}\mathbf{v} \wedge \mathbf{r}. \tag{2.2}$$

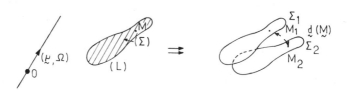

Figure 2.4. Surface section $(\Sigma)$ resting on a dislocation line; Volterra process

(c) Suppose that $\mathbf{b}$ and $(\Omega, \mathbf{v})$ are symmetry operations of a perfect crystal. (In an amorphous body, $\mathbf{b}$ and $(\Omega, \mathbf{v})$ can take any value). It is then possible to fill the space left free by the rigid displacement with perfect crystal which will adjust itself perfectly to fit the lips $\Sigma_1$ and $\Sigma_2$ re-forming the bonds which were cut, without vacancies or repetitions. Instead of a space left free by the displacement $\Sigma_1$ and $\Sigma_2$, doubly-occupied regions may occur. The slice of excess perfect crystal is then taken and the bonds are re-established in the same way.

(d) The foregoing operations are purely *topological* and the detailed way in which the crystal responds to the operations to which it is submitted is not important here. For the moment the constitutive law governing the medium can be ignored, and (if this proves useful) a particular deformation can be chosen in an arbitrary manner. It is sufficient for our purpose that the deformation conserves local symmetries (local existence of the Bravais lattice, smooth variations of angular relations). That being so, and the medium having an appropriate constitutive law, forces then develop, and stresses must be applied to maintain the crystal in the state of deformation which has been imposed upon it. If these stresses cease, a medium is obtained with a singularity along $(L)$ but in which all singularities have disappeared on $\Sigma_1$ and $\Sigma_2$.

This result deserves comment before it is clarified in the simple example relating to solid crystals which will be studied shortly. First of all, the final

---

† Too much importance must not be attached to the fact that (2.2) is valid only for an infinitely small rotation, which would commute with the translation $\mathbf{b}$. The non-commutability properties of rotations and translations do not introduce specific phenomena at this level. This can be verified from the examples given. Non-commutability changes the detail of the phenomenon, but not its nature. On the other hand, the interactions of lines characterized by non-commutable symmetries must be more closely analysed (cf. Chapters 8 and 10).

medium is no longer simply connected. This is expressed by equation (2.2). Consider a circuit ($\gamma$) closed around $L$ in the perfect material; if all the strains undergone by each point of ($\gamma$) between the perfect state and the final relaxed state are summed up, the deformation must be found not to depend on the path followed, in particular between the points of departure and arrival on ($\gamma$), and the summation must be found to be equal to $\mathbf{d(r)}$.

This condition can be explained in the simple case where the displacement is a pure translation. We then note the displacement $\mathbf{u(r)}$ from a point $\mathbf{r}$ of the perfect medium. We must have

$$\int_{(\gamma)} d\mathbf{u} = \mathbf{b} \tag{2.3}$$

which says that $d\mathbf{u}$ is not a total exact differential.

On the other hand, it is affirmed that the final result does not depend on the choices of surface to be cut, and that all singularities disappear on $\Sigma_1$ and $\Sigma_2$. It is understood by this that because the Volterra process conserved local symmetries on $\Sigma_1$ and $\Sigma_2$ (because $\mathbf{d}$ is a displacement which respects the symmetries of the crystal), any elasticity which, in the relaxed state, would affect $\Sigma_1$ and $\Sigma_2$ by causing the appearance of discontinuities of stress, would not have a physical meaning. From this point of view, linear elasticity is no problem; as demonstrated by Weingarten, $\mathbf{u(r)}$ is singular only on $L$ in this case.

Note again that the Volterra process indicates that a dislocation line is always a closed line, whether it ends at infinity, on the surface of the crystal, or on another dislocation. It also shows that two classes of lines can be distinguished:

(a)   translation dislocations $\mathbf{d(r)} = \mathbf{b}$, important in crystalline solids; $\mathbf{b}$ is known as the Burgers dislocation vector; and

(b)   rotation dislocations (disclinations) $\mathbf{d(r)} = \Omega\mathbf{v} \wedge \mathbf{r}$, important in liquid crystals, vector systems and to describe the effects of deformation due to Bloch walls in ferromagnets.

It should be noted that the Volterra process does not use symmetry operations with a reflection (centre of symmetry, plane of symmetry), known as symmetry operations of the second kind. In fact, a continuous deformation causing the transition from state I to a state II obtained by a reflection from state I cannot be produced in a crystal. In such a reflection the orientation of a reference trihedron is changed.

The Volterra process is topological in nature. However, it does not show the topological stability of the objects which it defines. This problem can only be tackled with the methods of topology outlined in Chapter 10, where a classification of lines using criteria other than those of crystalline symmetry will be proposed, and the relationship between the classification will be discussed. On the other hand, the Volterra process does not use energy considerations. It cannot therefore directly provide criteria of physical stability.

## 2.2 TRANSLATION DISLOCATIONS IN CRYSTALLINE SOLIDS. TOPOLOGY. WEINGARTEN'S THEOREM

In this section the now classic case of dislocations in crystalline solids is outlined in order to illustrate the concepts of the preceding section as simply as possible.

### 2.2.1 Edge dislocations and screw dislocations

Figure 2.5a represents a section of a cubic crystal along a plane perpendicular to a fourfold axis. A dislocation is introduced perpendicular to the plane of the figure along L, of Burgers vector **b** equal to the parameter of the cube. This is the smallest Burgers vector allowed. Figure 2.5b shows the result, which can be imagined to have been produced with the help of the Volterra process. The surface section is along the half-plane Σ and the displacement of the lips by a quantity **b** leads to the introduction of a supplementary atomic plane.

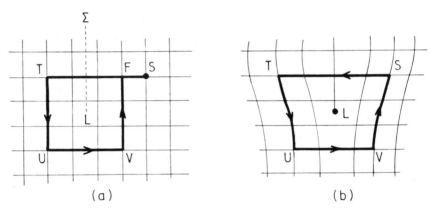

Figure 2.5. Burgers circuit. Two dimensional illustration: (a) perfect crystal; (b) crystal containing a dislocation

In the crystal deformed by the presence of dislocations, the regions of *good crystal* should be distinguished from regions of *bad crystal*. In the former, a local Bravais lattice can be defined, i.e. all the atoms have the same nearest neighbour atoms as in the perfect crystal. In fact, in all regions of good crystal in a deformed crystal there is an atom-by-atom correspondence with a region of perfect crystal. This is illustrated in Figure 2.5. Consider a closed, oriented circuit (γ) *STUVS*, situated entirely in the region of good crystal, and surrounding the line L (Figure 2.5b). It obviously corresponds to the circuit in the perfect crystal *STUVF*. The circuit *STUVF* is open. The vector **FS** constitutes the closure defect. This is obviously the Burgers vector of the dislocation surrounded by (γ) (Burgers circuit):

$$\mathbf{b} = \mathbf{FS}. \tag{2.4}$$

The sign of **b** depends on the orientation of the Burgers circuit. The following convention, *FS/RH* (final start, right hand), is currently used. Let there be a dislocation line *L*, oriented with an arbitrary direction; the direction of the Burgers circuit will then be defined by adopting the Maxwell cork-screw rule (right hand). The Burgers vector will then be the closure defect **FS** directed towards the image *F* in the perfect crystal from the end of the Burgers circuit. This convention has been adopted for the dislocation in Figure 2.5; it is therefore directed towards the observer.

Figure 2.5 represents a line perpendicular to its Burgers vector: this is known as an *edge dislocation*. If the line is parallel to its Burgers vector it is known as a *screw dislocation*. Figure 2.6 represents the two cases. Figures 2.6a and 2.6b, which both illustrate edge dislocations, are equivalent: the only difference between the two is the choice of the cut surface, which has no influence on the final result; Figure 2.6c illustrates a screw dislocation. The starting material in the perfect state clearly had the form of a cylindrical crown.

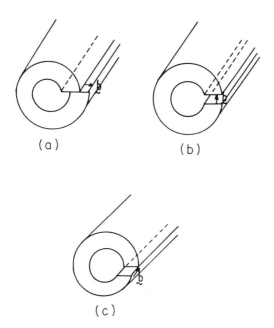

(a)

(b)

(c)

Figure 2.6. Schematic representation of rectilinear edge dislocations and of a screw dislocation

A screw dislocation is, by definition, rectilinear. An edge dislocation may take the form of a planar loop perpendicular to its Burger's vector. A dislocation is known as mixed if the Burgers vector possesses components parallel and perpendicular to the line.

## 2.2.2 Conservation rules

Consider a dislocation node (Figure 2.7) at a point $I$, and direct the lines either towards $I$ or away from $I$. It can then be easily shown that

$$\Sigma_i \mathbf{b}_i = 0. \tag{2.5}$$

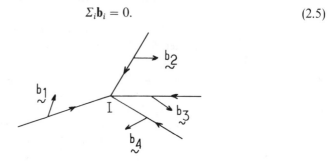

Figure 2.7. Dislocation node

## 2.2.3 Weingarten's theorem

In elasticity of small deformations: (a) a strain tensor can be defined:†

$$e_{ij} = \tfrac{1}{2}(u_{i,j} + u_{j,i}), \tag{2.6}$$

where the $u_i$ are the components of the displacement vector $\mathbf{u}(\mathbf{r})$; and (b) a symmetrical stress tensor linked to the strain tensor by the law of linear behaviour can also be defined:

$$\sigma_{ij} = C_{ijkl}e_{kl} = \sigma_{ji}. \tag{2.7}$$

In the absence of applied bulk forces, the equations of equilibrium are written:

$$\sigma_{ij,j} = 0, \qquad \text{equation in the bulk,} \tag{2.8a}$$

$$\sigma_{ij}v_j = f_i^{(e)}, \qquad \text{boundary conditions, the } f_i^{(e)} \text{ being the forces exerted per unit surface.} \tag{2.8b}$$

The stresses $\sigma_{ij}$ which result from the application of the forces $f_i^{(e)}$ on the surface are known as applied stress.

The equations of equilibrium (2.8a) in the presence of the constitutive law (2.7) satisfy a uniqueness theorem: when the forces applied to the surface are zero, the stresses $\sigma_{ij}$ are identically zero. To demonstrate this result, the medium must be simply connected. It is not the same if the medium contains dislocations. Dislocations create a field of non-zero stresses $\sigma_{ij}$ even in the absence of forces applied to the surface. These are known as *internal stresses* as opposed to *applied*

† The following notations will be used:

$$f_{,i} = \frac{\partial f}{\partial x_i}; \qquad u_{i,j} = \frac{\partial u_i}{\partial x_j}.$$

From equation (2.7) onwards Einstein's summation convention for repeated indices will be used.

*stresses.* The $\sigma_{ij}$ stresses are continuous throughout, except on the dislocation, because of Weingarten's theorem.

*Theorem.* For the strains $e_{ij}$ in a deformed medium to be finite and continuous throughout, except on a closed line $L$, the relative displacement $\Delta u_i(M)$ of the lips of the cut surface resting on $L$ must be a displacement which is possible in a non-deformable solid, i.e.

$$\Delta u_i(M) = b_i + \Omega_{ij}x_j(M), \qquad (2.9)$$

where $\Omega_{ij} = -\Omega_{ij}$, which indeed expresses the fact that the relative displacement of the two lips of the section is that of a non-deformable solid, since the term $\Omega_{ij}x_j(M)$ represents a pure rotation.

A demonstration of this theorem is to be found in Nabarro (1967). It depends on a hypothesis of small displacements. But, as indicated above (subsection 2.1.4), it is probably a particular case of a general theorem, independent of such a hypothesis and affirming that the Volterra process leads to fields of distortions discontinued only on a line, that is if the operations carried out on the surface section are symmetry operations of the medium. Here $b_i$ and $\Omega_{ij}$ are such operations, and are the only ones carried out (reflections are excluded).

(a) In an amorphous solid all displacements **b** and all rotations $\Omega$ are allowed. It appears, then, according to Weingarten's theorem, that by the effect of any one of these operations (or their product) on the surface section a field of internal stresses is created which is continuous throughout. Some discontinuities of the rotation tensor $\omega_{ij}$ remain, but here they have no physical importance since $\omega_{ij}$ rotations can only be measured on a lattice traced in the solid and this lattice is non-existent here.

(b) This is not the same for a solid crystal. If the surface section is not to be marked by orientation discontinuities of the crystal lattice, $\Omega$ must be a rotation allowed by the symmetries (obviously **b** must also be a symmetry translation). But even if $\Omega$ is not a symmetry rotation, i.e. even if the surface section is physically marked in the crystal, the $e_{ij}$ (and therefore the $\sigma_{ij}$) will be continuous. In fact, the cut surface will appear as a wall bounded by the dislocation line $L$, i.e. as a disorientation boundary.

### 2.2.4 Dislocation arrays

In a well-annealed crystal containing a small density of dislocation ($10^3$ to $10^8$ dislocations/cm$^2$) of different signs, these are organized in a relatively stable three-dimensional lattice, formed essentially by triple nodes. The array eliminates the elastic deformation by summation of terms of opposite sign. This is the so-called *Frank network* (Figure 2.8).

If the crystal contains an excess of dislocations of a certain sign, produced by plastic deformation, these, in the first stage, can form *stackings* which cause the hardening of metals; then after thermal treatment, they organise themselves into

22

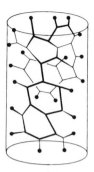

Figure 2.8. Frank network (Reprinted by permission of the Council of the Institution of Mechanical Engineers from Cottrell, 1957)

walls (grain boundaries) which relax the stress considerably, but introduce slight disorientations between the grains which they separate. This *polygonized* structure which is superimposed on the three-dimensional Frank network, causes what crystallographers have known for a long time as the *mosaic crystal*. Some of the geometric properties of these walls (cf. Nye, 1953) will now be discussed.

A perfect crystal is taken and a collection of its reticular parallel planes is considered. The crystal is then deformed by making the planes glide in relation to each other, while maintaining a constant distance between planes (Figure 2.9). Figure 2.9a shows the simple example where this glide is one-dimensional.

Suppose, moreover, that the distances between atoms on the layers remain constant. It is clear that in such a situation, where *strains are zero* ($e_{ij} = 0$;

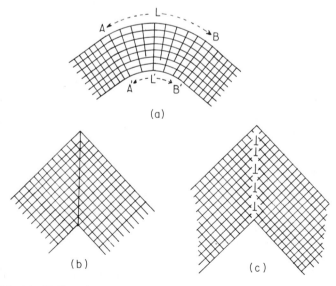

Figure 2.9. (a) Uniformly *polygonized* crystal; (b) tilt boundary; (c) schematic representation of tilt boundary

$\omega_{ij} = 0$) regularly distributed edge dislocations should be introduced (Figure 2.9a). Densities of dislocations can also be considered and a Burgers circuit ($\gamma$) traced which follows two parallel layers and joins them by two segments normal to these layers. The Burgers construction indicates that the total Burgers vector is distributed with a density $1/R$ per unit surface (in the plane of the diagram), where $R$ is the radius of curvature of the layers. Note that between the normals $AA'$ and $BB'$ the crystal has turned through an angle $\Delta\omega = L/R = L'/R'$, and that the figure can be extended to the right and the left with a non-distorted crystal structure. Now let $A$ tend towards $B$, and $A'$ towards $B'$, while $\Delta\omega$ and $d$ are constant. The density $1/R$ varies, but $\Delta\omega(L/R)$ does not; therefore the sum of the Burgers vectors $(1/R) \cdot Ld = \Delta\omega \cdot d$ remains constant. The process does not therefore require the introduction of any supplementary dislocations. All the dislocations are finally found to be on a wall $AA'$ (Figure 2.9b) with a linear density $\Delta\omega$.

In other words, a wall of edge dislocations all of the same sign separates the two disoriented grains. If these dislocations are at distance 1 from each other (Figure 2.9c) their individual Burgers vector is

$$\mathbf{b} = 1\Delta\omega.$$

Moreover, there are no long-range stresses.

Note that in this very detailed example the rotation from one crystallite to another occurs along an axis contained by the boundary. This is known as a *tilt boundary*. It consists of an array of parallel edge dislocations. The second typical case is known as a *twist boundary* (Figure 2.10), in which case the relative rotation between the two crystallites is about an axis perpendicular to the boundary. It can be shown that this disorientation can be produced without long-range stresses from two rectangular networks of screw dislocations of Burgers vector $\mathbf{b} = 1\Delta\omega$ situated on the boundary.

The process can be generalized (Read and Shockley, 1950; Frank, 1949) to apply to any disorientation boundary. It is always possible to invent such a geometry with networks made of dislocations of the same sign, and which are situated on the boundary. This is indeed what is observed, and there are no longer any long-range stresses, except in a zone of size $\sim 1$ in the region of the boundary each grain. For a boundary of any form, separating two grains disoriented by an angle $\Delta\omega$ about the rotation axis $\gamma$, the distribution of dislocations to be present in the boundary can be obtained as follows.

To simplify the demonstration, the angle $\Delta\omega$ is taken to be small. Take a perfect crystal and design the boundary in such a way, for example, that the right-hand side of the crystal has a definite orientation (cf. Figure 2.11). In the region of point $A$ on the boundary, the orientation of the left-hand side is obtained by submitting the (left side) crystal to a rotation about the $\gamma$ axis, through an angle $\Delta\omega$, from an arbitrary origin $O$. If $O$ is close enough to $A$ and $\Delta\omega$ is small, $OA$ can be made into a vector in a plane tangential to the boundary and we have:

$$\mathbf{AA'} = \Delta\omega\gamma \wedge \mathbf{OA}. \tag{2.10}$$

24

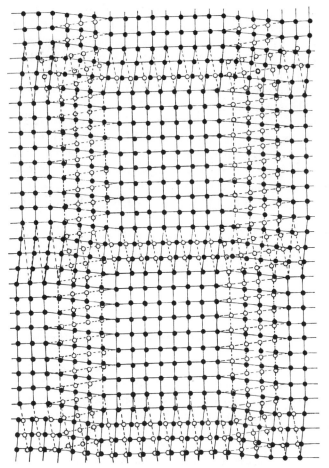

Figure 2.10. Twist boundary (from Read, 1953. Reproduced by permission of McGraw-Hill)

If $O$ is on the boundary, $\mathbf{AA'}$ is the closure defect of the closed circuit $(\gamma)$. It can then be considered to be the Burgers vector of the dislocation array perpendicular to this circuit and contained in the boundary. It is obvious that at each point $O$ of the boundary two circuits, $(\gamma)$ and $(\gamma')$, must be considered in two planes perpendicular to each other and perpendicular to the boundary in order to obtain all the necessary dislocations. It can also be seen that there is a certain arbitrariness in the choice of $(\gamma)$ and $(\gamma')$ and therefore in the choice of boundary dislocations. This arbitrariness is removed in a real crystal, because of energy considerations.

Suppose, for example, that there is a twist boundary (plane) perpendicular to $\gamma$. Consider a circuit $(\gamma)$ and the vector $\gamma \wedge \mathbf{OA}$ which is perpendicular to the plane of this circuit. A dislocation density can then be used traversing the circuit perpendicularly and carrying Burgers vectors in the same direction. These are

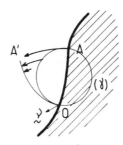

Figure 2.11. Disorientation boundary (see text)

screw dislocations. In a circuit where $(\gamma')$ is perpendicular to $(\gamma)$ there is the same arrangement.

Finally, note that because of Weingarten's theorem these dislocations do not cause discontinuities in the stress field. But this theorem is always *implicitly* supposed valid, as indicated in subsection 2.1.4. We will return to this question when discussing defects relevant to elasticities other than those of ordinary solids.

### 2.3   ROTATION DISLOCATIONS (*disclinations*). TOPOLOGY

#### 2.3.1   Solids

It is clear from the Volterra process that if the rotation axis which defines the line is not along the line itself strong distortions will be produced along the actual core of the line. This observation indicates that in solids the only disclinations capable of existing are probably rectilinear, and are on the rotation axis along the line. These are known as wedge disclinations (Figure 2.12).

Suppose that the crystal is limited to a cylindrical crown, centred on $L$, of radii $r_c$ and $R$. The stresses introduced by the wedge disclination have been calculated by Timoshenko (1951):

$$\sigma_{rr} = A\left\{\ln\frac{r}{R} - \frac{r_c^2}{r^2}\ln\frac{r_c}{R}\right\},$$

$$\sigma_{\theta\theta} = A\left\{\ln\frac{r}{R} + \frac{r_c^2}{r^2}\ln\frac{r_c}{R} + 1\right\}, \qquad (2.22)$$

$$\sigma_{zz} = v(\sigma_{rr} + \sigma_{\theta\theta}),$$

where $A = \mu\Omega^2/2\pi(1 - v)$.

The stresses (and distortions) increase with $r$ and the line energy per unit length is:

$$W = \frac{\mu\Omega^2}{16\pi(1 - v)}(R^2 - r_c^2)\left[1 - \frac{4r_c^2 R^2}{(R^2 - r_c^2)^2}\left(\ln\frac{R}{r_c}\right)^2\right], \qquad (2.12)$$

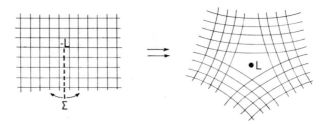

Figure 2.12. Wedge disclination

which is of the same order of magnitude as the energy of an edge dislocation of Burgers vector $\frac{1}{2}R\Omega$.[†] Wedge disclinations will not therefore be permitted except in exceptional circumstances, for example in lamellar crystals, where they can be located between two lamellae without their energy being prohibitive (Friedel). In fact, in solid crystals, they can only really exist in the form of pairs of disclinations of opposite signs situated a short distance apart (Figure 2.13).

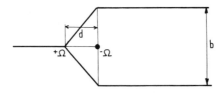

Figure 2.13. Pair of wedge disclinations of opposing signs

Their sum is in fact equivalent to an edge dislocation of Burgers vector $\mathbf{b} = 2d \tan \Omega/2$, where $d$ is the distance between the lines, measured normally to the rotation vector $(\Omega, \mathbf{v})$. At long distances, the elastic effects are therefore analogous to those of an edge dislocation.

Volterra had considered the possibility of the existence of rotation dislocations with vectors perpendicular to the line (Figure 2.14). It appears clearly in the figure that such objects, which for $\Omega = \pi$ are true Moebius volumes, can only exist with finite dimensions. A less restrictive definition of these objects, known as *twist disclinations*, is now given.

Consider a wedge disclination displaced parallel to itself from a position $L$ to a position $L'$, and suppose that $\Omega$ is displaced with the line. A point $M$ of the portion of the cut surface common to $L$ and $L'$ (it can always be imagined that $\Sigma$

---

† Remember that the line energy of a translation dislocation in a solid is of the form (Friedel):

$$W = \frac{\mu b^2}{4\pi} \ln \frac{R}{r_c},$$

where $R$ is of the order of the minimum distance between dislocations of opposite signs and $r_c$ a core radius $r_c \sim b$.

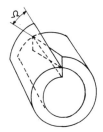

Figure 2.14. Schematic representation of a twist dislocation

and $\Sigma'$ have a common part) after this displacement of $L$, undergoes a supplementary displacement which reads:

$$\mathbf{b} = \mathbf{\Omega} \wedge \mathbf{A'A}, \qquad (2.13)$$

where $A$ and $A'$ are any points on $L$ and $L'$. Note that $\mathbf{b}$ is a constant vector, independent of $M$. On the other hand it will be a Bravais lattice vector, if the positions $L$ and $L'$ are separated by a vector of the Bravais lattice and $\mathbf{\Omega}$ is a permitted rotation. It can therefore be concluded that in the displacement of a wedge disclination a translation disclination is emitted (or absorbed).

Now suppose that a dislocation of Burgers vector $\mathbf{b}$ meets a wedge disclination $\mathbf{\Omega}$ on only one part of its length. A kink $AA'$ is produced on the wedge line (Figure 2.15) to which is attached a dislocation of Burgers vector $\mathbf{b}$. If the process is repeated by bringing in other $\mathbf{b}$ dislocations, it can be seen that $AA'$ can be lengthened as much as we wish, provided that $\mathbf{b}$ dislocations are attached to it with a density given by equation (2.13). $AA'$ is then a *twist disclination*. It has been shown at the same time that the Moebius dislocation can be lengthened infinitely by attaching translation dislocations to it at regular intervals.

The dislocations attached to $AA'$ form a wall, in the sense that has been indicated above in subsection 2.1.4. This comparison will be developed further when the magnetostriction of ferromagnets is examined (Chapter 9). It should be remembered that a dislocation wall which is limited in space behaves at its ends

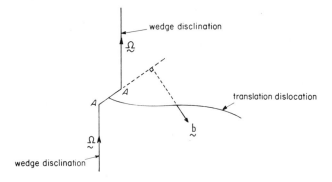

Figure 2.15. Twist line; operational definition

like a dislocation of rotation. If the wall is bounded by a line which intersects the dislocations which comprise it, this line has a twist component. If the wall is bounded by a line parallel to one of the dislocations which comprise it and does not intersect the others, the result still applies, but the line is a wedge disclination. The reader will be easily convinced by imagining that Figure 2.9c ends at $L$; this gives Figure 2.16.

Dislocations of rotation in solids will be studied in more detail in Chapter 8.

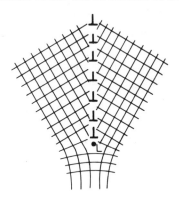

Figure 2.16. Finite tilt boundary

### 2.3.2 Directional medium

In a solid, the only currently admitted disclinations have small rotation vectors, which are in general unrelated to the symmetries of the medium. These are the ends of dislocation walls. Cosserat directional media of the 'liquid crystal' type, on the other hand, offer many examples of large angle disclinations. In fact, historically, these were the first line defects to be really observed (Lehmann, 1904; G. Friedel and Grandjean, 1910; G. Friedel, 1922). These observations were extremely easy to make (polarizing microscope) because of the large birefringence of these anisotropic liquids. But the theory behind them has only recently been formed. We limit ourselves here to the topological properties of these disclinations, leaving the study of their energetic properties in liquid crystals to another chapter. Also note that disclinations in spin systems are quite similar phenomena (cf. Chapter 9).

#### 2.3.2.1 *Wedge disclination*

Figure 2.17 represents the action of the Volterra process for a relative rotation of $+\pi$ of the two lips of $\Sigma$. The final relaxed state is represented in Figure 2.17c and necessarily has a ternary symmetry. Figure 2.18 represents the most commonly observed wedge disclinations. In all this topology it is supposed that the molecules remain in the plane perpendicular to the line. When the elastic properties of these media are given, the validity of this hypothesis will be discussed.

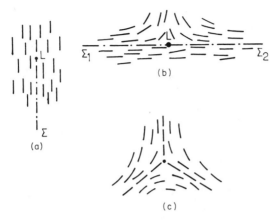

Figure 2.17. Volterra process in a directional medium. Creation of a wedge line

The strength $S$ is defined as follows: let there be a closed circuit $(\gamma)$ surrounding a wedge line; give this circuit an arbitrary direction and, starting from a point of origin on this circuit, trace the hodograph of the unit vectors tangential to the molecules which the circuit meets. If the hodograph and the circuit have the same direction, $S$ will be positive and equal to $\alpha/2\pi$, where $\alpha$ is the angle turned in the hodograph (Figure 2.19).

Figure 2.18. Wedge lines

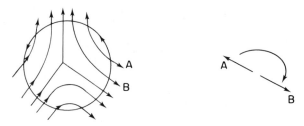

Figure 2.19. Hodograph of a wedge line

It will be noted that $S$ is a scalar, independent of the direction of the circuit ($\gamma$), while $\Omega$ depends (as does the Burgers vector) on the orientation of the dislocation line. We have $\Omega = 2\pi S\mathbf{v}$, where $\mathbf{v}$ is a unit vector along the line.

There are three possible configurations for $S = 1$. They correspond topologically to different possible cut surfaces. The physical transition from one to the other is achieved by cooperative rotation of the molecules at a constant angle about an axis parallel to the line.

### 2.3.2.2 Twist disclination

A twist disclination has been defined by a wall of edge dislocations. This definition was given for solids, is purely topological and can be extended to liquid crystals. But at least in the case of nematics, any translation is permitted. Thus, if a finite translation $\mathbf{b}$ is introduced, it is dissipated by viscous relaxation into infinitesimal translations $d\mathbf{b}$. This means that it no longer becomes necessary to be limited to rectilinear twist disclinations in a nematic. In fact, any curved line $L$ of rotation vector $\Omega$, can be accommodated in the nematic phase if a density of translation dislocations is attached to it:

$$d\mathbf{b} = \Omega \wedge \mathbf{t}\, ds, \qquad (2.14)$$

where $\mathbf{t}$ is the vector tangential to the line and $ds$ the arc length element. Equation (2.14) is a generalization of equation (2.13) and its demonstration is similar. In fact it is only necessary to attach parallel rotation vectors to two neighbouring points $P$ and $P'$ (Figure 2.20) and to consider the effects of these rotation vectors on the cut surface at $M$. The displacements

$$d(\mathbf{P}) = \Omega \wedge \mathbf{PM} \quad \text{and} \quad d(\mathbf{P'}) = \Omega \wedge \mathbf{P'M}$$

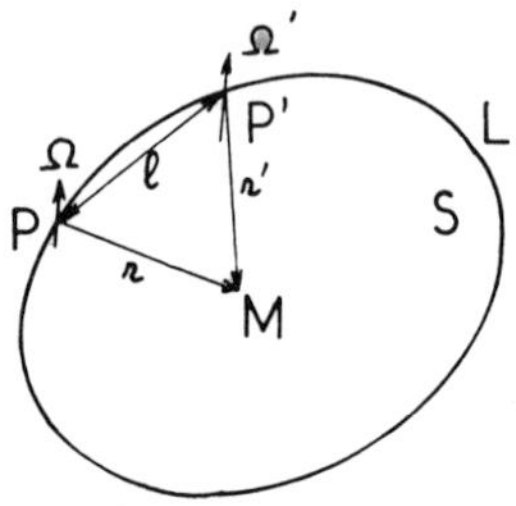

Figure 2.20. Dislocation density on a line of any given form

are only compatible with the Volterra process if translation dislocations

$$d(\mathbf{P'}) - d(\mathbf{P}) = \mathbf{\Omega} \wedge \mathbf{MM'} = \mathbf{\Omega} \wedge \mathbf{t}\, ds$$

are attached to the disclination between $P$ and $P'$.

### 2.3.2.3 *De Gennes–Friedel process*

The simplicity of the previous result, which in fact makes it possible to eliminate the role of translation dislocations in the case of a mixed line in a nematic, indicates the following generation process, simpler than the Volterra process for a nematic medium.

De Gennes has proposed the construction of disclination lines in nematics by carrying out the following operation ($P_G$) on the surface section: the molecules on either side of $\Sigma$ are submitted to pure rotations $\mathbf{\Omega}_1$ and $\mathbf{\Omega}_2$ such that $\mathbf{\Omega}_2 - \mathbf{\Omega}_1 = \mathbf{\Omega}$, around their centre of gravity. This is the equivalent to the Volterra process in nematics; the Volterra process indeed breaks down into a translation and a rotation. More precisely, the translation can be considered (Friedel) to displace a molecule (in the case of a nematic) or a crystallographic direction (in the case of a solid crystal) *parallel* to itself ($T$) and the rotation makes the molecule or direction turn *about its centre of gravity* ($P_G$). The process in Figure 2.17 must therefore be described somewhat differently: it can in fact be broken down into

$$P_v = P_G + T \tag{2.15}$$

according to the scheme of Figure 2.21 (where the breakdown for a rotation dislocation of angle $\alpha$ is represented). But it is known that translation dislocations are relaxed in a nematic. Therefore it remains true that

$$P_v = P_G \tag{2.16}$$

in nematics.

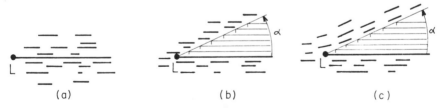

Figure 2.21. De Gennes–Friedel process

### 2.3.2.4 *Twist disclinations: schemes; strength*

Figure 2.22 represents schemes of pure twist disclinations. In Figures 2.22a and b it is supposed that the configuration remains planar; in Figure 2.22c it is radial. In each case it is supposed that $\Omega = \pm \pi$. The difference between Figures 2.22a and b is that in one case the rotation of the director follows a left-hand helix, then a right-hand helix, if a circuit is followed around the line; the other diagram shows the converse situation. Attach to each point of the circuit the molecule which meets it at that point and consider the ribbon formed by the circuit and these molecules (to obtain a well formed ribbon, it is necessary to consider only those circuits which are nowhere tangential to the molecule). Thus the circuit will display singular points, as in Figure 2.22a. These ribbons are Moebius ribbons (Bouligand); the strength of the twist line can be defined as the sign of the corresponding ribbon. It is a pseudo-scalar property, as a Moebian figure is inverted in its mirror image.

Figures 2.22a, b and c represent different topological possibilities relating to the same disclination. It is necessary to remember that physical reality imposed by *energy* considerations leads to a scheme which is intermediate among the various possibilities.

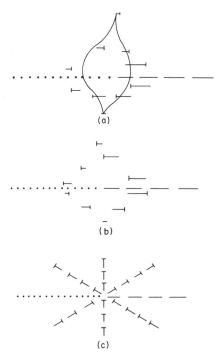

Figure 2.22. Twist lines

### 2.3.2.5 *Loops*

Figure 2.23 represents a twist loop of planar configuration: $\Omega$ is perpendicular to the plane of the loop. This is an experimental situation which is frequently encountered. One will recognize that the molecules turn in a right-hand helix inside the cylinder formed by the loop and the vertical. In the plane of the loop itself the molecules undergo an abrupt rotation from one side of $L$ to the other but in every other plane the rotation is continuous. Figure 2.23 is of a topological nature, and therefore constitutes an approximate representation of reality, as does Figure 2.22.

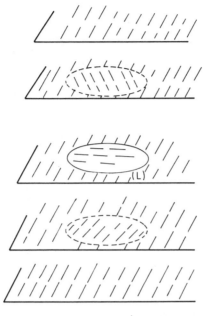

Figure 2.23. Twist loop

Loops consisting of wedge and twist parts are less common (Figure 2.24). They can be found, for example, in a plane containing $\Omega$. In fact if $\Omega$ is constant along the line, $S$ changes sign accordingly on the opposite wedge parts. There is therefore a transition from one to the other which has been discussed by Bouligand *et al.* (1973). It will be noted that in a plane perpendicular to $\Omega$ the configuration either forms a pinch† (Bouligand) or an inverted pinch. But in every case in the diagram the configuration is perfect at long distances.

### 2.3.2.6 *Conservation laws*

There are no laws of conservation for disclinations except in particular cases. The reason for this is that successive rotations are not commutable, except of course if they are parallel. In this case in a node of lines there is quite evidently the relation

$$\Sigma_i \Omega i = 0$$

by orienting all the lines towards the node or away from the node, with the same convention as for dislocations.

† *Translators' note:* In French, 'pincement'.

34

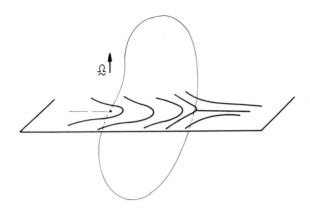

Figure 2.24. Closed loop and pinch

Attention must be drawn at this point to a curious but profound result (Harris, 1970b; Nabarro, 1970). Consider a closed surface on which is placed a field of directors. The point singularities of this field can be considered as wedge lines piercing the surface. The sum of the strengths of these lines is shown to be equal to Euler's surface characteristic. In the case of the sphere this gives

$$\Sigma_i S_i = 2$$

and the derivation is as follows.

First define the strength of a closed circuit ($\gamma$) traced on a surface (Figure 2.25) by expression (2.17)

$$S = 1 - \frac{1}{2\pi} \oint d\theta, \tag{2.17}$$

where $\theta$ is the angle between the arbitrarily but continuously oriented director and the tangent to the circuit oriented so that $S = 0$ if there is no singular point inside the circuit. $S$ is equal to the strength of the singularities inside the circuit (Figure 2.25b illustrates this for $S = 1$).

Now suppose that ($\gamma$) is drawn on the surface of a sphere (Figure 2.25c) and consider in turn the inside and outside of ($\gamma$). It is clear that to calculate the order of ($\gamma$) for its 'exterior' it is sufficient to change the sign of the integral. Therefore

$$S_{int} = 1 - \frac{1}{2\pi} \oint d\theta,$$

$$S_{ext} = 1 + \frac{1}{2\pi} \oint d\theta; \tag{2.18}$$

thus

$$S_{int} + S_{ext} = \Sigma_i S_i = 2.$$

In Figure 2.25c there are effectively two configurations $S = 1$.

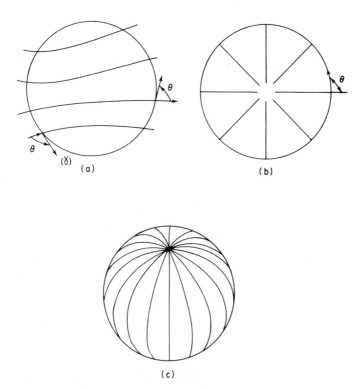

Figure 2.25. Singularities of a vector field on a sphere (see text)

# Chapter 3
# Nematics

At the end of Chapter 2 a topological description of disclinations was given using directors $\mathbf{n(r)}$. From this study the following points should be remembered.

(1) Disclinations of half-integral strength are permitted in a nematic because of the physical equivalence $\mathbf{n(r)} \sim -\mathbf{n(r)}$.

(2) The Volterra process leads to the definition of wedge disclinations ($\mathbf{\Omega}$ along the line) and twist disclinations ($\mathbf{\Omega}$ perpendicular to the line). The latter can only exist if they are coupled with (translation) dislocations terminating on the line with density

$$db\frac{db}{ds}ds = \mathbf{\Omega} \wedge \mathbf{t}, \qquad (3.1)$$

where $\mathbf{t}$ is the unit vector tangential to the line and $ds$ the element of arc length. If condition (3.1) is not met, then the phenomenon produced by the Volterra process can only be of finite length (see Figure 2.14 in Chapter 2).

(3) Relaxation theorem: the symmetry group of nematics allows the presence of translation dislocations of any Burgers vector. Condition (3.1) can therefore be met whatever the form taken by the line of disclination; wedge, twist and mixed lines will therefore be obtained. Moreover, since the nematic is liquid, the relaxation of deformations (to which only the forces of viscosity are opposed) allows all translation dislocations $\mathbf{b}$ to be dissipated into infinitesimal dislocations $\mathbf{db}$. The consideration of translation dislocations in a nematic is therefore pointless (this does not apply in a medium possessing the same symmetries, but solid, for example in a reticulated nematic) and the Volterra process can be replaced by:

(4) The de Gennes process, which consists of the (relative) local rotation of the molecules of the surface section through an angle $\mathbf{\Omega}$ about the axes passing through their centres of gravity.

In this chapter, only some aspects of the physical properties of dislocations in nematics will be discussed. First, wedge disclinations of integral strength will be studied and it will be shown how they differ fundamentally from dislocations at half-integral strength. In the former, the core may have no singularities, while this is impossible for the latter. Different physical properties result from this. Surface lines, i.e. singular configurations localized near the sample boundaries, will then be considered. This last problem is extremely important since it touches on the physics of the substrate on which the liquid crystals are deposited. The theoretical aspects will be seen to connect this problem with that of the spread of the core of a dislocation in a crystalline solid in its glide plane (Peierls–Nabarro model).

A very particular type of defect found in nematics, which has not yet been discussed, is singular points. These will be considered as part of the study of lines of integral strength and in certain applications of the Peierls–Nabarro model. The physical concept of singular points will be discussed at a deeper level in Chapter 10.

Before tackling these questions, some concepts of elasticity in nematics need to be explained. We will be restricted to static elasticity, which will be presented from a phenomenological point of view. The elasticity of director media offers an interesting example of elasticity of torques and rotations, but is at the opposite extreme from the usual elasticity of stresses and strains. However, certain analogies of a purely phenomenological nature between stresses and torques on the one hand and strains and rotations on the other, will be discussed and applied to dislocations.

## 3.1 CHEMICAL SPECIES

On this subject, Gray (1962, 1969, 1973) and Demus *et al.* (1974) should be consulted for new compounds and G. Friedel (1922) for old compounds.

The nematic materials most currently used all consist of two benzine rings associated with different functions and with short, flexible exterior chains, according to the following general scheme:

$$R - \langle O \rangle - A = B - \langle O \rangle - R'.$$

For a long time, PAA (paraazoxyanisole) has been the material most commonly used:

PAA:  $CH_3 - O - \langle O \rangle - N = N - \langle O \rangle - O - CH_3$
$\downarrow$
$O$

This is nematic between 116 and 135°C. More recently MBBA (methoxy p-benzylidene p-butylaniline) has been studied a great deal; it is nematic at more manageable temperatures (from 20 to 45°C):

MBBA:  $CH_3 - O - \langle O \rangle - CH = N - \langle O \rangle - CH_2 - CH_2 - CH_2 - CH_3$

These molecules are about 20 Å in length and 5 Å in width. They are fairly unstable, and currently a certain number of more stable compounds are beginning to be used, including tolanes (Malthète *et al.*, 1971) and biphenyls (Gray, 1973). The search for stable compounds is important for the industry of liquid crystals display systems.

An article by Malthète *et al.* (1976) discusses the molecular properties leading to nematogenic or smectogenic characteristics in a compound.

## 3.2   ELASTOSTATICS OF NEMATICS

Consider a *weakly* deformed nematic medium i.e. such that the variations in orientation of the director $n(r)$ only become perceptible above a distance $l$ much greater than $a$ (mean distance between molecules). Some of these variations of the director can be demonstrated by optical observation through a polarizing microscope. The nematic crystal is a positive uniaxial crystal, with the optical axis along $n$; if these variations are on the scale of the wavelength of light or even slower, the deformed crystal appears locally as a perfect crystal. In any case, since $l \gg a$, i.e. since the components of the gradient of $n$ obey the inequality

$$an_{i,j} \ll l, \tag{3.2}$$

the crystal can be treated s a continuous medium.

### 3.2.1   Fundamental deformations and free energy density (Frank, 1958; de Gennes, 1975)

Let $F_d$ be the free energy density of the deformed medium. Phenomenologically, $F_d$ can be constructed from the following observations:

(a) $F_d$ must be equal to zero as $n_{i,j} = 0$, i.e. for the non-deformed crystal;
(b) $F_d$ must be even in $n$ because of the physical equivalence $n(r) \sim -n(r)$; and
(c) $F_d$ must be invariant in the continuous group of rotations about the local $n(r)$.

If $F_d$ is limited to quadratic terms in the $n_{i,j}$ components of the gradient, and constructed in accordance with the rules of symmetry given above, three fundamental invariants corresponding to three fundamental modes of distortion of the perfect crystal appear. These invariants, which describe three independent types of deformation of the medium (Figure 3.1a, b, c), are as follows:

Figure 3.1. (a) Splay (Chatelain friction, planar anchoring); $(\text{div}\,n)^2 = 1/r^2$; (b) bend (homeotropic anchoring): $(n \wedge \text{rot}\,n)^2 = 1/r^2$; (c) twist (Mauguin twisted area): rotation of angle $\alpha$: $(n \cdot \text{rot}\,n)^2 = \alpha^2/h^2$

### 3.2.1.1 *Splay deformation*

This can be produced by placing a liquid crystal between two glass slides placed at a small angle such that the anchoring conditions on the surface of the slides favour an alignment perpendicular to the line of intersection of the two slides, and contained in the plane of these slides. Such an anchoring can be produced by rubbing the two glass plates in a well-defined direction with drawing paper (the long molecules of fatty acids which are always present on the paper or on the hands of the operator are spread in a unidirectional manner on the glass and invariably create conditions favourable to alignment). This is *Chatelain friction,* often replaced today by the fabrication of directional substrates (evaporation under vacuum of silicon monoxide (SiO) on a glass substrate, with the direction of evaporation at a small angle to the substrate; cf. Chapter 5, Figure 5.24).

### 3.2.1.2 *Bend deformation*

This can be produced by inserting liquid crystal between two glass slides at a small angle and treated in such a way as to favour anchoring of molecules perpendicular to the slides (homeotropic anchoring).† There is a natural tendency towards homeotropy in non-treated slides. This tendency will be accentuated if the slide is treated with detergent (see subsection 3.3.1).

### 3.2.1.3 *Twist deformation*

This occurs when a nematic is inserted between two parallel slides with planar anchoring, such that the anchoring directions are at an angle $\alpha$. Figure 3.1c represents a Mauguin 'twisted area'. The molecules at an angle to the diagram are symbolized by nails, with their points directed towards the observer and with a length proportional to their projection into the plane of the diagram.

A positive stiffness constant is associated with each of these fundamental deformations. The free energy density is written:

$$F_d = \tfrac{1}{2}K_1 (\operatorname{div}\mathbf{n})^2 + \tfrac{1}{2}K_2 (\mathbf{n}\cdot\operatorname{rot}\mathbf{n})^2 + \tfrac{1}{2}K_3 (\mathbf{n}\wedge\operatorname{rot}\mathbf{n})^2. \qquad (3.3)$$

The order of magnitude of the $K_i$ can be estimated by observing that if $F_d$ is an energy per unit volume, the $K_i$ are energies per unit length (erg/cm or dyne). The typical energy involved in this situation is the interaction energy $U$ between the neighbouring molecules, while the typical length involved is the molecular distance $a$; therefore

$$K_i \sim \frac{U}{a} \sim 10^{-6}\,\text{dynes.}$$

The intermolecular forces are essentially of the Van der Waals type i.e. $U \sim 0.1\,\text{eV} = 10^3\,\text{K} = 5 \times 10^{-14}\,\text{ergs.}$ They are weak forces. A detailed study of

† The term homeotropy is used to designate anchoring conditions perpendicular to the boundaries; it was introduced in this sense by G. Friedel (1922).

microscopic interactions in liquid crystals has still to be done. We also have $a \sim 5 \times 10^{-8}$ cm. This leads to the indicated order of magnitude, which is found by precise measurements of the stiffness coefficients. For example, in paraazoxyanisole (PAA) at 120°C (Zvetkov, 1937; Saupe, 1960):

$$K_1 = 0.7 \times 10^{-6} \text{ dynes}; \qquad K_2 = 0.43 \times 10^{-6} \text{ dynes};$$

$$K_3 = 1.7 \times 10^{-6} \text{ dynes};$$

in paramethoxybenzylidene p-butylaniline (MBBA) at 22°C (Haller, Léger):

$$K_1 = 0.58 \times 10^{-6} \text{ dynes}; \qquad K_2 = 0.33 \times 10^{-6} \text{ dynes};$$

$$K_3 = 0.7 \times 10^{-6} \text{ dynes}.$$

It will be noted that the twist constant $K_2$ is (experimentally) always smaller than the other two. Twist deformations are therefore favoured, if allowed by the boundary conditions.

Finally, it is observed that according to (3.3) the order of magnitude of energy per molecule associated with distortion is of the order of

$$\frac{K_i}{l^2} a^3 = U \left(\frac{a}{l}\right)^2$$

where $l$ is the length associated with the envisaged deformation. This energy is seen to be necessarily less than $U$, if $l \gg a$ (see beginning of section).

### 3.2.2 Effects of magnetic or electric field (de Gennes, 1975)

The molecules in nematics tend to be aligned parallel to a magnetic field; in fact benzine groups, which have many peripheral electrons, tend to repel any magnetic field perpendicular to them and therefore place themselves in a plane containing this field. This diamagnetism involves energies per molecule which are low in comparison with the thermal agitation energy $k_B T$, but becomes important since it is cooperative in nature, if a large number of molecules is involved. It can be described phenomenologically by introducing a supplementary term into $F_d$:

$$\Delta F_d = -\tfrac{1}{2}\chi_a(\mathbf{H} \cdot \mathbf{n})^2, \tag{3.4}$$

which is equal to zero for $H$ perpendicular to the director. However, when $H$ is parallel to $\mathbf{n}$ it gives a negative contribution to the free energy and therefore a more stable state, $\chi_a$ being positive. $\chi_a$ is the anisotropy of magnetic susceptibility ($\chi_a \sim 10^{-7}$ e.m.u. c.g.s. units).

The expression $F_d + \Delta F_d$ gives three characteristic lengths $\xi_i$ linked to the magnetic field:

$$\xi_i = \frac{1}{H} \left(\frac{K_i}{\chi_a}\right)^{1/2}, \tag{3.5}$$

which can be observed to become greater as $H$ becomes smaller; this expresses the cooperative effect mentioned above. The magnetic effects of a given field $H$ are felt in samples of typical dimensions greater than $\xi_i$. This is the basis of the measurement of the stiffness coefficients (Freedericks' transitions, cf. de Gennes, 1975) which it is not necessary to discuss here.

The effects of the electric field can be treated in an analogous manner, at least in a perfectly non-conductive material. There is therefore a term for dielectric anisotropy:

$$\Delta F_d = -\tfrac{1}{2}\varepsilon_a(\mathbf{E}\cdot\mathbf{n})^2 \tag{3.6}$$

similar to (3.4) ($\varepsilon_a \sim 0.1$ e.s.u, can be positive or negative). A weakly induced polarization effect called flexo-electricity was foreseen by Meyer (1969). Usual symmetry considerations show that this polarization can be written as an even function of $\mathbf{n}$:

$$\mathbf{P} = e_1\mathbf{n}\,\mathrm{div}\,\mathbf{n} + e_3\mathbf{n}\wedge\mathrm{rot}\,\mathbf{n}. \tag{3.7}$$

Meyer attributed this effect to the presence of permanent dipolar moments in molecules possessing a certain type of symmetry (Figure 3.2). This effect has been recently demonstrated experimentally (Schmidt *et al.*, 1972; Prost and Pershan, 1976).

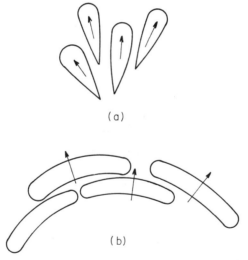

( a )

( b )

Figure 3.2. Molecular asymmetries leading to a flexoelectric effect; (a) preferential coupling with splay; (b) preferential coupling with bend

The effects of an electric field are often perturbed by movements of material linked to displacements of charge, if the material contains ionizable impurities. However, the effects of an electric field are of great interest, because of the low voltage values which they require ($\varepsilon_a$ small) and because situations occur where $\varepsilon_a < 0$ i.e. where the director, in the ground state, is situated arbitrarily in a plane perpendicular to the applied field. The effects of an electric field give rise to many defects of various kinds and in conditions where $\varepsilon_a < 0$ there arise quite particular point defects (umbilics) (Rapini, 1973; Rapini, Léger and Martinet, 1975; Dreyfus, 1979).

### 3.2.3 Minimization of free energy

Consider a virtual variation of the director consisting of a local rotation and manifesting itself by a variation $\delta\mathbf{n}(\mathbf{r})$. In this variation the virtual variation of energy $\delta\int F_d \, d^3\mathbf{r}$ is zero. The condition $\mathbf{n}^2 = 1$ is taken into account in introducing a Lagrange multiplier $\lambda/2$. The total energy therefore reads:

$$W = \int\left[F_d - \frac{\lambda}{2}(\mathbf{n}^2 - 1)\right]d^3\mathbf{r}$$

and its variation has the form:

$$\delta W = \int\left[\frac{\partial F_d}{\partial n_i} - \frac{\partial}{\partial x_j}\left(\frac{\partial F_d}{\partial n_{i,j}}\right) - \lambda n_i\right]\delta n_i \, d^3\mathbf{r} + \int\frac{\partial F_d}{\partial n_{i,j}}\delta n_i \, dS_j. \tag{3.8}$$

The equilibrium equations are therefore obtained in the form:

$$h_i = -\frac{\partial F_d}{\partial n_i} + \frac{\partial}{\partial x_j}\left(\frac{\partial F_d}{\partial n_{i,j}}\right) = -\lambda n_i, \tag{3.9}$$

where $F_d$ has the general form given by (3.3), (3.4) and (3.5).

The vector $\mathbf{h}$ is known as a molecular field. Equation (3.9) says that it is colinear to the director. The equilibrium equation is therefore written, eliminating $\lambda$:

$$\mathbf{n} \wedge \mathbf{h} = 0. \tag{3.10}$$

It is convenient to interpret $\mathbf{n} \wedge \mathbf{h}$ as a bulk torque. In fact suppose now that $\delta\mathbf{n}$ represents a *real* variation of the director and introduce the real rotation $\omega_i$ which transforms it from $\mathbf{n}$ to $\mathbf{n} + \delta\mathbf{n}$. By definition:

$$\mathbf{n} = \boldsymbol{\omega} \wedge \mathbf{n}. \tag{3.11}$$

By choosing $\boldsymbol{\omega}$ perpendicular to both $\mathbf{n}$ and $\mathbf{n} + \delta\mathbf{n}$, which does not at all restrict the generality, one gets:

$$\boldsymbol{\omega} = \mathbf{n} \wedge \delta\mathbf{n},$$

$$\omega_i = \varepsilon_{ijk}n_j\delta n_k, \tag{3.12}$$

$$\delta n_i = \varepsilon_{ijk}\omega_j n_k.$$

Substitute this expression of $\delta\mathbf{n}$ into equation (3.8). Denoting as $\Delta W$ the real variation of free energy in the field of infinitesimal rotations $\boldsymbol{\omega}(\mathbf{r})$, it becomes:

$$\Delta W = \int \boldsymbol{\omega} \cdot (\mathbf{h} \wedge \mathbf{n}) \, d^3\mathbf{r} + \int (\mathbf{C} \cdot \boldsymbol{\omega}) \cdot d\mathbf{S}. \tag{3.13}$$

$\Gamma = \mathbf{n} \wedge \mathbf{h}$ ($= 0$ in equilibrium) can then be interpreted as the (bulk) torque exerted by the medium on the director $\mathbf{n}$. The tensor $\mathbf{C}$ has been introduced:

$$C_{ij} = \frac{\partial F_d}{\partial n_{k,i}} \varepsilon_{kjl} n_l. \tag{3.14}$$

$C_{ij} \, dS_i$ represents the (surface) torque exerted by the exterior medium on the surface element $dS$. Note $\chi_j$ this exterior torque, measured per unit surface. When the medium is in equilibrium, the boundary conditions read:

$$C_{ij} \nu_i = \chi_j, \tag{3.15}$$

where $\mathbf{v}$ is the unit vector normal to the boundaries and directed towards the exterior.

### 3.2.4 One constant approximation

Equations (3.8) will not be discussed in the general case where $F_d$ is given by equation (3.3). These non-linear equations are difficult to deal with except in cases of high symmetry. It is usual to make a *one constant approximation* ($K = K_1 = K_2 = K_3$). The free energy density is then written:

$$F_d = \tfrac{1}{2} K \left[ (\operatorname{div} \mathbf{n})^2 + (\operatorname{rot} \mathbf{n})^2 \right] = \tfrac{1}{2} K (n_{j,j} n_{k,k} - n_{k,j} n_{j,k} + n_{k,j} n_{k,j}).$$

If we notice that

$$n_{j,j} n_{k,k} = (n_{j,j} n_k)_{,k} - n_{j,jk} n_k,$$

$$-n_{j,k} n_{k,j} = -(n_{j,k} n_k)_{,j} + n_{j,jk} n_k,$$

then

$$F_d = \tfrac{1}{2} K (n_{k,j} n_{k,j}) + \tfrac{1}{2} K (n_{j,j} n_k - n_{k,j} n_j)_{,k}. \tag{3.16}$$

The second term of equation (3.16) is of the type $g_{k,k}$. In the total free energy $\int F_d \, d^3 \mathbf{r}$ its contribution is reduced to a surface term in view of Stokes' theorem. It is therefore sufficient to minimize the reduced free energy, whose density reads:

$$F_d = \tfrac{1}{2} K n_{k,j} n_{k,j} = \tfrac{1}{2} K \left[ \left( \frac{\partial \mathbf{n}}{\partial x} \right)^2 + \left( \frac{\partial \mathbf{n}}{\partial y} \right)^2 + \left( \frac{\partial \mathbf{n}}{\partial z} \right)^2 \right] \tag{3.17}$$

to obtain the equilibrium equations for the one constant approximation. These are written:

$$K \, \Delta \mathbf{n} = -\lambda \mathbf{n}, \tag{3.18}$$

while the surface torque tensor takes the form:

$$C_{ij} = K \varepsilon_{kjl} n_{k,i} n_l. \tag{3.19}$$

### 3.2.5 One constant elasticity: planar case

In this section, situations will be considered where the director remains parallel to a fixed plane. This hypothesis corresponds to cases which we have already quoted:

    (a) wedge disclinations (Figure 2.18);
    (b) twist disclinations (Figure 2.21); and
    (c) planar loops (Figure 2.22).

Let $\phi(x, y, z)$ be the angle of the director with a fixed direction in the $xy$ plane. We have:

$$n_x = \cos \phi; \qquad n_y = \sin \phi,$$

which expressions, according to equation (3.16), lead to

$$F_d = \tfrac{1}{2}K(\nabla\phi)^2. \tag{3.20}$$

Minimizing $W$ leads directly to the equation

$$\Delta\phi = 0, \tag{3.21}$$

which is proved to be equivalent to equation (3.18) with

$$\lambda = (\nabla\phi)^2 = \frac{2F_d}{K}.$$

Equation (3.21) possesses a whole class of singular solutions in two dimensions of the type

$$\phi = S\theta + \phi_0, \tag{3.22}$$

where $\theta$ is the azimuthal coordinate in the $x,y$ plane. $S$ must be integral or half-integral, since $\phi$ must vary by an integral multiple of $\pi$ when a closed circuit is described around the origin. $S$ is therefore the strength of the wedge disclination with axis $Oz$. The schemes of Figure 2.18 are easily recognizable. Note that $S = 1$ is peculiar; $\phi_0 = 0$ gives the radial configuration and $\phi_0 = \pi/2$ the circular configuration, the intermediary values of $\phi_0$ giving spiral figures. For $S \neq 1$, changing $\phi_0$ means turning the figures through an angle equal and opposite to $\phi_0$.

Planar twist disclinations are obtained by taking $n_x = 0$; $n_y = \cos \phi$; $n_z = \sin \phi$; $\theta$ now designating an azimuthal angle in the $y,z$ plane, we have again:

$$\phi = S\theta + \phi_0. \tag{3.23}$$

This is the case of Figure 2.22b.

Finally, $\Delta\phi = 0$ has three-dimensional solutions (Friedel and de Gennes, 1969) of the type:

$$\phi = \frac{S}{2}\Omega\,(\mathbf{r}), \tag{3.24}$$

where $\Omega(\mathbf{r})$ designates the solid angle at which any closed line $(L)$ is seen from point $\mathbf{r}$ (Figure 3.3).

Along a closed circuit $(\gamma)$ surrounding $L$ once and passing through its interior, $\phi$ varied by a quantity $2\pi S$. $S$ is therefore the strength of $L$ and must be taken as equal to an integer or half-integer.

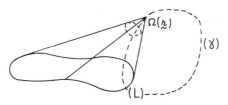

Figure 3.3. Dislocation line $L$. The director is placed on $\mathbf{r}$ at an angle $(S/2)\Omega(\mathbf{r})$ with respect to a fixed direction, while remaining parallel to a fixed plane. $(\gamma)$ designates a Burgers circuit

### 3.2.5.1   Free energy

The discussion will at first be restricted to the case where the line is entirely of wedge or twist character. According to (3.20) and (3.22) the free energy density is written:

$$F_d = \frac{K\,S^2}{2}\frac{1}{r^2}. \tag{3.25}$$

This quantity becomes infinite on the core. A core radius $r_c$ is introduced. By an easy calculation the total energy of the line (per unit length) is written in a way similar to that of a screw dislocation in a solid:

$$W = \pi K S^2 \ln R/r_c, \tag{3.26}$$

which can usefully be compared to the energy of a screw dislocation:

$$W = \frac{\mu b^2}{4\pi} \ln R/r_c. \tag{3.27}$$

The dependence on $S^2$ ($\sim b^2$) will be noticed. This indicates that the most stable planar rotation dislocations are of low strength ($S = \pm\tfrac{1}{2}$). It will also be noted that, on introducing numerical values, the orders of magnitude of (3.26) and (3.27) are comparable.

If the geometry is not planar, the preceding conclusion concerning stability is no longer valid. This will be discussed below.

Planar loops of the type shown in Figure 3.3 can easily be obtained in Mauguin twisted areas. They relax the twist when the angle of twist $q_0 h$ is large ($q_0 = 2\pi/p_0$, $p_0$ being the pitch of the twist). If this imposed twist is taken into account, the angle is written:

$$\phi = \frac{S}{2}\Omega(r) + q_0 z.$$

Friedel and de Gennes (1969) have shown that the total energy of the loop is then:

$$W = K\left(\frac{\pi}{8}L - \pi q_0 \Sigma\right),$$ (3.28)

where $L$ is the self-induction of the loop, using the analogy of an electrical circuit, and $\Sigma$ the surface of the loop projected on the plane perpendicular to the axis of twist. For a circular loop of radius $R$, we have $L = 4\pi R \ln R/r_c$ and $\Sigma = \pi R^2$. By minimizing (3.27) with respect to $R$ we obtain the critical radius

$$R_c \sim \frac{1}{4q_0} \ln \frac{1}{q_0 e_c}$$ (3.29)

above which the loop grows spontaneously. The activation energy of this process is given by:

$$E \sim \frac{K}{q_0}\left(\frac{\pi}{4}\ln\frac{1}{q_0 r_c}\right)^{1/2}.$$ (3.30)

$E$ would be very greatly reduced by making the dislocation diffuse, but this is only possible for integral values of $S$, and in this case the line almost certainly carries singular points (see below). It is possible that the nucleation of loops of half-integral values of $S$ is favoured by defects in the walls.

### 3.2.5.2   Core energy

The most probable hypothesis on the nature of core energy is that the core consists of an isotropic liquid. The molecular energy is therefore of the order of $k_B \Delta T_c$, where $k_B$ is Boltzmann's constant and $\Delta T_c$ the difference in temperature at the nematic–isotropic point of transition. Let $r_c$ be the radius of the core, $M$ the molecular mass, $N$ Avogadro's number and $\rho$ the density of the liquid. The total energy per unit line is written:

$$W = \pi K S^2 \ln \frac{R}{r_c} + k_B \Delta T_c \pi r_c^2 \frac{\rho N}{M}.$$ (3.31)

$W$ is minimized in relation to $r_c$; this gives:

$$r_c^2 = \frac{S^2}{2} \frac{M}{\rho N} \frac{K}{k_B \Delta T_c}.$$ (3.32)

Let us take: $\rho = 1$, $M/N = \rho a^2 l$ ($a = 5 \times 10^{-8}$ cm, $l = 20 \times 10^{-8}$ cm) $K = 10^3$ K, $k_B \Delta T_c = 10$ K ($10°$ below $T_c$). This gives:

$$r_c \sim 10^{-6} \text{ cm (i.e. 100 Å).}$$ (3.33)

It is a microscopic region far from $T_c$, but it quickly becomes predominant in the vicinity of $T_c$. In practice, it is possible that the properties of an isotropic core are linked to impurities which could preferentially diffuse in such regions.

Fan (1971) proposed a core model in which the amplitude of the order parameter would vary continuously and would tend towards zero (isotropic liquid) on the axis. The variation can be calculated using the Landau–Ginsburg expansion of free energy (Ginsburg, 1955).

### 3.2.6 One constant elasticity: general formulae of dislocation theory. Peach and Koehler force

One constant elasticity is a very particular case of elasticity of torques and rotations, but it is conceptually interesting as it has strong analogies with the usual elasticity of stresses and strains. First of all, note that $C_{ij,i} = 0$;

$$C_{ij,i} = K\varepsilon_{kjl}(n_{k,ii}n_l + n_{k,i}n_{l,i}) = 0. \tag{3.34}$$

In equation (3.34) the first term is zero because of the equations of equilibrium (3.18) and the second because it is the product of a symmetrical tensor and an antisymmetrical tensor. Equation (3.34) implies that a quantity of the type

$$\int_L C_{ij}\,dS_i,$$

integrated over a surface element $S$ only depends on the contour $L$ bounding $S$, and not on the choice of $S$. Now consider a contour $L$ in a perfect nematic and carry out the de Gennes process on the cut surface $S$ of $L$. A dislocation line $\Omega$ is obtained which creates a field of surface torques $C_{ij}(\mathbf{r})$. $C_{ij}(\mathbf{r})$ is proportional to $\Omega$. Let us indeed denote $d\omega_i$ the rotation which transforms $\mathbf{n}(\mathbf{r})$ to $\mathbf{n}(\mathbf{r} + d\mathbf{r})$. According to (3.12):

$$d\omega_i = \varepsilon_{ijk}n_j\,dn_k$$
$$= \varepsilon_{ijk}n_jn_{k,l}\,dx_l. \tag{3.35}$$

The tensor $K_{li} = \varepsilon_{ijk}n_jn_{k,l}$ is the analogue of the strain tensor of solids. Obviously:

$$C_{li} = KK_{li}. \tag{3.36}$$

This equation is quite similar to the one which links stress and strain:

$$\sigma_{ij} = C_{ijkl}e_{kl} \tag{3.37}$$

and possesses the same properties of linearity. Also, pushing the analogy further:

$$F_d = \tfrac{1}{2}Kn_{k,j}n_{k,j} = \tfrac{1}{2}K_{li}C_{li}. \tag{3.38}$$

$K_{li}$ is therefore the analogue of the $e_{ij}$; $C_{li}$ of the $\sigma_{ij}$.

The quadratic character of $F_d$ with respect to the angular gradients $K_{li}$ implies that the equilibrium equations $C_{ij,i} = 0$ are linear (equivalent to equation (3.18)) with respect to the same gradients, indicating that the solutions to these equations are proportional to $\Omega$, just as the $e_{ij}$ strains are proportional to $\mathbf{b}$.

The self-energy of a dislocation line $L$ can therefore be calculated by applying reversibly to the molecules of its cut surface an angular variation $s\Omega_i$, where $s$ is a parameter which varies from 0 to 1. The resulting surface torques are equal to $sC_{ij}$. The work done on the cut surface between the values $s$ and $s + ds$ is equal to:

$$\int_S sC_{ij}\,ds\Omega_j\,dS_i$$

which, integrated between $s = 0$ and $s = 1$, gives the formula:

$$W = \frac{1}{2}\int_S c_{ij}\Omega_j\,dS_i, \qquad (3.39)$$

which is similar to the formula giving the self energy a dislocation line in a solid:

$$W = \frac{1}{2}\int_S \sigma_{ij}b_j\,dS_i. \qquad (3.40)$$

In the same way, the interaction energy between two lines $L(C_{ij}, \Omega_j)$ and $L'$ $(C'_{ij}, \Omega'_j)$ is equal to:

$$W_1 = \int_{S'} C_{ij}\Omega'_j\,dS'_i = \int_S C'_j\Omega_j\,dS_i. \qquad (3.41)$$

Finally, the existence of a formula analogous to that of Peach and Koehler should be noted. The configurational force $\mathbf{F}$ acting on a line $L$ in the field of torques $C_{ij}$ (applied torques or internal torques due to the presence of other lines) is given by:

$$F_i = \varepsilon_{ikl}C'_{kj}\Omega_{jl}t_l, \qquad (3.42)$$

where $\mathbf{t}$ is the unit vector tangent to the oriented line $L$.

*Exercises*

(1) Calculate the interaction between two wedge parallel lines of opposite sign situated at distance $d$ from each other. Show that they attract each other with a force $F = 2\pi KS^2/d$. Use formula (3.41). Compare the calculation with that of de Gennes (1975).

(2) Calculate the image force being exerted on a wedge line situated at a distance $h$ from the surface. Show that the interaction with the surface is repulsive if the orientation of the director is fixed on the boundary and that it is attractive if the boundary conditions at the surface impose $C_{ij}\,dS_i = 0$ (de Gennes, 1975).

## 3.3 NON-SINGULAR DISCLINATIONS OF INTEGRAL ORDER

### 3.3.1 Stability of planar disclinations under three-dimensional perturbations

The examples in the preceding section were restricted to planar dislocations. The stability of the planar situation under more general distortions involving the three components of the director has not been discussed. This problem has been tackled by Dzyaloshinskii (1970) and Anisimov and Dzyaloshinskii (1972). Their results can be summarized as follows.

(a) If the stiffness constants are no longer supposed equal, the only stable planar $S = 1$ are those for which $\phi_0 = 0$ (radial configurations) and $\phi_0 = \pi/2$ (concentric circle configurations). Lines with spiral configurations are forbidden. (Dzyaloshinskii, 1970).

(b) In an infinite medium, lines of integral strength are in fact entirely unstable, and tend to disappear. This can be shown from Figure 3.4, where a cross-section of a radial $S = +1$ dislocation is represented. It is clear that the molecules can be aligned along the axis (3.4b). This phenomenon leads by degrees to the disappearance of the line, unless the boundary conditions prevent the director from following the initial direction of the line (e.g. the homeotropic conditions of Figure 3.4b). However, for $K_2 > 2K_3$, $K_1 > K_3$ the planar concentric lines $S = +1$ are stable.

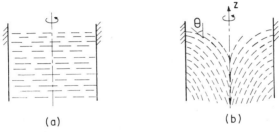

(a)    (b)

Figure 3.4. Instability of an $S = 1$ line under a three-dimensional perturbation

(c) Lines of half-integral strength are stable under three-dimensional perturbations. More precisely:

if $K_2 > \frac{1}{2}(K_1 + K_3)$, the planar wedge lines $S = \pm\frac{1}{2}$ are stable;
if $K_2 < \frac{1}{2}(K_1 + K_3)$, the twist lines $S = \pm\frac{1}{2}$ are stable but planar.

The latter case, which corresponds to values found in experiments, is the most interesting, but it has not yet been systematically studied.

In fact, as well as these arguments of *physical stability*, which are based on the examination of the variation of the free energy as a function of $K_1$, $K_2$ and $K_3$, there is also the consideration of *topological stability* which demonstrates a fundamental difference between the integral lines and half-integral lines. Consider a line $L$. Imagine a circuit traced around the line, and attach to each point on the circuit the molecule at that point. In this way a ribbon of width $a$ is constructed. (The circuit under consideration will be so chosen that it is nowhere tangent to a molecule.) In the case of a half-integral line, this ribbon has to be a Moebius ribbon, as it begins with a molecule **n** and returns to the same molecule, but of opposite sign, $-\mathbf{n}$. If this ribbon is reduced to zero, by leading it through transformations which conserve the topology to enclose $L$ more and more, an infinite number of molecular directions will remain on the ribbon (Bouligand).

The core of a half-integral line is therefore necessarily singular. The situation is quite different for an integral line: the ribbon constitutes an ordinary surface, which can be reduced while conserving its topology to a single molecule. This is illustrated by Figure 3.4b. The discussion of its topological stability will be taken up more generally in Chapter 10.

To summarize, the planar description is in general borne out for half-integral lines, and is insufficient only in the core zone, as described above. On the other hand, an entirely different theory is necessary for integral lines.

### 3.3.2 Lines ($S = +1$) without singular core (Cladis and Kléman, 1972a; Meyer, 1973a). One constant elasticity

Consider the cylindrical geometry of Figure 3.4b. Note $\theta$ the angle made by the director with the axis $Oz$, and $\phi$ the angle made by the projection of the director in the horizontal plane with a fixed direction (spherical coordinates). We shall be restricted to one constant elasticity. The free energy density (equation (3.17)) reads:

$$F_d = \frac{K}{2} [(\nabla\theta)^2 + \sin^2\theta\,(\nabla\phi)^2].$$  (3.43)

By minimizing the total energy $\int F_d\,d^3\mathbf{r}$ in the usual way, we obtain the equations of equilibrium:

$$\Delta\theta = \sin\theta\cos\theta\,(\nabla\phi)^2,$$
$$\Delta\phi + 2\cot\theta\cdot\nabla\theta\cdot\nabla\phi = 0.$$  (3.44)

Suppose that the symmetry is cylindrical and the projections of $\mathbf{n}$ radial, then $\theta = \theta(r)$, $\phi = \theta_r$ (where $\theta_r$ is the polar angle of point $(x, y)$ in the horizontal plane) and we have

$$r\frac{d}{dr}\left(r\frac{d\theta}{dr}\right) - \sin\theta\cos\theta = 0.$$  (3.45)

Equation (3.45) gives the three-dimensional behaviour we were looking for. By putting $R(d\theta/dr) = u(\theta)$, it is easy to obtain the first-order integral:

$$r^2\left(\frac{d\theta}{dr}\right)^2 = A^2 - \cos^2\theta$$  (3.46)

where $A$ is a constant of integration. The integration of the equation must also take into account the boundary condition $\theta(R) = \pi/2$, where $R$ is the radius of the cylinder. The nature of the solutions depends on the value of $A$. Figure 3.5 shows the variation $\theta = \theta(r)$;

*For $A > 1$:* the molecule turns indefinitely, and always in the same direction, approaching the axis $r = 0$. It can be shown that the rotation becomes faster and faster:

$$r(\theta)/r(\theta + 2\pi) = \exp[2\sin\alpha K(\alpha)],$$  (3.47)

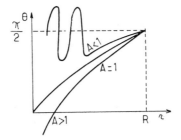

Figure 3.5. Different solutions of equation (3.46)

where $(\sin \alpha)^{-1} = A$, $K(\alpha)$ being the complete elliptical integral of the first kind. The details of the calculations have no great interest here; this solution is obviously to be rejected, as it leads to infinite energy on the axis.

*For A < 1:* the molecule turns indefinitely, changing direction each time $\theta$ attains the value $\arcsin A$. Here also the speed of rotation becomes faster and faster, and this solution is to be rejected for the same reasons as the former.

*For A = 1:* We have the non-singular solution

$$r = R \cot \left| \frac{\pi}{2} - \frac{\theta}{2} \right|, \tag{3.48}$$

where $r$ tends towards zero as $\theta \to 0$.

The calculation of the energy from equation (3.48) leads to

$$W = 2\pi K. \tag{3.49}$$

This expression does not involve any core radius as the line is not singular, and does not involve $R$. Equation (3.48) indicates moreover that the different solutions are homothetic to each other. In a term borrowed from the theory of dislocations in solids, equation (3.48) can be said to represent a core *spread* over all the available space.

The foregoing results are extended to any integral line of strength $S$. Generally it is found that

$$\phi = S\theta_r + \phi_0,$$

$$\left( \frac{r}{R} \right)^{|S|} = \cot \left| \frac{\pi}{2} - \frac{\theta}{2} \right|, \tag{3.50}$$

$$W = 2\pi K |S|.$$

Note also that the energy is proportional to $|S|$, and not $S^2$ as in the planar case.

Figure 3.6, represents in projection on the $x,y$ planes, integral $S$ configurations with an entirely spread core.

This type of calculation can be extended to twist lines (Meyer, 1973a). Thus, the configuration in Figure 3.7 represents a twist line $|S| = 1$ of energy $2\pi KS$, which does not have core singularity.

52

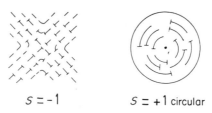

$S = -1$        $S = +1$ circular

Figure 3.6. Configuration of integral wedge lines with spread core

Figure 3.7. $|S| = 1$, twist line

### 3.3.3 Lines ($S = +1$) without singular core. Three constants elasticity (Cladis and Kléman, 1972a)

The radial $S = +1$ involves splay ($K_1$) and bend ($K_3$). It is not difficult to show that there is a non-singular solution which takes into account the difference between $K_1$ and $K_3$. We obtain:

$$K_1 < K_3 : W = \pi\left(K_1 + K_3\frac{k}{\tan k}\right); \qquad \tan^2 k = \frac{K_3 - K_1}{K_1},$$

$$K_1 > K_3 : W = \pi\left(K_1 + K_3\frac{k}{\tanh k}\right); \qquad \tanh^2 k = \frac{K_1 - K_3}{K_3}. \tag{3.51}$$

The circular $S = +1$ involves bend and twist deformations ($K_2$). Here also there are non-singular configurations, whose energy reads:

$$K_2 < K_3 : W = \pi\left(K_2 + K_3\frac{k}{\sin k}\right); \qquad \sin^2 k = \frac{K_3 - K_2}{K_3},$$

$$K_2 > K_3 : W = \pi\left(K_2 + K_3\frac{k}{\tanh k}\right); \qquad \sinh^2 k = \frac{K_2 - K_3}{K_3}. \tag{3.52}$$

The expressions of the deformations can be found in the article quoted. Differences in the energy values between this article and equations (3.51) and (3.52) will be noted. These are due to the fact that the free energy terms which are integrated into surface terms have been neglected here.

### 3.3.4 Singular points

An examination of Figure 3.4b indicates that the tilting of the director towards the axis can be made either towards the $z > 0$ or towards the $z < 0$ (as shown). If the two cases in the diagram are present on the same line, singular points of two types are obtained (Figure 3.8).

Other types exist for other integral lines. These singular points are very frequently observed. In fact, it now appears that G. Friedel's nuclei (1922), which are obtained in great numbers when a thin sample of nematic is placed between two glass slides with degenerate anchoring, and which have been described up to now as lines of integral strength, always carry singular points.

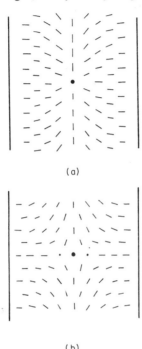

(a)

(b)

Figure 3.8. Singular points on a wedge $S = +1$

Saupe (1973) has shown that, in the one constant approximation, equation (3.44) possesses solutions representing singular points (for integral values of $S$ only):

$$\phi = S\theta_r + \phi_0$$

$$\tan\frac{\theta}{2} = \left(\tan\frac{\lambda}{2}\right)^{\pm|S|}$$

(3.53)

where $\lambda$ is, in spherical coordinates, the angle between the $O_z$ axis and the point where the director makes an angle $\theta$ with $O_z$. This solution obviously contains the two configurations of Figure 3.8a and b, in the immediate vicinity of the singular point. In fact for $S = 1$:

$$\theta = \pm\lambda$$

(3.54)

i.e. for $\theta = \lambda$, the 'radial' singular point, and for $\theta = -\lambda$, the 'hyperbolic' singular point. These qualifications relate to the configuration in a meridian section. The projection of the configuration in the equatorial plane is that of a planar $|S| = 1$.

For $S = -1$ the two configurations $\theta = \pm\lambda$ are equivalent. In fact, consider two rectangular meridian sections, one for $\theta_r = 0$, the other for $\theta_r = \pi/2$, and suppose that $\phi_0 = 0$; we obtain:

(a) for $\theta_r = 0$: $n_x = \pm\sin\lambda$; $\qquad n_z = \cos\lambda$,

$$ \tag{3.55} $$

(b) for $\theta_r = \pi/2$: $n_y = \pm\sin\lambda$; $\qquad n_z = \cos\lambda$.

If the first configuration $(\theta_r = 0)$ represents a radial point, the second $(\theta_r = \pi/2)$ represents a hyperbolic point, and vice versa.

In practice singular points are observed in the vicinity of free surfaces (growth of nematic spherolites in isotropic phase; free drops, described by Meyer, 1972a). Their existence is in this case fundamentally linked to the isotropic nature of surface tension (see below).

They have also been observed in capillary tubes, in which a nematic liquid crystal is introduced with homeotropic anchoring (Williams *et al.*, 1972). Melzer and Nabarro (1977) have studied the motion of these points in the same geometry and concluded that the equilibrium configuration of a singular point may be a small disclination loop of cylindrical symmetry, and of strength $S = \pm\frac{1}{2}$. These authors have also been able to produce circumferential surface orientation in the capillary. In that case a circular $S = +1$ appears along the axis of the tube. Singular points are visible at the junction of two regions of opposite helicities; their configurations are not so easily described as in the former case: they are alternatively *noeuds* and *cols*, terminology originating in a classification (which was first given by Poincaré, 1886) of the singularities of a velocity field.

## 3.4 SURFACE DISCLINATIONS

The study of mutual interactions between liquid crystals and the surfaces on which they are deposited is of primary importance, both in fundamental theory and in applications. First, a classification will be given of the different types of substrate surfaces. This will be done as a function of the nature on the anchoring of the nematic and *anchoring energy* will be defined which will be distinguished from surface tension. This energy can be reached through a study of *surface defects*, which will be defined precisely below. The elastic theory of these dislocations, which generally have their cores spread out across the surface, bears a fundamental analogy with the theory of splitting of ordinary dislocations (Peierls–Nabarro). These studies will be followed by the discussion of some particular cases (effect of a magnetic field on a free surface, singular points on a surface...) which use the same methods. Finally, two interesting phenomena linked to anchorings of very weak energy will be discussed: the observation of walls and the existence of quasibidimensional anchoring transitions.

### 3.4.1 Preparation of surfaces

The substrate–nematic interaction is defined first by the direction which it imposes on the molecules in the vicinity of the substrate, and secondly, by the nature of the substrate. There have been experimental attempts to obtain homogeneous substrates which permit uniform orientations over macroscopic distances, and are therefore capable of supporting nematic single crystals.

The orientation of the director on the substrate is defined by two angles, for example, using spherical coordinates, the angle $\phi$ ($0 < \phi < \pi$) with the normal to the surface and the azimuth $\theta$ measured in the plane of the surface in relation to a fixed direction in the substrate. It should be noted that the angles $\phi$ and $\pi - \phi$, and $\theta$ and $\theta + \pi$ designate the same molecular direction. Anchoring will be said to be *homeotropic* if $\phi = 0$ (molecules normal to the surface) and *planar* if $\phi = \pi/2$. If the surface imposes a definite direction $\theta_0$, $\phi_0$ on the nematic, then the term *easy direction* will be used; if $\theta_0$ is not fixed, then the terms *conic anchoring* and *degenerated easy direction* will be used.

The principal substrates are the following.

*Organic supports.* Monomolecular layers of soaps or lipids are deposited on glass in the following way: the polar heads of the molecules become attached to the glass (strong dipolar interaction) with the paraffinic tails perpendicular to the layer (Figure 3.9). In the case of long paraffinic chains, with high densities of polar heads, these layers determine a *homeotropic* anchoring, which can be attributed essentially to steric factors.

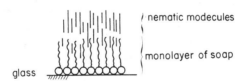

Figure 3.9. Homeotropic anchoring on a monolayer of detergent (or phospholipid)

If the density of these polar heads is lower (the area occupied by a polar head is of the order of 25 Å), and if the paraffinic tails no longer form a compact assemblage, then the latter can become deformed. Thus, the introduction by capillary action of a nematic between the two layers lays down the tails in the direction of flow, the angle obtained not depending on the speed of flow, as an equilibrium is attained. *Chatelain* anchoring, which has already been mentioned, corresponds to the situation where the density is sufficiently low for the organic molecules to spread themselves completely over the substrate.

A detailed study of the influence of the density of molecules of hexadecyltrimethylammonium bromide (HTAB), with discussion of the nature of the forces coupling the nematic molecules to those of detergent has been made by Proust *et al.* (1972, 1978). According to the density of detergent and the nature of the nematic, anchoring can be either planar or homeotropic. Porte (1976), on the other hand, has studied the oblique contact of **MBBA** with a short-chained surface active agent. Such sorts of study have also developed because of applications to display systems of increased longevity (for a review see Guyon and Urbach, 1976).

Let us mention here two other types of anchoring which are not understood.

(1) Ryschenkow's (1975) amorphous layers, obtained by evaporation on glass of the products of the breakdown of heated paper. An extremely *weak* degenerated anchoring is obtained.

(2) Friedel and Grandjean's surface films (G. Friedel, 1922): if a nematic is introduced by capillary action onto a very clean glass plate, the direction of anchoring is linked to that of flow and remains so even after heating of the nematic above the clearing point.†

*Crystalline substrates.* Grandjean (1916) made a very detailed study of the various types of liquid crystals on cleavage layers of many crystals. The simple crystallographic directions are the determining factors here.

Thus on a cleavage [001] of rock salt, PAA is directed along the two diagonals (110) and (1̄10). On the same cleavage, anisalzadine‡ makes a small angle ($\sim 10°$) with these directions. The electrical effects are probably dominant: the electron rich centres of the two molecules (benzene rings, free doublet of nitrogen) can match spatially with the $Na^+$ cations. The same applies to MBBA (benzene rings).

On micas (cleavage [001] of phlogopite, for example) nematics of the PAA group (different from PAA by simple substitution at the end of the chains) are oriented along the three directions of the ternary system. Anisalzadine is oriented at 30° from these directions. The distances between cationic sites measured along ternary directions are 5.4 Å (cf. PAA 5–8 Å for distances between benzene sites); at 30° from the ternary directions these distances are of the order of 10 Å (cf. anisalzadine: 8–11 Å).

Another interaction mechanism (Rapini, 1972a) is the coupling between the vibrations of the surface lattice (oscillating dipole) and the dipole induced into the molecule of the nematic (this involves the high electrical polarizability of the very mobile $\pi$ electrons).

*Mechanically prepared substrates.* Substrates on which a film of Au or SiO is evaporated at oblique incidence orient many liquid crystals very strongly along unidirectional planar directions or at an angle to the surface (Guyon and Urbach, 1976). The same applies to mechanically diamond polished surfaces.

---

† The nematic–isotropic transition temperature is called the *clearing point* indicating that it is characterized by the sudden disappearance of the opalescence of the nematic.

‡ Anisalzadine (see G. Friedel, 1922)

$$CH_3-O-\bigcirc -CH=N-N-CH-\bigcirc -O-CH_3$$

solid $\longleftrightarrow$ nematic $\longleftrightarrow$ isotropic
160°C 180°C

Chemical polishing gets rid of organic debris of all kinds; on a glass prepared in this way and dried, a weak homeotropic anchoring can be obtained. However, surface hydrolysis generally destroys it very rapidly and this leads to a degenerate planar anchoring (Friedel and Grandjean's films).

*Contact with air.* Experiments indicate that contact with air causes a strong anchoring, often homeotropic or at a small angle to the normal at the boundary.

*Contact with water.* Thin films of MBBA on water have been studied by Proust *et al.* (1978).

### 3.4.2 Distinction between surface tension and anchoring energy

Consider a zone of surface $S$ where the nematic covers a certain substrate, and suppose that thermodynamic equilibrium has been attained. A reversible variation $dS$ of the surface $S$ is now made. The variation in energy of the nematic includes a bulk term and a surface term. The variation of the latter can be written:

$$dW = \gamma \, dS. \tag{3.56}$$

$\gamma$ is the surface tension coefficient of the liquid–substrate contact. It is a quantity which depends on these two elements.

$\gamma$ can be interpreted as a configurational force per unit line, which is exerted on the contour bounding $S$. We can in fact write $dS = ds \cdot dl$, where $ds$ is the element of arc of the contour and $dl$ its perpendicular displacement to $ds$ (Figure 3.10a).

This leads to the principle of measurement of surface tensions by measurement of angles of contact (Figure 3.10b). We have

$$\gamma_b \cos \theta = \gamma_a \tag{3.57}$$

where $\gamma_a$ designates the surface tension at liquid substrate contact and $\gamma_b$ the surface tension at liquid–air contact.

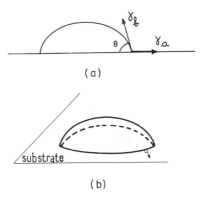

(a)

(b)

Figure 3.10. (a) principle of the measurement of surface tension; (b) surface tension is exerted on the contour of the drop. If $\gamma < 0$ its direction is along the outward normal

58

Must it be considered that surface tension in a nematic depends (for a given substrate) on the orientation of the nematic on the substrate? G. Friedel in his classic article of 1922 stresses the fact that surface tension in a nematic is an isotropic quantity. This is its essential difference from surface tension in an ordinary crystal, which depends on the orientation of the faces. Friedel used the following experimental argument in support of this: if a nematic is grown in its isotropic phase, the globules which appear are *spherical*. On the other hand, the growth of an ordinary crystal is characterized by the appearance of connecting faces, the relative size of which depends on the anisotropy of the surface tension, faces with weak tension having a larger surface.

If this tendency for ordinary crystals to grow along well-defined faces is known as the anisotropy of surface tension (which supposes that the unique orientation of a crystal can be defined) it is quite clear that this concept is not true for a nematic, as surface energy depends on the *relative* orientation of the nematic in relation to the substrate. For the tendency to form spheres in the above case is indicated by the very fact that the orientation which the nematic takes spontaneously is that which is imposed *locally* by the substrate, i.e. that of minimal surface energy. Thus, for homeotropic orientation, the globule will necessarily be a *deformed* nematic crystal, which also possesses a singular point at its centre (Figure 3.11a). However, in the case of planar orientation, the globule of deformed nematic will necessarily possess singular points, the sum of whose strengths will be Euler's characteristic of the sphere (cf. end of Chapter 2). In this second case, the surface energy of a sphere $R$ leads to a natural separation of the contribution of surface tension $\sim 4\pi R^2 \gamma$, from the contribution of singular objects $W_{\text{sing}}$.

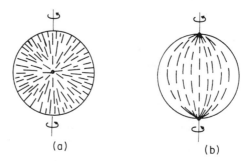

Figure 3.11. Nematic globules: (a) with central singular point; (b) with singular points at polar positions

The positive variation of surface energy which appears when the molecule takes a direction $\theta$, $\phi$ will be referred to as the *anchoring energy function* $W(\theta, \phi)$. This is different from the direction of minimum surface energy. By definition $W(\theta_0, \phi_0) = 0$.

The term *anchoring energy* will be given to the maximal value of $W(\theta, \phi)$ i.e. the energy which must be expended for the molecule to go along the most difficult direction. Whereas (isotropic) surface tensions are of the order of 20 to 100 erg/cm$^2$, the anchoring energies vary between $10^{-4}$ and $10^{-2}$ erg/cm$^2$, i.e. between about $10^{-6}$ eV/molecule and $10^{-4}$ eV/molecule.

Planar anchorings will be dealt with in detail and a first approximation to $W$ will be in the form:

$$W = W_s \sin^2 \theta, \tag{3.58}$$

where $W_s$ is the anchoring energy.

### 3.4.3  Definition of a surface disclination (Vitek and Kléman, 1975)

Examples have been given of singular points on the surface. The existence of points dislocations in general requires quite a strong degeneracy of the anchoring. If this is not the case, surface defect *lines* are more likely to be formed and may be marked in places by point defects.

The following experiment (which is not just conceptual) clarifies this concept. A *solid* crystal of a material which has a nematic phase and which possesses a *grain boundary* adjoining the substrate is placed on a glass slide (Figure 3.12a). The molecules in each grain have a well defined orientation; let $\Delta\theta$ be their disorientation from one grain to another. The crystal is heated until it reaches the mesomorphic phase. If the substrate has not been specially treated, it will very probably form Grandjean and Friedel's films, and the final configuration is shown in Figure 3.12b, where the boundary has disappeared through viscous relaxation of the orientation of the molecules. Meanwhile, a *line* remains on the substrate, which can be considered as the intersection of a wall (now spread throughout the mass) and the substrate.

(a)                              (b)                              x

Figure 3.12. Surface lines as intersection of a disorientation wall and a surface

These surface lines display the following characteristics, which differentiate them from bulk lines:

(a) They can have any order $\Sigma = \Delta\theta/2\pi$, as Figure 3.12 indicates. A complementary line of order $\Sigma_c = S - \Sigma$, where $S$ is an integer or half-integer, can be traced in the virtual medium on the outside of the sample. As $\Delta\theta$ can be infinitely small, it can be seen that it is possible to define a density of surface lines $\partial\theta/\partial x$, while such a concept has little sense in the bulk.

(b) As a result of the existence of this density, the cores of surface dislocations may always be spread out, thus causing the disappearance of all singularities in the director. This fact can be explained by saying that $S = \Sigma_c + \Sigma$ can always be chosen with integral strength. In particular, the bulk lines of half-integral strength can, in certain circumstances, be attached to the surface and thus cause the disappearance of their core singularity (which is necessary in the bulk).

(c) There is therefore a *topological barrier* to the detachment of a singularity from the surface. This barrier is infinite when $\Sigma$ is neither integral nor half-integral (as in the diagram) and such a line can only be detached by the nucleation of other singularities, such as singular points.

Like bulk lines, surface lines obey summation rules at their intersections with each other or with bulk lines. These rules were recognized by G. Friedel. Figure 3.13 gives an example.

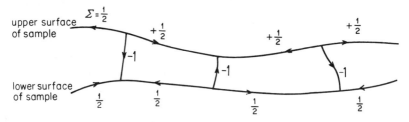

Figure 3.13. Relations of strength between surface lines and *oriented* bulk lines

The spreading of the core is a property similar to the Peierls–Nabarro model of the spreading of dislocations in their glide plane. This question can be dealt with by considering that the structure of the core results in a competition between the surface terms (anchoring) and bulk terms, such that infinite (core) energies do not appear anywhere. But first a simplified calculation of the energy and stability of surface lines will be given.

### 3.4.4 Lines or walls?

Let there be (Figure 3.14) a layer of nematic with parallel faces and planar anchoring ($W = W_s \sin^2 \theta$). The most obvious surface defects are of order $\Sigma = \frac{1}{2}$ which correspond to a distribution of $\mathbf{n}(\mathbf{r})$ in the vicinity of the surface itself and which essentially display bend and splay. If this configuration is extended throughout the material, a wall is obtained (Néel wall, in terminology borrowed from magnetism). This will lead to another surface defect on the opposite layer. This wall only exists if the interactions between defects on the opposing layers are strong enough for two defects of opposite sign to have a tendency to become parallel to each other due to attractive forces. If the layer is thick enough, it is to be expected that the defects would exist independent of each other on the opposing faces (thus the presence of the defect in Figure 3.14b necessarily leads to the presence of another defect, but this one may just as well be on one layer as on the other). This means therefore that their line energy should be independent of the

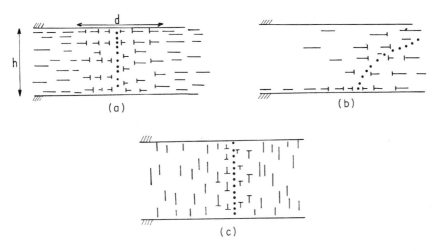

Figure 3.14. Relations between surface defects and walls; (a) Néel wall; (b) surface defect and diffuse wall; there necessarily remains a surface where **n** is perpendicular to the plane of the diagram; (c) Bloch wall

thickness $h$. Let $K$ be this energy of the order of magnitude of the stiffness of the nematic.

Where a wall exists, there are additional terms: a surface term of the order of $dW_s$ ($d$ = width of wall, $W_s$ = anchoring energy) and a bulk term of the order of $(K/d^2)hd = K(h/d)$, where the denominator $1/d^2$ is quite obviously of the order of magnitude of $(\text{div } \mathbf{n})^2$ (splay) or of $(\mathbf{n} \cdot \text{rot } \mathbf{n})^2$ (bend). Therefore, per unit length of the wall:

$$E_p = dW_s + K\frac{h}{d}, \tag{3.59}$$

where $d$ must take a value such that $E_p$ is a minimum, i.e.:

$$d = \left(h\frac{K}{W_s}\right)^{1/2}. \tag{3.60}$$

The total energy is then obtained by substituting this value of $d$ into the preceding expression:

$$E_p = 2(hKW_s)^{1/2} = 2K\left(\frac{2h}{b}\right)^{1/2}, \tag{3.61}$$

where we have introduced the length:

$$b = K/2W_s. \tag{3.62}$$

This length (known as the extrapolation length) plays an essential role in the physics of surface lines.

Compare $E_p$ to the case of the diffuse wall ($E_d \sim 2K$). It can be seen immediately that:

(1) if $h \leqslant b$, walls will be favoured; and
(2) if $h \geqslant b$, the wall will be diffuse.

Therefore walls will only be obtained for very thin samples. Note immediately that for $W_s \sim 1\,\mathrm{erg/cm^2}$, we have $b \sim 50\,\text{Å}$: case (1) can never be observed experimentally. Case (2) has been observed with Ryschenkow's anchoring ($b \sim 50\,\mu$).

Figure 3.14c shows a topology of surface lines in the case of homeotropic anchoring; the corresponding wall is called a Bloch wall.

### 3.4.5  Peierls–Nabarro model (Vitek and Kléman)

Let us take the $Oz$ axis as perpendicular to the substrate and suppose the problem is planar and one-dimensional. The azimuthal angle $\theta$ of the director depends only on $x$ and $z$. It will be supposed for maximum simplification that the sample is infinite in the direction $z > 0$. The configuration minimizes the energy:

$$F = \tfrac{1}{2}K \int_0^\infty \mathrm{d}z \int_{-\infty}^{+\infty} \mathrm{d}x \left[ (\nabla\theta)^2 + \frac{W(\theta)}{K}\,\delta(z) \right], \tag{3.63}$$

and after an easy calculation we obtain:

$$\nabla^2\theta = 0 \qquad \text{(in the bulk)},$$

$$\frac{\partial W}{\partial \theta} = K \left(\frac{\partial \theta}{\partial z}\right)_{z=0} \qquad \text{(on the surface)}, \tag{3.64}$$

with the boundary conditions:

$$\theta(x = +\infty, z) = 0; \qquad \theta(x = -\infty, z) = 2\pi\Sigma, \tag{3.65}$$

which characterize the presence of a defect along the $Oy$ axis on the surface. It is shown that on the surface $\theta$ obeys the integral equation:

$$-\frac{\mathrm{d}W}{\mathrm{d}\theta} = \frac{K}{\pi} \int_{-\infty}^{+\infty} \left(\frac{\partial \theta}{\partial x'}\right)_{z=0} \frac{1}{x - x'}\,\mathrm{d}x', \tag{3.66}$$

i.e. that the experimental data of the variation of the director allows us to obtain the surface energy function $W(\theta)$. $\mathrm{d}W/\mathrm{d}\theta$ is Hilbert's transform of $K\,(\partial\theta/\partial x')_{z=0}$.

*Demonstration* (Peierls). Put $\phi(x) = \theta(x,0)$ and expand $\phi(x)$ as a Fourier integral:

$$\phi(x) = \int_{-\infty}^{+\infty} \mathrm{d}k\, a_k \exp(\mathrm{i}k\,x).$$

If $\theta(x,z)$ is expanded in the same way:

$$\theta(x, z) = \int_{-\infty}^{+\infty} dk a_k g(z) \exp(ik\,x),$$

the bulk equation $\nabla^2 \theta = 0$ leads immediately to the relation:

$$\frac{d^2 g}{dz^2} - k^2 g(z) = 0,$$

i.e.

$$\theta(x, z) = \int_{-\infty}^{+\infty} dk a_k \exp(ikx - |k|\,z),$$

where the minus sign is chosen in order to ensure that $\theta(x, z)$ does not diverge for $z \to +\infty$.

Inverting the Fourier series, we have:

$$a_k = \frac{1}{2\pi} \int dx' \phi(x') \exp(-ikx').$$

Let:

$$\theta(x, z) = \frac{1}{2\pi} \int_{-\infty}^{+\infty} dk dx' \phi(x') \exp\{ik(x - x') - |k|\,z\}$$

$$= \frac{1}{2\pi} \int_0^{\infty} dk \int_{-\infty}^{+\infty} dx' \exp\{ik(x - x') - kz\} \phi(x') + \text{c.c.}$$

Integrating over $k$, we obtain:

$$\theta(x, z) = \tfrac{1}{2}\delta(x - x' + iz) + \text{c.c.} + \frac{1}{2\pi} \int_{-\infty}^{+\infty} dx' \phi(x') \frac{1}{z - i(x - x')} + \text{c.c.}$$

from which:

$$\frac{\partial \theta}{\partial z} = \frac{i}{2\pi} \int_{-\infty}^{+\infty} dx' \phi(x') \frac{\partial}{\partial x'} \left[ \frac{1}{z - i(x - x')} - \frac{1}{z + i(x - x')} \right]$$

where $\partial/\partial z$ is replaced by $\pm i \partial/\partial x'$.

This integral is transformed by an integration by parts, to obtain finally:

$$\frac{\partial \theta}{\partial z} = \frac{i}{2\pi} \int_{-\infty}^{+\infty} dx' \frac{\partial \phi}{\partial x'} \frac{1}{z - i(x - x')} - \text{c.c.}$$

which for $z = 0$ leads to the result we are seeking:

$$\left( \frac{\partial \theta}{\partial z} \right)_{z=0} = \frac{1}{\pi} \int_{-\infty}^{+\infty} \left( \frac{\partial \theta}{\partial x'} \right)_{z=0} \frac{1}{x - x'} dx'.$$

Let us compare this with the Peierls–Nabarro analysis for ordinary dislocations. $W(\theta)$ is the analogue of the fault energy $\gamma(u)$. $dW/d\theta$ is a restoring torque, while $d\gamma/du$ is a restoring force, $u$ being a displacement. The shear modulus $\mu$ is replaced by $K$, the stiffness constant of the nematic. $2\pi\Sigma$ is analogous to $\mathbf{b}$, the Burgers vector. Remember that the equations of equilibrium for a dislocation take the form:

$$\nabla^2\mathbf{u} = (1 - 2\nu)\,\text{grad div }\mathbf{u},$$

$$\frac{d\gamma}{du} = \mu\left(\frac{\partial u}{\partial z}\right)_{z=0}.$$

If we are dealing with a screw dislocation, we have only to consider that the component $u$ is along $y$, and put $\text{div }\mathbf{u} = 0$. The analogy is then complete.

As $dW/d\theta$ is a restoring torque in $x$, the fundamental integral equation can be interpreted in the following manner: $(\partial\theta/\partial x')_{z=0}$ is the surface (rotation) dislocation density on $x'$ and $(K/\pi)[1/(x - x')]$ is the torque exerted in the molecule at $x$.

Two important formulae must be added.

(a) The solution of the equation $\nabla^2\theta = 0$ which satisfies the boundary conditions can be written in the integral form:

$$\theta(x, z) = \frac{1}{\pi}\int \frac{z}{z^2 + (x - x')^2}\,\theta(x', z = 0)\,dx'. \tag{3.67}$$

(b) The total surface energy, due to variations in anchoring, does not depend on $W(\theta)$. We have:

$$E_s = 2\pi K\Sigma^2. \tag{3.68}$$

In fact, by definition:

$$E_s = \int_{-\infty}^{+\infty} W(\theta)\,dx = \int_0^{2\pi\Sigma} W(\theta)\frac{dx}{d\theta}\,d\theta.$$

Integrating by parts:

$$W(\theta)\frac{dx}{d\theta} = \frac{d}{d\theta}(xW(\theta)) - x\frac{dW}{d\theta},$$

from which:

$$E_s = -\int_0^{2\pi\Sigma} x\frac{dW}{d\theta}\,d\theta.$$

Replacing $dW/d\theta$ by its value given by the fundamental integral equation, we have:

$$E_s = \frac{K}{\pi}\int \frac{x}{x - x'}\left(\frac{\partial\theta}{\partial x}\right)_{z=0}\left(\frac{\partial\theta}{\partial x'}\right)_{z=0}\,dx\,dx'.$$

Changing $x$ into $x'$ and adding the two expressions of $E_s$ obtained in this way, yields:

$$E_s = \frac{K}{2\pi} \int \int \left(\frac{\partial \theta}{\partial x}\right)_{z=0} dx \left(\frac{\partial \theta}{\partial x'}\right)_{z=0} dx' = 2\pi K \Sigma^2.$$

### 3.4.6 Applications

Taking $W = W_s \sin^2 \theta$, the solution for $\Sigma = \frac{1}{2}$ reads:

$$\theta = \arctan \frac{z+b}{x}, \tag{3.69}$$

or again, by taking the origin back to the coordinate $z = -b$:

$$\theta = \phi,$$

where $\phi$ is the polar angle in these new axes: this is the equation for a *wedge line of strength S* = 1, situated at point $x = 0, z = -b$. This point being virtual, there are no singularities in the distribution $\theta(x)$ for $z = 0$. The calculation of line energy is not difficult, and we obtain:

$$W = \frac{\pi K}{2} \left(\ln \frac{R}{b} + 1\right), \tag{3.70}$$

where $R$ is the radius of the half-cylinder in which the energy density has been integrated. These results were obtained by Meyer (1972b).

*Commentary on the symmetry of the solution.* We have calculated the distribution $\theta(x, z)$ in a semi-infinite medium, by supposing a symmetrical anchoring energy function. The solution is therefore symmetrical due to both the symmetry of $W(\theta)$ and the symmetry of the boundary conditions. The following distinctions will be made:

An intrinsic symmetry (or asymmetry) which displays itself experimentally in the close vicinity of the core, due to the symmetry properties of $W(\theta)$. Ionic crystals should be good candidates for substrates with intrinsic asymmetry;

An extrinsic symmetry (or asymmetry) which appears over long distances due to the boundary conditions. Figure 3.14b gives an example.

### 3.4.7 Methods of measuring $W(\theta)$

The fundamental integral equation shows the method to be followed in the planar case. $\theta(x, z = 0)$ must be measured. Note that $\theta(x, 0)$ varies significantly in value on a distance which can be estimated by writing:

$$E_s = dW_s,$$

where $E_s$ is the surface energy. According to equation (3.68),

$$d \sim 2\pi \Sigma^2 \frac{K}{W_s} = 4\pi \Sigma b; \tag{3.71}$$

$d$ is therefore of the order of $b$. Only a small $W_s$ will therefore be easily measurable. In fact the lines always have greater width than that given by this estimation. This is due to the interaction between layers and the image force. Thus, for a layer of thickness $h$ and possessing a wall, we have (cf. above):

$$d \sim (hb)^{1/2} \tag{3.72}$$

a quantity which can clearly be much larger than the former. It is therefore generally convenient to take the image force into account. For various experimental results, see Guyon and Urbach (1976).

### 3.4.8 Mutual influence of two parallel substrates

The influence of the interactions between lines situated on two parallel substrates can be estimated if the thickness $2L$ of the layer is greater than $b$. The boundary conditions take the form:

$$K\frac{\partial \theta}{\partial z} = -\frac{dW}{d\theta} \qquad (z = -L),$$

$$K\frac{\partial \theta}{\partial z} = +\frac{dW}{d\theta} \qquad (z = -L). \tag{3.73}$$

Note $\phi_+(x)$ and $\phi_-(x)$, the distributions of the director at the boundaries:

$$\phi_+(x) = \theta(x, L); \qquad \phi_-(x) = \theta(x, -L). \tag{3.74}$$

By using the Peierls method, the integral form of the distribution $\theta(x, z)$ is written:

$$\theta(x, z) = \frac{1}{2\pi} \int_0^\infty dk\, dx' \frac{e^{-ik(x'-x)}}{\sinh 2\,kL} [\phi_+(x')\sinh k(L+z)$$

$$+ \phi_-(x')\sinh k\,(L-z)] + \text{c.c.} \tag{3.75}$$

and the boundary conditions (equation (3.73)) must be rewritten using this expression. The width of the surface dislocations is of the order of $b$ (or larger is the effects of walls are dominant). Therefore only the values of $k$ in the vicinity of $2\pi/b$ (or $2\pi/(Lb)^{1/2}$) count, but in any case, the inequality $kL \gg 1$ is satisfied. Equation (3.75) then takes the simplified form:

$$u_-(x) = \left(\frac{\partial \theta}{\partial z}\right)_{z=-L} = -\frac{1}{\pi}\int_{-\infty}^{+\infty} dx' \frac{\partial \phi_-}{\partial x'}\frac{1}{x-x'}$$

$$+ \frac{1}{\pi}\int_{-\infty}^{+\infty} dx' \frac{\partial \phi_+}{\partial x'}\frac{x-x'}{(x-x')^2 + 4L^2},$$

$$\tag{3.76}$$

$$u_+(x) = \left(\frac{\partial\theta}{\partial z}\right)_{z=+L} = \frac{1}{\pi}\int_{-\infty}^{+\infty} dx' \frac{\partial\phi}{\partial x'} \frac{1}{x-x'}$$

$$- \frac{1}{\pi}\int_{-\infty}^{+\infty} dx' \frac{\partial\phi_-}{\partial x'} \frac{x-x'}{(x-x')^2 + 4L^2}.$$

In each of these equations the first integral represents the contribution of a given boundary of the sample to the torque at the point $x$ of the same boundary. The second integral represents the contribution of the other boundary.

In the same approximation we have:

$$\theta(x,z) = \frac{1}{\pi}\int_{-\infty}^{+\infty} dx' \left\{\frac{\phi_-(x')(z+L)}{(z+L)^2 + (x-x')^2} + \frac{\phi_+(x')(z-L)}{(z-L)^2 + (x-x')^2}\right\}. \tag{3.77}$$

Let us consider only the particular situation where $\phi_+(x') = \phi_-(x')$, i.e. where we have the same distribution at the two boundaries. This is important as it can easily be carried out by slowly cooling the isotropic phase. Two nematic spherolites then appear with diametrically opposed singular points, often along the vertical. As they grow, these spherolites finally touch the surfaces and leave on them linear defects which are identical on both faces. It can be seen that their interaction energy is great enough to explain their stability. We will put $\phi(x) = \phi_+(x) = \phi_-(x)$.

### 3.4.8.1  Surface energy

On each of the boundaries the surface energy reads: $F_s = -\int_0^\pi x(dW/d\theta)\,d\theta$ (cf. subsection 3.3.5). We therefore have, for the two boundaries:

$$F_s = \frac{4K}{\pi}L^2 \int_{-\infty}^{+\infty} dx\,dx' \frac{\partial\phi}{\partial x}\frac{\partial\phi}{\partial x'} \frac{1}{(x-x')^2 + 4L^2}. \tag{3.78}$$

Let us substitute into $F_s$ the non-perturbed value of $\partial\phi/\partial x$, i.e. the value relative to the semi-infinite medium ($L = \infty$). This yields:

$$F_s = \pi K\left(1 - \frac{b}{L+b}\right). \tag{3.79}$$

A perceptible reduction in the surface energy therefore takes place if $b$ is not negligible compared with $L$. Note that for $W_s \sim 10^{-2}$ erg/cm$^2$, $b$ is of the order of a micron.

### 3.4.8.2  Bulk energy

This is written here as:

$$F_v = \frac{K}{2\pi}\int_{-\infty}^{+\infty} k\,dk \tanh kL\, \phi(x)\,\phi(x')\mathrm{e}^{-ik(x-x')}\,dx\,dx'. \tag{3.80}$$

This quantity does not significantly depend on $L$ for $kL > 1$. Therefore only the surface energy is involved. Furthermore it is easy to separate out the (negative) interaction term in the surface energy. Let us write this, supposing first of all that the two lines are not at the same abscissa. This yields, in the most symmetrical form possible:

$$F_{SI} = -\frac{K}{4\pi} \int dx\, dx' \frac{(x - x')^2}{(x - x')^2 + 4L^2} \left( \frac{\partial \phi_-}{\partial x'} \frac{\partial \phi_+}{\partial x} + \frac{\partial \phi_+}{\partial x'} \frac{\partial \phi_-}{\partial x} \right). \quad (3.81)$$

Suppose that $\phi_+$ is centred at $a$ and $\phi_-$ is centred at $-a$. We have:

$$\frac{\partial \phi_-}{\partial x'} \frac{\partial \phi_+}{\partial x} + \frac{\partial \phi_-}{\partial x'} \frac{\partial \phi_+}{\partial x} = b^2 \left\{ \frac{1}{b^2 + (x' - a)^2} \frac{1}{b^2 + (x' + a)^2} + \text{sym} \right\}. \quad (3.82)$$

$F_{SI}$ tends towards zero for infinite $a$. The interaction energy is therefore the most negative possible for $a = 0$, i.e. when the lines are placed above one another.

### 3.4.9  Dislocation near a surface (Meyer, 1973a)

Meyer studied the behaviour of a twisted line $S = \frac{1}{2}$ situated at a distance $d$ from the surface. Such a line possesses a singular core of radius $r_c$, and is pushed to the surface by its image force if it is situated at a distance $d \gg b$. It can then indeed be considered that the surface molecules conserve their easy orientation: the image of the line is then a dislocation of the same sign at distance $-d$. On the other hand, when $d$ diminishes, the surface molecules are no longer fixed. If we are limited to a distance $d \gg r_c$ (so that the question of the disappearance of the core does not have to be considered) but which still remains small, the line energy takes the form:

$$W = W_{\text{core}} + \frac{\pi K}{4} \left[ \ln \frac{R^2}{br_c} - \ln \frac{2d}{b} - 2 \exp \frac{2d}{b} \cdot E_1 \left( \frac{2d}{b} \right) \right], \quad (3.83)$$

where

$$E_1(x) = \int_x^\infty \frac{e^{-u}}{u}\, du$$

is the exponential integral function.

We then see that there is attraction of the line at a short distance from the surface which is essentially the case of $d < b$, and there is a subsequent transformation into surface lines.

### 3.5  EFFECT OF A MAGNETIC FIELD ON A FREE SURFACE. INFINITE ANCHORING

This problem has been examined by de Gennes (1970). His results are presented here in a slightly different way, dealing in particular with the boundary conditions and using the methods given above.

It is supposed that the anchoring energy is infinite and that the director makes an angle $\Phi_0$ with the plane tangential to the surface. The long-range boundary conditions of the surface are imposed by the magnetic field $H$ which is taken as horizontal. The deformation $\zeta$ of the free surface will now be derived (Figure 3.15).

Figure 3.15. Deformation of a nematic free surface under the influence of a magnetic field

Let $\theta$ be the angle of the director with $H$, $\xi = (1/H)(K/\chi_a)^{1/2}$ the characteristic length linked to $H$. The dependence of $\theta$ on $z$, according to de Gennes and using the one constant approximation, obeys the law:

$$\xi \frac{d\theta}{dz} = \sin\theta. \tag{3.84}$$

This equation approximately satisfies the bulk equilibrium equation:

$$\nabla^2\theta - \frac{1}{\xi^2}\sin\theta\cos\theta = 0. \tag{3.85}$$

The dependence on $x$ is not fully satisfied, but it will be supposed that the variations in $x$ are slow. Let us now consider only surface terms, and write the boundary conditions by using a method of virtual work for the surface. With virtual displacements $\delta\theta$ and $\delta\zeta$, the free energy undergoes a variation:

$$\delta E_s = \int K\frac{\partial\theta}{\partial z}\delta\theta\,dx + \int A\left(\frac{\zeta}{\lambda^2} - \zeta''\right)\delta\zeta\,dx$$
$$+ \int \left\{p + \frac{K}{2}\left[\left(\frac{\partial\theta}{\partial z}\right)^2 + \frac{\sin^2\theta}{\xi^2}\right]\right\}\delta\zeta\,dx, \tag{3.86}$$

which must be zero whatever the variations $\delta\theta, \delta\zeta$. In equation (3.86), the first integral term represents the work done by the restoring torque on the surface (cf. equation (3.13)), the second the work done by surface tension $A$ and gravity $A/\lambda^2 = \rho g$. The third integral term expresses the fact that the total volume is unchanged in the variation $\delta\zeta$. $p$ is the associated Lagrange multiplier. But it is known in nemato-statics (cf. de Gennes, 1975) that the quantity $p + (\frac{1}{2})K(\nabla\theta)^2 + (\frac{1}{2})K/\xi^2\sin^2\theta$ is a constant, which here only has the effect of altering the origin of $\zeta$. The third integral term can therefore be taken as zero.

Integrating (3.86) over a period $L$, the following notations and relations are used:

$$\zeta = \Sigma A_n \cos (2n + 1)\frac{\pi x}{L}, \tag{3.87}$$

$$\theta = \pm \psi_0 + \frac{\partial \xi}{\partial x}, \qquad \begin{array}{ll} +\psi_0 & \text{for } -L < x < 0, \\ -\psi_0 & \text{for } 0 < x < L, \end{array}$$

$$\delta\zeta = \Sigma \delta A_n \cos (2n + 1)\frac{\pi x}{L},$$

$$\delta\theta = -\Sigma (2n + 1)\delta A_n \sin (2n + 1)\frac{\pi x}{L}. \tag{3.88}$$

The actual virtual variations are thus expanded in a Fourier series. The coefficients of $\delta A_n$ are equal to zero. It then becomes:

$$A_n = \frac{4K}{L\xi} \frac{\sin \psi_0}{\frac{A}{\lambda^2} + \frac{\pi^2}{L^2}(2n + 1)^2 \left(A + \frac{K}{\xi}\cos \psi_0\right)}, \tag{3.89}$$

$$\xi = aH \frac{K}{A} \frac{\lambda^2}{L\xi} \sin \psi_0 \frac{\cosh a\pi(1 - x/L) - \cosh a\pi x/L}{\sinh a\pi}, \tag{3.90}$$

where

$$a = \frac{L}{\pi\lambda}\left(\frac{A}{A + K/\xi \cos \psi_0}\right)^{1/2}. \tag{3.91}$$

The maximum slope at the peak of the singularities is given by:

$$\varepsilon = \left(\frac{\partial \zeta}{\partial x}\right)_{x=0} = \frac{\pi}{L} \frac{Ka}{A} \frac{\lambda}{\xi} \sin \psi_0, \tag{3.92}$$

which is reduced to the value obtained by de Gennes for $a \sim L/\pi\lambda$. This approximation is valid for $K/\xi \ll A$, which is generally true.

The energy reads:

$$F = -\frac{2K^2}{\pi A \xi^2} \frac{1}{a} \sin^2 \psi_0 \cdot \tanh \frac{a\pi}{2} \tag{3.93}$$

to which must be added a line energy (of the singularities) of the order of $K/L$ which takes into account the fact that in this region there is a strong variation of $\theta$ with $x$.

*Validity.* This calculation supposed an infinite anchoring energy. But if $W_s$ is small in comparison to the energy of the system and, in fact, if the inequality

$$W_s < \tfrac{1}{2}A\varepsilon^2 \tag{3.94}$$

occurs, the approximation of the infinite anchoring is no longer correct. For $W_s \sim 5 \times 10^{-3}\,\mathrm{erg/cm^2}$ ($A = 10\,\mathrm{erg/cm^2}$; $\xi \sim 10^{-4}\,\mathrm{cm}$), the approximation is correct: obviously it is no longer so for a large enough $\xi$, i.e. for weak magnetic fields. We recall that these anchoring energies on the free surface are generally large.

## 3.6 SURFACE SINGULAR POINTS (FRIEDEL'S NUCLEI)

A nematic placed between two plates of degenerate planar anchoring gives numerous nuclei (subsection 3.2.4) which are essentially vertical lines of integral strength ($S = \pm 1$). In fact these lines can be shown to carry singular, but virtual, points, i.e. they are situated outside the sample.

Let us begin with equation (3.53) which is written again in cylindrical coordinates:

$$\tan\theta = \pm r/z \qquad (3.95)$$

where the origin of the coordinates is taken to be at the singular point itself. Let $z_0$ be the coordinate of the surface. Note that:

$$\left(\frac{\partial\theta}{\partial z}\right)_{z=z_0} = -\frac{r}{r^2 + z_0^2}. \qquad (3.96)$$

If the anchoring energy is written $W = W_s\cos^2\theta$, we have:

$$\frac{\partial W_s}{\partial\theta} = -\frac{2rz}{r^2 + z^2}. \qquad (3.97)$$

The boundary conditions can be satisfied with this expression. They are written:

$$\frac{\mathrm{d}W}{\mathrm{d}\theta} = \pm K\frac{\partial\theta}{\partial z} \qquad (3.98)$$

(a plus sign for the lower coordinate, a minus sign for the other case). Taking a line $S = 1$, one obtains:

$$\tan\theta = r/z, \; z_0 = -\frac{K}{2W_s}, \quad \text{radial singular point,}$$

for the upper plate; and

$$\tan\theta = -r/z, \; z_0 = +\frac{K}{2W_s}, \quad \text{hyperbolic singular point,}$$

for the lower plate.

In both these cases we are dealing with virtual points. Similar results are obtained for the line $S = -1$. The two points are again virtual but of the same kind, and are turned around each other at $\pi/2$ about the vertical.

## 3.7 QUASI-TWO-DIMENSIONAL NEMATICS

We have discussed in subsection 3.3.4 the conditions in which walls can be observed in a nematic: the sample must be sufficiently thin, or the anchoring energy sufficiently low. Ryschenkow's anchoring (1975) which is of the order of $10^{-4}$ erg/cm$^2$, makes it possible to attain such a situation with samples of the order of 10 $\mu$m thick, the extrapolation length $b$ being of the order of 50 $\mu$m. Walls are in fact observed whose topological relations with nuclei, which are numerous with a degenerate conic anchoring of this type, are described in Ryschenkow and Kléman (1976).

A curious transition of the anchoring takes place, which is observed within the nematic range of MBBA, on samples of thickness $h < b$ (Figure 3.16).

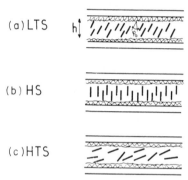

Figure 3.16. Variation of anchoring with temperature

(a) At low temperatures, the anchoring is conic, and the angle $\phi_0(T)$ measured between the director and the normal to the substrate decreases as the temperature increases. This phase is the one which possesses the walls and nuclei texture mentioned above. Of particular interest is the fact that the defect density seems remarkably constant over the whole area, and that the mean distance between defects (walls and nuclei) is of the order of 100 $\mu$m. We will call this phase LTS (low temperature structure).

(b) When $T$ increases, a homeotropic structure (HS) is gradually reached, which is characterized by the presence of Bloch walls resembling those of the LTS.

(c) A second, more abrupt transition, appears at a temperature quite close to the clearing point. A new structure (HTS) appears which is quite different to the others (no walls) and it is obvious that the extrapolation length $b$ becomes smaller than $h$.

We shall not dwell upon the origins of these structural transitions (cf. de Gennes and Dubois-Violette, 1975) but rather discuss a specific feature of the HS and LTS. This is their quasi-bidimensional character. Everything occurs as if the variations in $z$ in the samples were reduced to zero ($\partial/\partial z = 0$). The LTS phase therefore appears as the analogue of a smectic C in two dimensions, the HS phase as the analogue of a smectic A in two dimensions. More precisely, the thermodynamic fluctuations occur on two very different scales:

(a) small scale bulk fluctuations correspond to the usual strong elastic forces; and
(b) large scale fluctuations involving the entire thickness of the sample correspond to weak restoring forces due to the anchoring. It is these which are going to be discussed and it will be supposed that they are entirely unconnected with the first.

### 3.7.1 LTS phase

Let us call $\varepsilon$ the small fluctuations in $\phi$ of the director $\phi = \phi_0 + \varepsilon$, and introduce the azimuth $\theta$. The free energy can be written in the form:

$$F_s = 2\,W_s(\phi) + \tfrac{1}{2}Kh[\sin^2\phi(\nabla\theta)^2 + (\nabla\varepsilon)^2], \tag{3.99}$$

where the factor 2 refers to the two limits, and $h$ is the thickness of the sample. Let us expand $F_s$ to second order in the fluctuations $\varepsilon$ and $\theta$, expanded in the Fourier series:

$$\varepsilon = \frac{1}{(2\pi)^2}\Sigma\varepsilon_{\mathbf{q}}\exp{(\mathrm{i}\,\mathbf{q}\mathbf{r})},$$

$$\theta = \frac{1}{(2\pi)^2}\Sigma\theta_{\mathbf{q}}\exp{(\mathrm{i}\,\mathbf{q}\mathbf{r})}, \tag{3.100}$$

$$\varepsilon_{\mathbf{q}}^* = \varepsilon_{-\mathbf{q}}; \qquad \theta_{\mathbf{q}}^* = \theta_{-\mathbf{q}}; \qquad \mathbf{q}^2 = q_x^2 + q_y^2.$$

We obtain, after integration over the $x, y$ plane:

$$\int F_s\,\mathrm{d}x\,\mathrm{d}y = \frac{1}{(2\pi)^2}\Sigma_{\mathbf{q}}\left[2\,W_s\varepsilon_{\mathbf{q}}\varepsilon_{\mathbf{q}}^* + \frac{Kh}{2}q^2(\sin^2\phi_0\theta_{\mathbf{q}}\theta_{\mathbf{q}}^* + \varepsilon_{\mathbf{q}}\varepsilon_{\mathbf{q}}^*)\right]. \tag{3.101}$$

In this expression, the fluctuations in $\theta$ and $\varepsilon$ are entirely unconnected. The fluctuations in $\theta$ are typical of a two-dimensional nematic. Following de Gennes' (1971) argument which refers to the fluctuation–dissipation theorem, (see, for example, Landau and Lifshitz, 1967) the mean quadratic fluctuation of this quantity reads:

$$\langle(\theta(\mathbf{r}_1) - \theta(\mathbf{r}_2))^2\rangle = \frac{k_B T}{h\pi K\sin^2\phi_0}\ln\frac{|\mathbf{r}_1 - \mathbf{r}_2|}{a} \tag{3.102}$$

where $a$ is a molecular length. This indicates that the fluctuations of the projection of the director in the $xy$ plane destroy the nematic order above a length of coherence $\xi_\theta(T)$:

$$\xi_\theta(T) = a\exp\frac{\pi Kh\sin^2\phi_0}{k_B T}. \tag{3.103}$$

This quantity is large, and it can be seen that the nematic order is conserved over long distances, except for $\Phi_0 = 0$ (homeotropic). We will therefore reach the same conclusion as de Gennes for the *two-dimensional nematic* of thickness $h$ a little greater than the molecular dimension. There is no true instability of the two-dimensional order because $\pi Kh \sin^2 \phi_0$ is large compared to $k_B T$.

The fluctuations of $\varepsilon$ are more original. The same fluctuation–dissipation theorem allows us to write:

$$\langle |\varepsilon_q|^2 \rangle = \frac{k_B T}{4W_s + Khq^2}. \tag{3.104}$$

These fluctuations *do not depend* on $\phi_0$. Calculate $\langle \varepsilon^2(\mathbf{r}) \rangle$:

$$\langle \varepsilon^2(\mathbf{r}) \rangle = 2 \int \frac{dq_x\, dq_y}{(2\pi)^2} \langle |\varepsilon_q|^2 \rangle = \frac{k_B T}{\pi Kh} \ln \frac{\xi_\phi}{a}, \tag{3.105}$$

where the factor 2 comes from the fact that $|\varepsilon_q|$ appears twice in the free energy ($\pm q$), and where we have introduced the length of coherence,

$$\xi_\phi^2 = \frac{bh}{2}, \tag{3.106}$$

independent of $\Phi_0$. Taking $h \sim 10\,\mu$m, $b \sim 100\,\mu$m, we obtain $\xi_\phi \sim 30\,\mu$m. This is the length of correlation of the fluctuation in $\phi$. We have $\xi_\phi \ll \xi_\theta(T)$ in the LTS phase. It may therefore reasonably be thought that the defect structure (Walls and nuclei) is essentially linked to the existence of a macroscopic $\xi_\phi$, and that the defects would somehow freeze these fluctuations by bringing in a favourable entropy term. The energy of the unfrozen fluctuations is in fact, from (3.99):

$$F_s \sim ck_B T \ln \frac{\xi_\phi}{a} \tag{3.107}$$

where $c = 1/\xi_\phi^2$ is the density of fluctuations.

The free energy of the defects of the same density is written:

$$F_{def} = c[hK - k_B T \ln c]. \tag{3.108}$$

It can be seen that $F_{def} < F_s$ for $\xi_\phi \gg a$, i.e. for low densities, which happens to be the case. There is therefore a natural tendency for the fluctuations to freeze. However, to complete the picture, fluctuations on a smaller scale would have to be taken into account.

This type of order is approached at long range and consists of equilibrium defects which recalls the predictions of Kosterlitz and Thouless (1972) on the nature of two-dimensional order.

### 3.7.2  HS phase

For $\Phi_0 = 0$ the fluctuations in $\theta$ are great and $\xi_\phi$ small. It then becomes necessary to take into account the coupling between the fluctuations in $\phi$ and $\theta$, and to add to the free energy the supplementary term (of order 3 and 4):

$$\Delta F_s = \tfrac{1}{2} Kh [\varepsilon \sin 2\phi_0 \cdot (\nabla\theta)^2 + \varepsilon^2 \cos 2\phi_0 (\nabla\theta)^2], \tag{3.109}$$

the expansion of which in the Fourier series after summation over the $xy$ plane reads:

$$\int \Delta F_s \, dx \, dy = \frac{1}{16\pi^4} Kh \sin 2\phi_0 \, \mathbf{p} \cdot \mathbf{s} [\varepsilon_q \theta_p \theta_s^* \delta(\mathbf{p} + \mathbf{q} - \mathbf{s}) + \text{c.c.}]$$

$$+ \frac{1}{16\pi^4} Kh\varepsilon_q \varepsilon_r^* \theta_s \theta_p^* \delta(\mathbf{s} + \mathbf{q} - \mathbf{r} - \mathbf{p}) \cos 2\phi_0 \cdot \mathbf{p} \cdot \mathbf{s}. \tag{3.110}$$

This expansion is justified in the domain where $\xi_0 \sim \xi_\phi$, i.e. for $\Phi_0 \leqslant 10^{-2}$ rad. The linear term $\varepsilon_q$ in $\Delta F_s$ can be neglected as its thermal mean is zero in equilibrium. In the same way the term $\frac{1}{2}Kh \sin^2 \phi_0 (\nabla\theta)^2$ (in $F_s$, equation (3.99)) can be neglected with respect to $\frac{1}{2}Kh\varepsilon^2(\nabla\theta)^2$ in the domain of values of $\phi_0$ under consideration. We are going to show that the coupling between fluctuations in $\theta$ and $\phi$ leads to the equality of the lengths of correlation of these two fluctuations.

To calculate the fluctuations in $\theta$, values of $\varepsilon$ of the LTS phase are used in perturbation, i.e.:

$$\langle \varepsilon_q \varepsilon_r^* \rangle = \frac{k_B T}{4W_s + Khq^2} \delta(\mathbf{q} - \mathbf{r}). \tag{3.111}$$

The thermal mean $\langle E \rangle = \langle (F + \Delta F_s) \, dx \, dy \rangle$ at zero fluctuations of $\theta$ is written:

$$\langle E \rangle = \frac{1}{2}W_s \langle \varepsilon^2 \rangle + \frac{1}{2}Kh\Sigma q^2 \langle \varepsilon_q \varepsilon_q^* \rangle + \frac{1}{2}Kh\langle \varepsilon^2 \rangle (\nabla\theta)^2, \tag{3.112}$$

which leads, for the mean quadratic values of fluctuations of $\theta$, and by using the fluctuation–dissipation theorem, to:

$$\langle \theta_p^2 \rangle = \frac{k_B T}{Kh\langle \varepsilon^2 \rangle p^2} = \frac{\pi}{p^2} \frac{1}{\ln \xi_\phi/a}. \tag{3.113}$$

In this expression the divergence of fluctuations for $\phi_0 = 0$ no longer appears. We have adopted for $\xi_\phi$ the value given by equations (3.105) and (3.106). From (3.113) we take $\xi_\theta$, and obtain:

$$\xi_\theta = \xi_\phi = \left(\frac{bh}{2}\right)^{1/2}. \tag{3.114}$$

This result, which demonstrates that the HS phase is stabilized by quasi-degenerate fluctuations (therefore having large entropy) of $\phi$ in the vicinity of $\Phi_0 = 0$, leads to the renormalization of $W_s$. In fact, let us take the thermal mean of $\langle E \rangle$ over the fluctuations of $\theta$. This reads:

$$\langle\langle E \rangle\rangle = \frac{2}{(2\pi)^2} W_s \langle \varepsilon^2 \rangle + \frac{1}{2(2\pi)^2} Kh \sum_q q^2 \langle \varepsilon_q^2 \rangle + \frac{1}{2\pi} \frac{h}{\xi_\phi^2} \left(\ln \frac{\xi_\phi}{a}\right)^{-1} \langle \varepsilon^2 \rangle \tag{3.115}$$

where we have used the relation

$$\langle (\nabla\theta)^2 \rangle = \left(2\pi \ln \frac{\xi_\phi}{a}\right)^{-1} \int_0^{2\pi/\xi_\phi} dq_x \, dq_y.$$

The integration in the plane $q_x, q_y$ is limited to wavelengths greater than $\xi_\phi$. We recognize in $\langle\langle E \rangle\rangle$ an effective anchoring energy

$$W_s^* = W_s\left[1 + 2\pi\frac{bh}{\xi_\phi^2}\left(\ln\frac{\xi_\phi}{a}\right)^{-1}\right] = W_s\left[1 + 4\pi\left(\ln\frac{\xi_\phi}{a}\right)^{-1}\right]. \qquad (3.116)$$

These effects may be important since the anchoring energy brought into effect is of the same order of magnitude as the terms of thermal agitation $Nk_B T$, where $N$ is the number of molecules per unit surface.

# Chapter 4

# Cholesterics

## 4.1 SYMMETRY PROPERTIES. GENERAL CONSIDERATIONS

The cholesteric phase is characterized by a helicoidal arrangement of molecules (Chapter 2, Figure 2.3). These are all parallel to a fixed plane direction (called the cholesteric plane) and turn about the normal to this plane with a pitch $p_0 = 2\pi/q_0$. This arrangement is liquid, i.e. the molecular directions which have just been defined are local mean directions. There also exist liquid type fluctuations of $p_0$ and of the direction of the local cholesteric axis $\chi$ (the normal to the cholesteric plane). Because of the equivalence $\mathbf{n} = -\mathbf{n}$, the periodicity along $\chi$ is equal to $p_0/2$.

In each cholesteric plane, the molecules are parallel to each other as in a nematic phase. When the pitch $p_0$ tends towards infinity, the cholesteric phase becomes more closely identified with the nematic phase. These two phases must be considered as different aspects of the same thermodynamic state of matter (G. Friedel, 1922). The cholesteric phase is due to the presence of optically active non-enantiomorphic (chiral) molecules, determining a left or right hand screw arrangement. The nematic phase is formed either of achiral molecules, or of a balanced mixture of left or right chiral molecules (racemic mixture). It is therefore understood that a pure cholesteric does not have a phase transition to a nematic phase, just as a pure nematic cannot have a phase transition to a cholesteric. On the other hand, there is nothing to prevent either of these phases undergoing a transition to a smectic phase.

The pitch of the different cholesteric compounds is very variable. It is of the order of several thousand Å in pure cholesterics such as cholesterol esters, or the chitinous shell of certain insects which have cholesteric pseudomorphoses (see end of section 4.1). These dimensions are responsible for remarkable optical properties in these substances. The rapid variation of pitch with temperature in cholesterol esters has led to their use as temperature indicators. However, in the study of defects by optical microscopy, it is important to have even greater periodicities (of the order of several microns). This is achieved by diluting weak concentrations of active molecules in a nematic phase. The pitch of the mixture $p_0$ is inversely proportional to the concentration $c$.

$$q_0 = \frac{2\pi}{p_0} = 4\pi\beta c. \tag{4.1}$$

$\beta$, the macroscopic twist power of the active molecules, is a pseudoscalar parameter and is independent of $c$ for weak concentrations. This power affects the direction and angle of the rotation $\omega$ which is induced by *one* active molecule placed parallel to the director in a perfect nematic (Brochard and de Gennes, 1970).

$$\omega = \beta \nabla \left(\frac{1}{r}\right). \tag{4.2}$$

The sign of the pitch is determined by the sign of the optical activity.

By mixing cholesterics of different kinds and opposing chiralities, the chiralities can be balanced at a given temperature $T^*$. The medium then behaves like a nematic. Above or below $T^*$, we have either the right-hand cholesteric or the left-hand cholesteric (G. Friedel, 1922; also Baessler and Labes, 1970; Adams and Haas, 1971).

Concentrated solutions of certain polypeptides (PBLG: Polybenzyl l-glutamate is a molecule of helicoidal conformation) in organic solvents, have been used to obtain cholesteric phases of large pitch (Robinson, 1966).

As for nematic phases, *frozen* mesophases can be created by reticulation of polymers whose monomers have a cholesteric phase (Strzelecki and Liébert, 1973). On the other hand nature offers many examples of biological materials which have helicoidal arrangements: organic matrices of the cuticle of certain crabs, DNA of the chromosomes of dinoflagellates etc.) (Bouligand 1969, 1975). It is possible that defects in these arrangements have a biological role.

In this chapter we refer essentially to observations of defects in liquid cholesterics, but the topological description given is valid for all types of cholesteric arrangements. However, different germination conditions and elastic effects bring about different textures. We will refer to Bouligand (1975, 1980) for the study of biological materials having cholesteric pseudomorphoses.

## 4.2 OPTICAL PROPERTIES

The extraordinary optical properties of cholesterics under circularly polarized light conditions will not be described here. The interested reader may refer to Chystiakov's (1967) review article and de Gennes' (1974) work. But for the interpretation of these optical observations it is important to know that a cholesteric behaves like an optically active medium of considerable rotatory power ($\sim 10^4$ deg/cm). The optical behaviour depends on the order of magnitude of the relation $\lambda/p_0$ (where $\lambda$ is the wavelength of light) and has a divergent anomaly, with the rotatory power changing sign close to the Bragg reflection which occurs at $\lambda = p_0$. The behaviour of small light wavelengths ($\lambda \ll (n_e - n_o)p_0$, $n_e$ extraordinary index, $n_o$ ordinary index) is particularly simple to visualize (Mauguin, 1911). It is as if the wave (extraordinary or ordinary) is guided by the helix and rotates along it with a velocity $c/n_e$ or $c/n_o$. The optical activity proper to the active molecules is therefore negligible under Mauguin's regime. This also happens in a nematic displaying small twists. These observations make the interpretation of optical observations of cholesterics of very large pitch particularly easy. Other regimes have been studied in detail by de Vries (1951), Cano (1967), Billard (1968), etc.

## 4.3  FREE ENERGY

The non-perturbed state of a cholesteric can be imagined as a twisted nematic:

$$\mathbf{n} \cdot \mathrm{rot}\, \mathbf{n} = -q_0, \tag{4.3}$$

splay (div $\mathbf{n}$) and bend ($\mathbf{n} \wedge$ div $\mathbf{n}$) being zero. The free energy can be written in the form

$$F = \tfrac{1}{2}K_1(\mathrm{div}\,\mathbf{n})^2 + \tfrac{1}{2}K_2(\mathbf{n}\cdot\mathrm{rot}\,\mathbf{n} + q_0)^2 + \tfrac{1}{2}K_3(\mathbf{n}\wedge\mathrm{rot}\,\mathbf{n})^2 - \tfrac{1}{2}\chi_a(\mathbf{H}\cdot\mathbf{n})^2, \tag{4.4}$$

where an eventual magnetic field has been taken into account. The minimization of equation (4.4) describes *static* deformations, to which the present discussion will be restricted.

Many experiments under magnetic field on *planar* samples, i.e. with the cholesteric axis perpendicular to the supporting slides, have shown either effects of instability under the field (Helfrich, 1970; Rondelez and Hulin, 1972) or effects of unwinding of the helicoidal arrangement (Sackmann *et al.*, 1967; Durand *et al.*, 1969; Meyer, 1969). These two types of effects are described below.

### 4.3.1  Helfrich instability

Where $\chi_a < 0$, the molecules tend to be perpendicular to $\mathbf{H}$ to minimize (4.4). If $\mathbf{H}$ is along the helicoidal axis, there is no effect of distortion.

On the other hand, for $\chi_a > 0$, an instability of *layers* appears in the same geometry, which is very similar to that which appears in smectics (see subsection 5.3.5). The threshold of the instability can be calculated by substituting for the free energy density (4.4) the density (see de Gennes; Lubensky, 1972):

$$F_{cg} = \tfrac{1}{2}K_2 q_0^2 \left(\frac{\delta p}{p_0}\right)^2 + \frac{3}{16}K_3(\mathrm{div}\,\mathbf{d})^2 + \tfrac{1}{2}\chi_a(\mathbf{H}\cdot\mathbf{d})^2, \tag{4.5}$$

where $\mathbf{d}$ is a unit vector along the local cholesteric axis. The expression (4.5) is a coarse grained version of the energy (4.4) and is only valid when the pitch of the layers is large enough compared to the distortions undergone by them. A local cholesteric axis of unit vector $\mathbf{d}$ and a relative variation of local pitch $\delta p/p_0$, where $p_0$ is the pitch in equilibrium, can therefore be defined without difficulty. The $K_2$ term is analogous to an energy of compression of the layers, and the $K_3$ term to an energy of curvature of the layers.

Helfrich (1970) has calculated the instability in one dimension. Delrieu (1974) showed that the instability thresholds in square lattices are lower than the one-dimensional instabilities, which anyway only appear for $\mathbf{H}$ fields oblique to the vertical. The reader should refer to the articles quoted and to subsection 5.3.5.

80

## 4.3.2 Unwinding of a cholesteric under magnetic field
(Meyer 1968; de Gennes, 1968a)

Let us suppose that $\chi_a > 0$, and that **H** is applied perpendicularly to the cholesteric axis which is in the plane of the sample. This last condition is necessary so that the boundary conditions do not impose a finite pitch on the cholesteric which is produced under a magnetic field (Figure 4.1). Above a critical field $H_c$, the cholesteric is completely unwound.

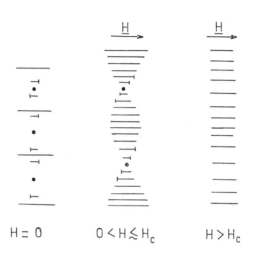

$$H = 0 \qquad 0 < H \lesssim H_c \qquad H > H_c$$

Figure 4.1. Unwinding of a cholesteric under a magnetic field $\chi_a > 0$

It is convenient to analyse the energy of the system by introducing the concept of wall, where the director rotates by an angle equal to $\pi$ (i.e. a half-pitch). In fact the transition towards the high field phase $(H > H_c)$ is not accompanied by the complete disappearance of the cholesteric. It always leaves behind inversion walls, which are traces of cholesteric pitches and cylindrical zones with generatrices parallel to the field, all metastable, and across which the director turns from $\pi$. However, they cannot be made to disappear by a continuous inverse rotation of the director (topological barrier). Conversely, when the field is reduced to below $H_c$, it is essentially the increase in the surface of the zones by terminal lengthening, folding up and filling up of the space, which returns it to the cholesteric phase. Meyer (thesis, 1969) has studied this phenomenon in detail and compared it to a two-dimensional dendritic growth.

For a given field **H**, we have a pitch $p(\mathbf{H})$ and the twist is essentially limited to a region of magnitude

$$\xi = \left(\frac{K_2}{\chi_a}\right)^{1/2} H^{-1}, \qquad (4.6)$$

a formula analogous to that of Freedericks' transition (see equation (3.5)). The pitch becomes infinite for $\xi = p_0/\pi^2$. The demonstration goes as follows:

Let $z$ be the direction of the cholesteric axis. Let us suppose that the only relevant term is twist $k_2$. Minimization of equation (4.4) yields:

$$\xi^2 \frac{d^2\theta}{dz^2} = \sin\theta\cos\theta, \tag{4.7}$$

where $\theta$ is the angle of the director with the direction of the field $H$ in the cholesteric plane. Equation (4.7) has as a first integral

$$\xi^2 \left(\frac{d\theta}{dz}\right)^2 = \sin^2\theta.$$

The free energy of a wall, decreased by the energy $\frac{1}{2}Kq_0^2$ of the configuration of the homogeneous state reads:

$$F_p = \int \left[ \frac{1}{2}K_2\left\{ \left(\left(\frac{d\theta}{dz}\right) - q_0\right)^2 - q_0^2 \right\} + \frac{1}{2}\chi_a H^2 \sin^2\theta \right] dz,$$

$$F_p = \chi_a H^2 \int_0^\pi \xi^2 \left(\frac{d\theta}{dz}\right) dz - q_0\xi\chi_a H^2 \int_0^\pi d\theta \tag{4.8}$$

$$= \xi\chi_a H^2(2 - \pi q_0 \xi).$$

Conditions become favourable to wall formation when this quantity becomes positive, i.e. for $\xi = 2/\pi q_0$; we therefore have a critical field:

$$H_c = \pi^2 \left(\frac{K_2}{\chi_a}\right)^{1/2} \frac{1}{p_0}. \tag{4.9}$$

## 4.4 GENERAL THEORY OF DEFECTS
### (Friedel; Kléman, 1969, 1970)

There are several possible classifications of defects in cholesterics. If only the symmetries of the medium are considered, translation dislocations corresponding to the periodicity of the arrangement and rotation dislocations can be distinguished (subsection 4.4.1). But the former can still be imagined as resulting from the pairing of two rotation dislocations a certain distance apart. This provides a core model (of translation dislocations) which has been confirmed by experiments. Each of the rotation dislocations then appears *imperfect*, and the core is a defect zone which it is convenient to analyse by using the concept of infinitesimal dislocation (subsection 4.4.2). The very numerous isolated screw dislocations can be given a particular description (subsection 4.4.3). We will also show that translation dislocations can again be imagined as particular forms of rotation dislocations.

Finally, we will discuss climb and glide (subsection 4.4.5) and kinks and singular points on defect lines (subsection 4.4.4).

82

This section does not deal with focal conic arrangements, similar to those observed in smectics (Chapter 5). These arrangements will be only mentioned, in the following section (4.5) in which we study cholesteric textures.

### 4.4.1 Perfect rotation dislocations (disclinations)

Perfect disclinations correspond to two types of quantifiable rotation, one along the cholesteric axis $\chi$, the other in the cholesteric plane along the molecule ($\lambda$) or perpendicular to the molecule ($\tau$).

#### 4.4.1.1 *Axis of rotation in the cholesteric plane ($\lambda$ and $\tau$ lines)*

We will designate $\tau^+(z)$ and $\tau^-(z)$ the dislocations with axes perpendicular to the molecules and situated in the cholesteric plane; $\lambda^+(z)$ and $\lambda^-(z)$ dislocations with axis parallel to the molecules and situated in the cholesteric plane. Figures 4.2 and 4.3 show these four types of defects. The similarity of the schemes to those found in nematics make our notations obvious. Moreover, they can be generalized $\tau^{+n}$, $\lambda^{-n}$, etc. where $n = 2S$, and $z$ varied in a discontinuous manner and by quantities equal to the half-pitch.

Defects of the $\lambda$ type do not possess core singularities: they are certainly of lower energy than the defects of $\tau$ type. In nematics, the core of defects $\tau$ where $S$ is half-integral cannot be made to disappear, but the $\tau$ with integral $S$ can be

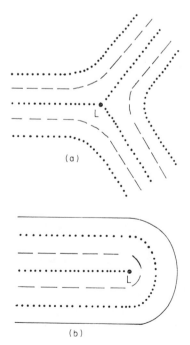

Figure 4.2. Rotation dislocations without core singularities: (a) $\lambda^+$; (b) $\lambda^-$

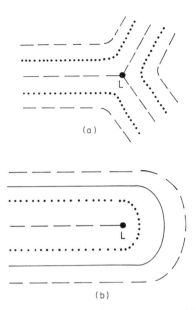

Figure 4.3. Rotation dislocations with core singularities: (a) $\tau^+$; (b) $\tau^-$

transformed in a continuous manner into $\lambda$ lines. But a $\lambda^\pm$ defect, which does not have a singular core, does not have a nematic analogue.

Moreover, a $\lambda$ or $\tau$ line is necessarily rectilinear and along the direction of the rotation axis which produced it by the Volterra process. It is in fact easy to ascertain by using formula (2.14) that curving such a line would necessitate the emission of translation dislocations impossible in a cholesteric.

The lines $\lambda^\pm$ and $\tau^\pm$ are not equivalent. It can be shown, for example, that a $\lambda^+$ can be transformed into a $\tau^\pm$, and vice versa, by emission or absorption of a line $\chi$. This algebra is of course equivalent to the fact that the product of a rotation $\pm\pi$ (along an axis $\lambda$ or $\tau$ of the cholesteric plane), with a rotation $\pm\pi$ along $\chi$, gives a rotation $\tau$ or $\lambda$.

### 4.4.1.2 *Rotation axis parallel to the cholesteric axis* ($\chi$ *lines*)

All the vectors in the cholesteric plane constitute allowed translations, dissipated viscously in a liquid cholesteric. The $\chi$ lines can therefore take any shape because of the relaxation theorem. From this point of view they are *identical* to the lines in nematics. Thus it is not difficult to see that the wedge $\chi$ lines obey the equation

$$\phi = S\theta + q_0 z \qquad (4.10)$$

where $\theta$ is the polar angle in the cholesteric plane and $\phi$ the azimuth of the director in this same plane. The configuration of the molecules is therefore that of a wedge line in a nematic, with a helicoidal cut surface having the pitch of the cholesteric. But this point of view is not the one with the most possibilities; the $\chi$

lines can also be described as *translation* lines. Consider a rotation $\Omega = \alpha$ of any angle, along the cholesteric axis, followed by a translation $T = -\alpha/q_0$ along the same axis. Taken together, these two operations constitute a symmetry operation of the cholesteric. We write symbolically:

$$(\Omega, T) = (\alpha, -\alpha/q_0) \simeq 0. \tag{4.11}$$

We have in the same way:

$$(\alpha, \alpha/q_0) \simeq (2\alpha, 0) \simeq (0, -2\alpha/q_0). \tag{4.12}$$

Let us take $\alpha = \pi/2$. The last expression indicates the equivalence between the rotation $\Omega = \pi$, the translation $T = -\pi/q_0$, and the translation–rotation $(\pi/2, \pi/2q_0)$. Harris (1970a) has given the name *dispiration* to a perfect dislocation combining the operations of translation $T$ and rotation $\Omega$ which are not symmetry operations separately.

Because of viscous relaxations of translation–rotation operations along the $\chi$ axis in a liquid cholesteric, there is an equivalence here between all dispirations $\alpha$. Taking this further, therefore, there is an equivalence between rotation dislocations $\chi$ ($= \pm n\pi$) and translation dislocations $T$ ($= \pm n\pi/q_0$). We can then use the terms screw dislocation $\chi$ ($\sim$ wedge) and edge dislocation ($\sim$ twist).

The twist $\chi$ model has been specifically used and calculated by de Gennes (1968b) to explain the *first* fine line observed in the *planar texture* of a Cano's (1968) wedge. We will return to this texture later. The model proposed by de Gennes is represented in Figure 4.4. It is essentially the arrangement of a twist nematic line $S = |\frac{1}{2}|$ on which is superimposed cholesteric twist. The identity of this line with an edge dislocation ($b = p/2$) is obvious; the core model of Figure 4.4 has been proved experimentally for the first line as Bouligand (1974) has shown. But the core model consisting in the pairing of two lines $\lambda$ or $\tau$ is more frequently found.

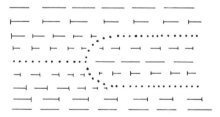

Figure 4.4. De Gennes' model of a twist $\chi$ line

### 4.4.2 Coupling of the $\lambda$ and $\tau$ lines

An isolated or perfect $\lambda$ or $\tau$ whose rotation axis is parallel to the cholesteric plane can be curved by associating with it a density of infinitesimal translation dislocations $\Omega \mathbf{v} \wedge \mathbf{ds}$ (see equation (2.14)). But these densities are topologically forbidden.

Another way of curving a line is to associate at each point of it a variable rotation vector, parallel or perpendicular to the local direction of the molecules. Let there be two positions $A_1$, $A_1'$ close to each other on such a line, $\mathbf{v}_1$ and $\mathbf{v}_1'$ being the corresponding rotation axes, $\mathbf{r}_1$ and $\mathbf{r}_1'$ the distances from $A_1$ and $A_1'$ at a point running over the surface section of the line (Figure 4.5). The relation

$$2\sin\frac{\Omega_1}{2}(\mathbf{v}_1' \wedge \mathbf{r}_1') = 2\sin\frac{\Omega_1}{2}(\mathbf{v}_1 \wedge \mathbf{r}_1)$$

$$+ 2\sin\frac{\Omega_1}{2}(\mathbf{dv} \wedge \mathbf{r}_1 - \mathbf{v}_1 \wedge \mathbf{ds}_1) + O_2 \qquad (4.13)$$

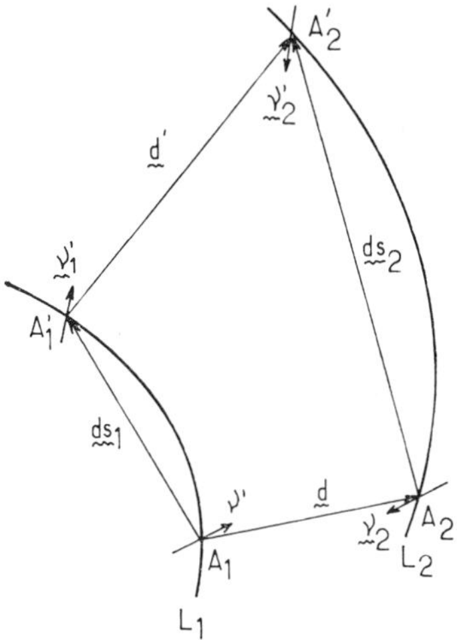

Figure 4.5. Pairing of $\lambda$ or $\tau$ lines of opposite sign (schematic)

shows that there may be a transition from the rotation $(\Omega_1, \mathbf{v}_1)$ at $A_1$ to the rotation $(\Omega_1, \mathbf{v}_1')$ at $A_1'$, by adding along $A_1 A_1'$ a continuous distribution of infinitesimal dislocations of two types:

(a) a density of rotation dislocations $\mathbf{dv}_1$, different from zero only if the line is not planar (in a cholesteric plane); and
(b) a density of translation dislocations $-\mathbf{v}_1 \wedge \mathbf{t}_1 \, ds$.

These operations are not of the types allowed in a cholesteric and cannot be compensated by a viscous relaxation of the crystal; isolated dislocations of this type are impossible. On the other hand, it is possible to imagine *pairs* of dislocations a short distance apart, of opposing rotations, which would compensate at long range for these continuous distributions of infinitesimal dislocations. The existence of these distributions then contributes to the total energy by a term limited to the core energy of the pair.

The simplest configurations involve two lines $\lambda^{\pm}$ or $\tau^{\pm}$. Let us consider two such perfect lines, i.e. rectilinear and situated in the cholesteric plane. Figure 4.6a

86

is obviously equivalent at long range to Figure 4.4, but the nature of the core is essentially different. Figure 4.6b and Figure 4.6c correspond, respectively, to the pairing of two $\lambda$ and of two $\tau$. It is clear that configuration 4.6b is of lower energy than configuration 4.6c. Different situations can be imagined (Figure 4.6d).

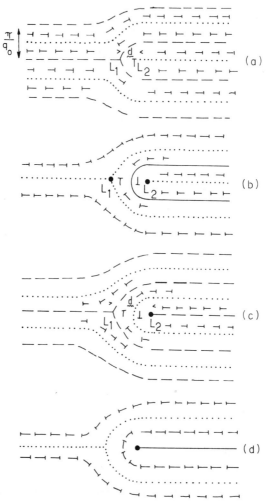

Figure 4.6. Pairing of parallel lines $\lambda$ or $\tau$ in perfect position

If $r_1$ and $r_2$ are the distances at $L_1$ and $L_2$ from the running point on the cut surface, the total effect of the pair $(L_1L_2)$ is written:

$$2\sin\frac{\Omega_1}{2}(v_1 \wedge r + v_2 \wedge r_2) = 2\sin\frac{\Omega_1}{2}(v_1 \wedge d), \qquad (4.14)$$

if **d** is the distance between the two lines. The arrangement $(L_1, L_2)$ therefore has at long range the same effect as a translation dislocation, of Burgers vector $2d$ parallel to the cholesteric axis. Such a dislocation is only acceptable for the crystal if $2d$ is a period of the lattice. We therefore have:

$$d = n \frac{\pi}{2q_0}. \tag{4.15}$$

The arrangement $(L_1 L_2)$ is also equivalent to a $\chi$. Note that this pairing of opposing rotation lines can also be produced in solid crystals (J. Friedel, 1964) with an analogous condition on the distance $d$. It can be produced in nematics, but without restriction on $d$; there is no (meta)stable solution and the two lines $L_1$ and $L_2$ attract and annihilate each other.

Let us look for the general conditions which must be obeyed by two *curved* lines $L_1$ and $L_2$. It may be thought in the first place that such a problem has a solution, for the coupling of $L_1$ and $L_2$ leads to a translation dislocation which is known to have no topological reason not to curve.

Let $A_1$, $A_2$ and $A_1'$, $A_2'$ be two pairs of points of opposing rotation axes (Figure 4.5). The relation:

$$2 \sin \frac{\Omega_1}{2} (\mathbf{v}_1' \wedge \mathbf{r}_1' + \mathbf{v}_2' \wedge \mathbf{r}_2') = 2 \sin \frac{\Omega_1}{2} (\mathbf{v}_1 \wedge \mathbf{r}_1 + \mathbf{v}_2 \wedge \mathbf{r}_2)$$

$$+ 2 \sin \frac{\Omega_1}{2} \{ \mathbf{v}_1 \wedge (\mathbf{ds}_2 - \mathbf{ds}_1) + d\mathbf{v} \wedge \mathbf{d} \} + O_2 \tag{4.16}$$

leads to the relation:

$$\mathbf{v}_1 (\mathbf{ds}_2 - \mathbf{ds}_1) + d\mathbf{v} \wedge \mathbf{d} = 0, \tag{4.17}$$

or again

$$\mathbf{v}_1 \wedge \mathbf{d} = \text{const.} \tag{4.18}$$

Finally the pairing of two lines can be described in the following manner: define at each point of a line $L_1$ an acceptable rotation axis (in the cholesteric plane perpendicular to the molecule) and a line $L_2$, so paired to $L_1$ that the homologous points $A_1$ (on $L_1$) and $A_2$ (on $L_2$) are situated at distance $d = n\pi/2q_0$ apart, measured in the same cholesteric plane, and so that the axis $\mathbf{v}_2$ is antiparallel to $\mathbf{v}_1$, the molecule at $A_2$ being parallel or perpendicular to the molecule at $A_1$. The ribbon between $L_1$ and $L_2$ is covered with a density of translation and rotation dislocations given by equation (4.13). The pair thus defined behaves at long range like a translation dislocation.

Two simple cases responding to this definition can be distinguished.

(1) *Translated planar loops.* $L_1$ and $L_2$ are two loops situated in the same cholesteric plane and related to each other by a translation $d = n\pi/2q_0$. $\mathbf{v}_1$ and $\mathbf{v}_2$ have fixed directions, and the molecules $A_1 A_2$ are related to each other by the same translation. However, it must be noted that such lines do not seem to occur

88

experimentally. The loops which are observed (Rault, 1972b; Bouligand, 1974) are more like the helicoidal lines described in the case which follows, and here reduced to a whorl closed by a kink.

(2) *Helicoidal concentric loops.* These loops are acceptable if their pitch is equal to the cholesteric pitch and the vectors $v_1$ and $v_2$ are antiparallel for points $A_1$ and $A_2$ having the same projection on the axis of the helix. They correspond to a uniform density of rotation dislocations radially crossing the ribbon $L_1L_2$. The axis of these rotations is radial, and their density is equal to the curvature of the projection of the helices on the cholesteric plane. It can be shown that the cholesteric has a different pitch inside and outside these helices. The effect of the pair is to introduce $n$ half-pitches (for $d = n\pi/2q_0$) inside the cylinder for each pitch of the pair. These configurations are understood better by using a mode of generation of helicoidal pairs which will be explained in the following subsection.

### 4.4.3 Helicoidal generation of pairs (Bouligand and Kléman, 1970)

Consider a pattern of dimension $p_0/2$ along an axis $pp'$, this pattern being semi-infinite in the direction transverse to $pp'$ and containing a molecular arrangement analogous to that of a perfect crystal (Figure 4.7a). Turn this rigid pattern about $pp'$ with a pitch $P = (\frac{1}{2})mp_0$.

This operation fills the space without recovery or vacancies and creates a singularity along the axis $pp'$. By convention we take $m > 0$ if the helicity of the

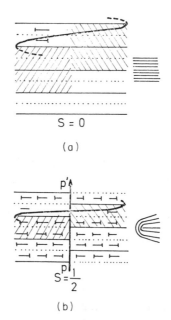

Figure 4.7. Helicoidal generation of $\chi$ lines. (a) $P = p_0$ $(m = 2)$ is the identity operation; (b) $P = p_0/2$ $(m = 1)$ a line $S = \frac{1}{2}$ is obtained

operation $P$ is the same as that of the cholesteric, $m < 0$ if it is not the same. Bouligand and Kléman have shown that the operation results in a wedge line $\chi$ along $pp'$ of order

$$S = 1 - \tfrac{1}{2}m. \tag{4.19}$$

It will be noted that for $m = 2$ no singularities are created. This will appear obvious on reference to Figure 4.7a. Figure 4.7b represents the helicoidal generation of a line $S = (\tfrac{1}{2})\,(m = 1)$; on the right the configuration of molecules in a plane perpendicular to $pp'$ is represented.

The planes perpendicular to $pp'$ can be considered as the transforms of the planes of the perfect cholesteric. This point of view, which describes the $\chi$ lines as rotation dislocations, suggests wrongly that the cholesteric is distorted to infinity. But this is not the case. The reader will in fact easily demonstrate that the final configuration will be obtained again by submitting each cholesteric plane, beginning with the perfect cholesteric, to a displacement $u = b\theta/2\pi$ along the direction $pp'$ with

$$b = -S \qquad p_0 = \left(\frac{m}{2} - 1\right)p_0, \tag{4.20}$$

i.e. by creating a screw dislocation along $pp'$, according to the usual Volterra process.

The core of the $\chi$ can be dissociated in the previously suggested form of a pair $L_1 L_2$; this can be understood by taking a modified pattern. Thus, Figure 4.8 represents a pattern formed by a pair $\lambda^+ \tau^-$, turning with a pitch equal to the half-pitch of the *exterior medium*, i.e. $m_e = 1$. The interior medium is of different pitch: for this $m_i = 2$. There is therefore no singularity on the axis (equation (4.19)). Finally, we have therefore created a singularity $S = S_e = \tfrac{1}{2}$, corresponding to a screw dislocation of Burgers vector $\mathbf{b} = -p_0/2$, and dissociated into two rotation

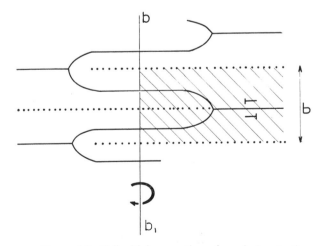

Figure 4.8. Helicoidal generation of a pair (see text)

90

dislocations $\lambda^+$ and $\tau^-$. It must be emphasized here that the pitch of the *exterior* medium is that of the cholesteric, while the pitch of the *interior* medium is that of the core zone distorted by the presence of the dislocation. The diameter of this core zone is established in order to minimize the free energy.

Two extreme cases are important:

(a) Where the exterior medium takes an infinite pitch: this is the situation of Figure 4.9a, which can be interpreted as the scheme of a cholesteric rod-like cell plunged into an isotropic medium. It will be seen that this pattern, wound with a pitch $m_i = 2$, allows the nature of the boundary between the rod and the medium surrounding it to be understood.

(b) Where the interior medium takes an infinite pitch (Figure 4.9b): here the interior is in fact replaced by a line $\lambda^{+2}$ inserted in a cholesteric *orthogonal* to the exterior cholesteric. This scheme corresponds to observations made by Rault (1973) on free planar drops of mixture MBBA–cholesterol propionate. Many helicoidal lines $\chi$ can be seen ($S = -1$) winding around their axis which consists of a $\lambda^{+2}$.

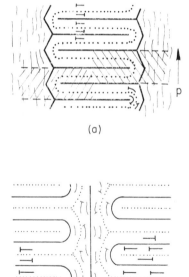

(a)

(b)

Figure 4.9. (a) Cholesteric rod-like cell† in an isotropic medium, created helicoidally; (b) helicoidal line ($S = -1$) surrounding a line $\lambda$ ($+1$): these lines are of opposite sign and the medium is not distorted at infinity

† *Translator's note:* In French, 'bâtonnet'. This word was used primarily by G. Friedel (1922) to describe the germs of smectic A phase nucleating in the isotropic phase. The configuration of Figure 4.9 bears some resemblance to them.

#### 4.4.4 Observations: lines, kinks, singular points

Many observations of helicoidal lines and edge lines have been made by Bouligand (1969, 1974) and Rault (1971, 1974a). We refer to these authors for the description of the principal phenomena observed. These are interpreted in accordance with the preceding topological models, and with the natural tendency of the cholesteric to form lines of the type $\lambda$ rather than $\tau$, i.e., for example, $\chi$ dissociated into $\lambda^+ \lambda^-$ lines, therefore of integral strength. Bouligand (1974) has also made many observations of kinks along the length of edge lines which cause their transition from one cholesteric plane to another and has shown that kinks without singularities can be distinguished from kinks which are analogous to singular points. These articles carry an iconography of cholesteric arrangement which is particularly careful and convincing.

Rault (1973, 1974a) made a detailed study of helicoidal lines and their different modes of glide and climb across the cholesteric.

#### 4.4.5 Line displacement

##### 4.4.5.1 *Isolated rectilinear $\tau$ or $\lambda$ lines*

Let us consider first the displacement of a $\lambda$ or a $\tau$ in an otherwise fixed cholesteric. Figure 4.10a represents, for example, the surface section of a $\tau^-$ being displaced along the cholesteric axis by a quantity $x < \pi/4q_0$. In the intermediate state of Figure 4.10b, the molecules in contact with the vacuum on the upper part of the surface section are not parallel to the molecules on the lower part. Therefore the left-hand side of the crystal must be deformed by a rotation $2xq_0$ of the molecules of a half-plane, or by opposed rotations $\pm xq_0$ on the two half-planes of the cell surface, to enable it to join together with the perfect crystal on the right. The result is shown in Figure 4.10c. The $S$ plane is obviously a plane of symmetry of the figure. It is clear that this is an excited state of $\tau^-$, as a simple reorganization of the core brings back $\tau^-$.

Note that the left-hand side of the crystal could also have been submitted to a rotation of $-\pi + 2xq_0$, which comes to the same thing as attaching a dislocation $\chi$ to $\tau^-$. The result is then an excited state of $\lambda^-$.

An analogous analysis can be made for the $\lambda^+$ and $\tau^+$. A displacement of $x$ along the cholesteric axis corresponds to a rotation of $xq_0$ of each molecule around the local helicoidal axis, and in the end to a rearrangement of higher energy of the molecules of the core. Figure 4.11a shows this rotation and Figure 4.11b describes the isoclinic lines of the same situation.

##### 4.4.5.2 *Absorption or emission of a $\chi$ line* (Frank)

The analysis above shows clearly that a $\lambda$ (or $\tau$) is transformed into a $\tau$ (or $\lambda$) by the absorption or emission of a $\chi$ parallel to the line. More precisely, Frank (1969,

92

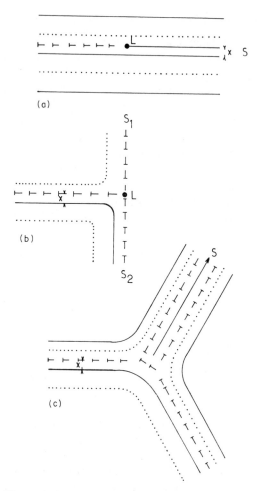

Figure 4.10. Displacement of a line $\tau^-$ (see text)

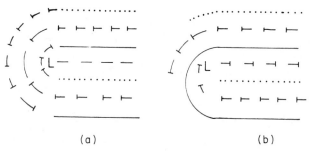

Figure 4.11. Displacement of a line $\lambda^+$ (see text)

oral communication to the Montpellier conference on liquid crystals) proposed the following reactions:

$$
\begin{array}{ll}
\tau_-(0) + \chi_+ \rightleftarrows \lambda_-(a); & \tau_+(0) + \chi_- \rightleftarrows \lambda_-(a), \\
\tau_-(0) + \chi_- \rightleftarrows \lambda_-(-a); & \tau_+(0) + \chi_- \rightleftarrows \lambda_-(-a), \\
\lambda_-(0) + \chi_+ \rightleftarrows \tau_-(a); & \lambda_+(0) + \chi_- \rightleftarrows \tau_-(a), \\
\lambda_-(0) + \chi_- \rightleftarrows \tau_-(-a); & \lambda_+(0) + \chi_+ \rightleftarrows \tau_+(-a),
\end{array}
\tag{4.21}
$$

where $a = \pi/2q_0$ is the quarter pitch.

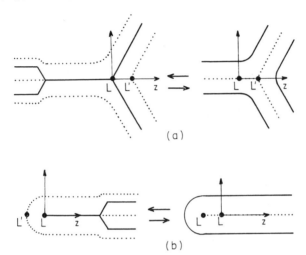

Figure 4.12. Absorption or emission of a line $\chi$ (Frank)

Frank has also made the following observation. Imagine a $\chi^+$ (Figure 4.13) turning anticlockwise around a $\tau^-(0)$ by an angle $2\pi$. In this process the $\chi^+$ is transformed into a $\chi^-$, which can again be described as the absorption of the $\chi^+$ by the $\tau^-$ on an axis of symmetry, and the emission of the $\chi^-$ on another axis of symmetry. According to reaction (4.21), the $\tau^-$ is displaced by a distance $-2a$ during the movement of the $\chi$. A line $\lambda$ or a line $\tau$ can therefore be displaced by turning a $\chi$ around it.

### 4.4.5.3 Climb and glide of a pair

Every movement of a $\chi$ or a pair can be decomposed into a glide along a direction parallel to the Burgers vector and a climb along a direction of the cholesteric plane, perpendicular to the Burgers vector.

The glide of a pair takes place by progressive emission of one line of the pair and its progressive absorption by the other line. This follows on from the discussion subsection in 4.4.5.2 and is represented in Figure 4.14. This process is probably difficult as it demands a core modification variable with the distance of the glide, a perfect situation being obtained at every quarter pitch. The glide of a

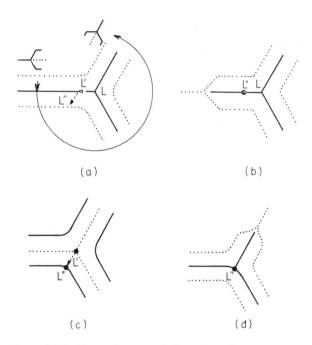

Figure 4.13. Absorption or emission of a $\chi$ line (see text)

non-dissociated $\chi$ has the same anisotropic character, this time because of the anisotropy of Frank's elasticity coefficients (the language of Peierls–Nabarro forces can be used: Caroli and Dubois-Violette, 1969, have calculated the energy of a twist $\chi$ as a function of its position on the cholesteric axis).

The process of climb is more favoured, both for non-dissociated $\chi$ and for the pair, and is observed particularly well in preparations of *planar textures* of the Cano wedge type (see below). Rault's observations, which indicate that the helicoidal pairs $\chi(-1)$ spread easily in large edge segments when they are obliged to take an oblique direction in relation to the cholesteric planes, must undoubtedly also be attributed to this facility for climb. It is interesting to note that this phenomenon does not exist for the helicoidal pairs $\chi(+1)$, which moreover have a core structure which is quite different from that of $\chi(-1)$ (Rault, 1974a).

## 4.5 TEXTURES

### 4.5.1 Classification of textures

There are essentially three types of texture distinguishable in cholesterics (G. Friedel, 1922; Bouligand, 1972b, 1973a, b, c, 1974; Rondelez, 1973): fan textures,[†] polygonal fields and planar textures. They differ from each other in the nature of

† *Translators' note:* In French, 'plage à éventails';

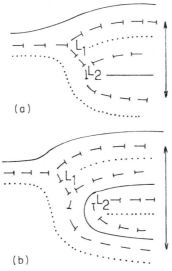

Figure 4.14. Glide of a pair

the defects of which they consist. They can be obtained in one sample between slide and coverslip, in a weak horizontal temperature gradient (or in a gradient of molecules chiralising a nematic) (Figure 4.15). There is a continuous transition from a texture with strata perpendicular to the isotropic liquid (fan) to a structure

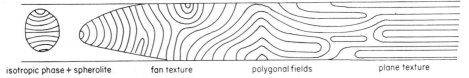

isotropic phase + spherolite      fan texture      polygonal fields      plane texture

Figure 4.15. Texture of a cholesteric between slide and coverslip: left to right: isotropic phase containing nucleated spherolites, fan texture, polygonal fields and planar textures

with planes parallel to the slides, which certainly produces a minimum of surface energy. Fan textures are complex arrangements of rotation dislocations and translation dislocations, as well of lines of flare[†] (Bouligand, 1973) which are analogous to focal lines in smectics. Polygonal fields are often regular arrangements of lines of flare and translation dislocations. Planar textures essentially contain edge dislocations.

### 4.5.2 Fan textures

For reasons not yet elucidated, the growth of a cholesteric in the isotropic phase takes place in the form of spherolites whose surface is perpendicular to the strata (Figure 4.9a). In all probability, the molecules are tangential to the surface

_____

† _Translators' note:_ In French, 'lignes d'évasement'.

96

and are arranged as a helicoidal rotation dislocation of the type shown in Figure 4.9a.† When two such spherolites meet, at variable incidences, they fuse, creating either an isolated rotation dislocation line (Figure 4.16b) (if the size of the objects is not much greater than the pitch, or if one of the objects is much smaller than the other) or a wall of edge dislocations of the same sign, if the two spherolites being paired are large compared to the pitch (Figure 4.16c). The role played by the helicoidal dislocation in this fusion has never been studied. In any case, this process of fusion is at the origin of the many different facies of the fan textures formed from subvertical cholesteric layers. Some of them are analysed here. The fan textures of the cholesterics seen on a large scale are similar to the fan textures of smectics studied by G. Friedel (1922).

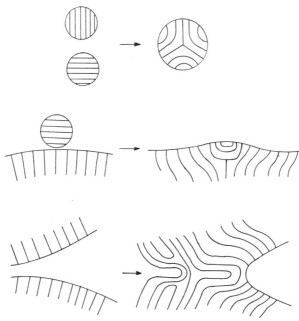

Figure 4.16. Fusion of two cholesteric areas (a) small spherolites; (b) spherolites of very different sizes; (c) wall of dislocations

### 4.5.2.1 Lines of flare

The coarse-grained version of the elasticity of cholesterics (equation (4.5)) suggests that the essential defects of cholesterics of small pitch are of the same type as those observed in smectics, i.e. *focal conics*. As a detailed study of these objects in cholesterics is not given here, it is advisable to read first the exhaustive description in Chapter 5. We remember here that the term for the compression of layers $K_2 q^2 (\delta p/p_0)$ is minimized (in fact reduced to zero) when the deformation of

---

† PBLG behaves in a very different manner from other cholesterics; it grows in the form of spherolites having a dislocation *radius* (see Robinson, 1966; Bouligand and Kléman, 1970).

the cholesteric takes place at *constant pitch*. Parallel surfaces perpendicular to the local cholesteric axis can therefore be defined in the deformed cholesteric. The normals to this axis envelop the two sheets of a *focal surface*, which is a singular locus for the molecular arrangement. If this focal surface is broken down into lines, we end up with an ellipse and a hyperbola which are confocal and situated in orthogonal planes. The term *conjugated lines* will be used.

The coarse-grained version introduces a small compression coefficient $K_2 q_0^2$ (with $K_2 = 10^{-6}$ dyne and $q_0 = 10^{+4}$ cm$^{-1}$, one has $K_2 q_0^2 \sim 10^2$ dynes/cm). The preceding scheme must therefore be modified by allowing large variations in thickness of the layers. On the other hand, it ignores the curvature terms of the envelope lines of **d** which become important for a strongly distorted situation. The analysis of these terms requires the use of methods which will be introduced in Chapter 6 and 7. For all these reasons, *the smectic scheme is only convenient in the first approximation*. In fact, there are thought to be considerable differences. Without the knowledge of a quantitative theory of deformed cholesterics, we will be restricted to a morphological description (simplified) inspired by Bouligand's work.

Conjugated lines are not necessarily conics, but may be line segments $L,L'$ such that by joining $L$ to any point on $L'$ in a straight line a congruence of straight lines normal to the cholesteric plane is obtained (see Figures 5.22 and 5.23). As we will show in Chapter 5, a relation of conjugation possessing such properties of layer–normal transversality can only be true for short orthogonal segments $L$ and $L'$. If $L$ and $L'$ are large compared to the pitch, the normals to the layers must be imagined not to be rectilinear. Furthermore, $L$ and $L'$ are not necessarily orthogonal. This situation occurs frequently: it was noted by Bouligand that the conjugated lines are not in focal positions. For convenience, anyway, the term focal domain will be used to designate the region of space described by these two conjugated lines. Experiment has shown that in a focal domain, the cholesteric layers locally have the form of a horse's saddle.† It will be seen in Chapter 5 that this geometry minimizes the term $(\mathrm{div}\,\mathbf{d})^2$ of the coarse-grained version.

In the vicinity of these lines, the layers form a boxed conical texture with cones axes along the line (Figure 4.17) which justifies the term *ligne d'évasement* in French; there is a strong maximum‡ in the pitch on the line and a maximum of the mean curvature $|1/R_1 + 1/R_2|$ of the layers, as on a focal line (hence the term of line of flare).

$\lambda^+$ disclinations are often the source of half lines of flare whose rotational geometry is compatible with the arrangement of the layers in the vicinity of a $\lambda^+$ (or $\tau^+$). Conversely, an entire splay line may be divided into a pair of disclinations of the same sign (Figure 4.18). The $\lambda^+$ disclinations therefore often play a focal role. The same may not apply to the $\lambda^-$ if one of the three sectors in which a $\lambda^-$ is

---

† Never a sphere except in the unorthodox case of PBLG.

‡ This experimental fact is certainly important. If the elasticity is treated in the same way as the compression and dilation of the layers, the essential defects would be screw dislocations (see section 5.8). As the coarse-grained version does not contain a linear term in $p/p_0$ it may be inferred that the version is insufficient.

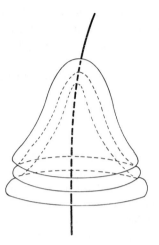

Figure 4.17. Line of flare

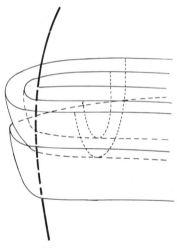

Figure 4.18. Relation between a line of flare and a $\lambda^+$ (or $\tau^+$) disclination. This configuration is found either on the surface (it must be imagined that the part in front of the diagram is virtual, see Figure 4.21) or in the bulk (Figure 4.22); it may then play a focal role

Figure 4.19. A $\lambda^-$ (or $\tau^-$) cannot play a focal role (after Bouligand)

spatially divided contains layers of opposite curvatures. It obviously is not the same with the two others (Figure 4.19). Furthermore, $\lambda^-$ lines which are involved in a relation of conjugation have never been found experimentally. On the other hand, it is observed that focal lines or lines of flare can be attached to them at singular points. This property is found in most of the facies of fan textures now being outlined.

### 4.5.2.2 *Facies with ellipses* (Figures 4.20a, b)

This facies is illustrated in Figure 4.20 (after Bouligand): it consists of focal ellipses situated on one of the plates, conjugated with a hyperbola crossing through the mass and continuing in a straight segment playing a focal role on the other plate. One part of the ellipse is replaced by a $\lambda^+$ linked to the centre of the ellipse by $\lambda^-$. On the surface itself, this arrangement is similar to that of the elementary pinch described in the chapter on nematics. It is obtained by the shearing of a sample with vertical layers (Bouligand 1973b). Figure 4.21 (after Bouligand) represents another possibility of conjugation between lines situated on opposing plates.

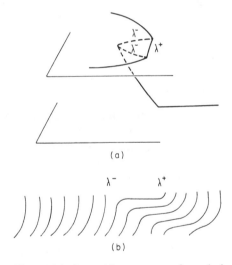

(a)

(b)

Figure 4.20. The top ellipse (a) is formed by two arcs of parabolas linked by a $\lambda^+$. The hyperbola consists of two segments, one in the bulk, the other on the lower surface. Figure 4.20b represents a vertical cut along the large axis of the ellipse (after Bouligand)

The *finger print texture* is an elliptical facies with a different origin. It has frequently been observed in the region of high temperature, near the transition (Rault and Cladis, 1971). These ellipses are formed by lines of flare in the form of hyperbolic segments which connect the vertical orientation of the variable layers. (cf. Bouligand, 1973).

100

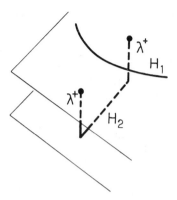

Figure 4.21. Conjugation of two surface hyperbolas $H_1$ and $H_2$. The focal domain is bounded in space by the presence of focal segments in $\lambda^+$

### 4.5.2.3 Zigzags and quadrilaterals

Vertical walls separating two zones of different orientation are frequently in the form of zigzags due to the splitting of translation dislocations on the boundary into pairs of rotation dislocations to which lines of flare are often linked. More general arrangements of vertical dislocations can lead to a typical aspect such as lozenges and polygons whose sides, marked by lines of flare, link neighbouring rotation dislocations of opposite signs.

### 4.5.2.4 Chevrons (Bouligand, 1973c)

In certain circumstances, lines of flare attached on the surface establish themselves into parallel lines situated at regular intervals and bound two by two by straight lines which are along the cholesteric axes between the top and bottom of the prepared sample. Thus, a line on one surface is connected to two lines on the other (Figure 4.22). A slightly analogous situation has been observed in *smectics*

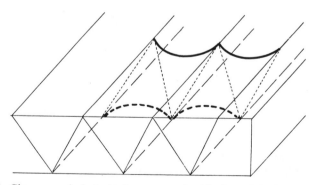

Figure 4.22. Chevrons: cholesteric planes organized in 'boxed' half-cones of revolution whose axes are alternatively on one lamella then on the other

(section 5.8). It should be noted that the lines of flare *conjugated* between the top and bottom are not in a focal relationship. In fact, each triangular domain, bound by a line $L$ on one face and two lines $L'$ and $L''$ on the other, can be constructed as a focal domain whose focal lines are $L$ and a circle at infinity in the plane perpendicular to $L$.

These *chevrons* have been observed by the application of a magnetic field along the helicoidal axis on a sample with vertical layers (Rault and Cladis, 1971).

### 4.5.2.5 *Blue phase*

Cholesterol myristate in the very vicinity of the isotropic–cholesteric transition has the optical aspect of a milky phase strongly diffracting in the visible region. X-Ray studies have shown it to have a cubic structure, and Saupe (1969) proposed a structural model in which this cubic phase consists in a lattice of disclinations of the cholesteric phase ($\lambda(S = 1)$) at a distance apart of the order of the pitch. Since then, the 'blue phase' has been recognized in a number of cholesterol derivatives (Coates and Gray, 1973; Stegemeyer and Bergmann, 1980) and the stability of this phase has attracted the interest of theoreticians. It is now recognized that many 'blue phases' (three in some cases, one being described as a blue fog and possibly amorphous (Marcus, 1981)) can be thermodynamically stable in a narrow range of temperature below the isotropic clearing point ($\sim$ a few degrees), displaying various cubic structures. The standard theory relies on a Landau expansion of the free energy (Brazovskii and Dmitriev, 1975; Hornreich and Shtrikman, 1980), where the order-parameter describing the transition between the isotropic to the mesomorphic state is written as a linear combination of spherical harmonies, and a set of basic wave vectors. The structure obtained in this manner is very similar to that one proposed initially by Saupe (1969). A very recent model makes a return to Saupe's ideas and describes the blue phase as a phase of defects (Meiboom *et al.*, 1981).

### 4.6.2.6 *Surface phenomena* (Cladis and Kléman, 1972b)

Fan textures, which correspond to a subvertical position of the layers, are incompatible with a uniform anchoring of the molecules on the surface. Cladis and Kléman have analysed and observed surface defects with homeotropic and planar anchoring conditions. Their results are shown in Figure 4.23. It would be useful to extend their calculations (done for infinite anchoring) explicitly to take into account the anchoring energies. On the other hand, no analysis has ever been made of the role of these surface defects in the nucleation and topology of the fan texture and polygonal fields.

### 4.5.3 **Polygonal fields** (Bouligand, 1973a)

This texture is defined by two disjointed and connected arrays of lines of flare. Each array is formed from (a) hyperbolic *surface* lines $H_i$ situated on one or other

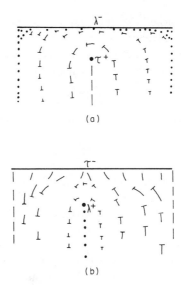

(a)

(b)

Figure 4.23. Distribution of the director **n** in the vicinity of the surface: (a) planar anchoring; (b) homeotropic anchoring

of the faces, forming a polygonal system and (b) *vertical* bulk lines projected down the vertices of the hyperbolic polygon towards the centres of the polygonal system conjugated on the other plate (Figure 4.24). The two conjugated segments $H_i$ and $H_i'$ are orthogonal. The projections of vertices $V_1'$ and $V_2'$ are the foci of the hyperbolic segment $H$. We therefore have here a tiling of space with focal domains of the type in Figure 4.21. The association between focal domains is governed by particular rules. The conjugated ellipses of the hyperbolic segments are formed from a hyperbolic segment and two straight segments. Each of these is vertical and belongs to several focal systems at the same time. Each is therefore the conjugate of all the hyperbolic segments of the polygon surrounding it, from which it is disjointed. But each hyperbolic segment is only conjugated with an 'ellipse'. This description is borrowed from the terminology used for focal conics

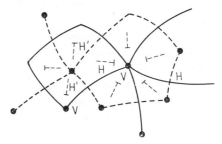

Figure 4.24. Polygonal field: theoretical scheme. The arrays $H_i$ and $H_i'$ are situated at different levels and their mutual projections are orthogonal. The vertical lines from the vertices of each polygon are lines of flare of the structure. In each sub-polygon the layers have a mean direction indicated here by the nail symbol

in smectics. But it will be found that the rules of association here authorize a total filling of the space with an infinite number of domains of analogous size, above all at the cost of considerable variations in the thickness of the layers and sometimes of the presence of screw dislocations. The symmetry between the two arrays of lines of flare is total.

The chevrons described above are a particular case of a polygonal field where the thickness of the layers remains constant. Quadrilaterals, the most symmetrical polygonal system, are illustrated in Figure 4.25.

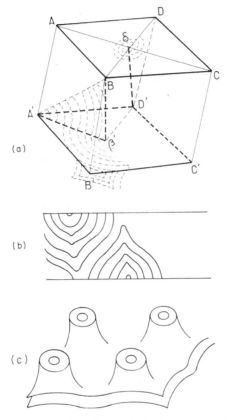

Figure 4.25. (a) Arrangement of domains and placing of cholesteric layers in a unit parallelopiped, extracted from a quadratic lattice; (b) placing of the cholesteric layers of a quadratic lattice cut vertically along the diagonal BD; (c) bird's eye view of the layers (after Bouligand)

Polygonal fields can be obtained by rocking the texture in a magnetic or electric field (Helfrich, 1970; Rondelez, 1973; Rondelez and Arnould, 1971; Rondelez et al., 1972) or by mechanical dilation (Rondelez; Clark and Meyer, 1973; Delrieu, 1974). The cholesteric storage mode, a non-persistent deformation obtained electrohydrodynamically, is of the same type (Heilmeyer and Goldmacher, 1968, 1969).

104

### 4.5.4 Planar textures

The defects in planar textures contained between two parallel plates are *edge dislocations*, which display numerous *kinks*, where the dislocations change plane. The frequent merging of edge dislocations in an accumulation of parallel lines gives the *oily streaks*[†] described by G. Friedel. The rotation dislocations into which the edge dislocations decompose are sometimes decorated with a line of flare, and focal domains then appear (see Bouligand, 1973b, 1974).

In a free droplet, the planar configurations intersected by the droplet give rise to characteristic configurations (domes and basins) which are observed as spirals (under natural light) and as concentric rings (between crossed nicols (Bouligand)). Note that if the molecules are projected on a plane cut obliquely from a cholesteric, the force lines of these projections form a series of arches[‡] (Bouligand, 1969) which are frequently observed in the section of the shell of certain crabs. This observation (and its correct interpretation) allowed Bouligand to propose a cholesteric pseudomorphosis for shells (Figure 4.26).

Figure 4.26. Bouligand arches: an array of cholesteric planes of diminishing dimensions have been stacked together. The overall aspect is of a double spiral described by the arches (after Bouligand)

These arches have been demonstrated by a decoration technique (Cladis *et al.*, 1971). The action of a magnetic field perpendicular to the cholesteric axis on the configuration of the free surface has been studied by Rault (1972a, 1974c): when the anchoring conditions on the free surface favour an orientation almost the same as the planar orientation, the layers have an instability of orientation called *commas*[§] (G. Friedel) which were recently interpreted by Rault (1974b). The asymmetry of the commas is linked to the sign of chirality of the cholesteric.

The texture observed in the *Grandjean–Cano wedge* corresponds to a simple and important physical situation. A cholesteric is inserted into a dihedron at a small angle, formed either from a fresh cleavage of mica (which will impose planar orientation) or two glass plates (Figure 4.27) rubbed in a direction perpendicular to the intersection of the dihedron. In practice, the cholesteric is introduced between a flat slide and a cylindrical glass, rubbed in a direction perpendicular to the generatrix of contact. A lattice of edge dislocations is then formed parallel to

[†] *Translators' note:* In French, 'stries huileuses'.

[‡] *Translators' note:* In French, 'arceaux'.

[§] *Translators' note:* In French, 'virgules'.

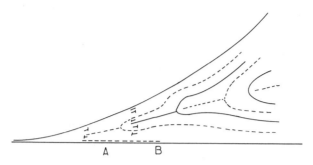

Figure 4.27. Grandjean–Cano wedge. Region A is of thickness $p/2$, region B is of thickness $p$

the edge, observed by Grandjean (1921). Grandjean had described these optical singularities as walls which he attributed, although in an unclear manner, to the intersection of the cholesteric layers, which are supposed as being stacked perfectly, and the substrates. The work of Cano (1968), Kassubeck and Meier (1969) and de Gennes (1968b) caused the idea that edge dislocations marked the discontinuous passage from an integral number $n$ of half-layers to another integral number $n + m$. The Orsay Liquid Crystal group (1969a, b) then showed that for the lines nearest to the edge, $m$ equals 1 (thin threads, also called single lines), but following that, essentially $m$ equals 2 (thick threads, also called double lines). The action of a magnetic field applied in the plane of the sample perpendicularly to the lines has remarkable effects. It is now known that under a field applied perpendicularly to the helicoidal axis the cholesteric pitch is expanded according to the law proposed by de Gennes:

$$\frac{q_0}{q(k)} = \left(\frac{2}{\pi}\right) E(k), \tag{4.22}$$

with $k = (2/\pi)H/H_c$, the functions $E(k)$ and $K(k)$ being the classic elliptical functions (see Abramovicz and Stegun), and becomes infinite for $H = H_c$ given by equation (4.9). The lines are therefore displaced towards the outside of the dihedron. When $H$ reaches a value of the order of $H_c/2$, the double lines take on a zigzag configuration while the single lines remain unaffected by the field. These different observations bring back the dissociation model $\lambda^+ \lambda^-$ of double lines (J. Friedel and Kléman, 1970). In fact the cholesteric axis tends to become perpendicular to $H$ ($\chi_a > 0$): the zigzag rotation of the line then leads to the diminution of the magnetic energy of the core of the $\lambda^+ \lambda^-$. By adopting a simplified model of this core (Figure 4.28) Friedel and Kléman estimate that the field necessary for the formation of zigzags is of the order of $H_c/m$. This effect is therefore observable for $m = 2$ (double lines) but not for single lines. Their explanation is as follows.

The line energy depends on the pitch of the medium $p(H)$ and on the angle made by the line with the field. Let us call $\theta$ the angle of the normal to the line $H$. The line energy $W(\theta)$ must then be distinguished from the *line tension* $\tau$, i.e. from

106

the configurational force which must be exerted along the tangent to the line, if the line is cut at any point, so as to maintain it at equilibrium. This issue is discussed in J. Friedel (1964). We have the relation:

$$\tau = W(\theta) + \frac{d^2 W}{d\theta^2}. \tag{4.34}$$

When the line is perpendicular to the field, $\tau$ may be reduced to zero and *the line may become unstable.* It will be supposed that in the line energy the dependence on $\theta$ comes essentially from the core. The surface of this region is of the order of (see Figure 4.28):

$$\frac{m^2 \pi^2}{4q(H)q(H \sin \theta)}.$$

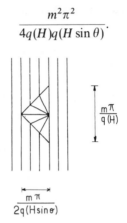

Figure 4.28. Schematic distribution of the cholesteric axes in the core of the pair

The term independent of $\theta$ can be estimated at double the energy of a dislocation $\pm\frac{1}{2}$ of dimension $m\pi/2q_0$,

$$\frac{\pi K}{2} \ln \frac{m\pi}{2q_0 r_c}.$$

We therefore have:

$$W(\theta) = \frac{\pi K}{2} \ln \frac{m\pi}{2q_0 r_c} + \frac{m^2 \pi^2}{4q(H)q(H \sin \theta)} f(H'), \tag{4.24}$$

where $H' = \langle H \sin \theta \rangle$ is the mean value of the effective field in the core, and $f(H)$ is the energy density of a cholesteric submitted to a field $H$ perpendicular to the helicoidal axis. This quantity has been calculated by de Gennes (1968a). Let us suppose that $\theta$ is small. We then have $H' \sim H\theta \times 2/\pi$; the approximation of the small fields may be made in de Gennes' equations. This yields, by putting $q(H) \sim q_0$:

$$f(H') \sim -K \frac{q_0^2}{8} h^2 \tag{4.25}$$

with:

$$h = \frac{\pi}{2} \frac{H}{H_c} \langle \theta \rangle, \tag{4.26}$$

and finally:

$$W(\theta) \sim \frac{\pi}{2} K \ln \frac{m\pi}{2q_0 r_c} - \frac{m^2\pi^2}{4} \frac{K}{8} \left(\frac{H}{H_c}\right)^2 \theta^2,$$

$$\tau = \left[ W(\theta) + \frac{d^2 W}{d\theta^2} \right]_{\theta=0} = \frac{\pi}{2} K \ln \frac{m\pi}{2q_0 r_c} - \frac{m^2\pi^2}{16} K \left(\frac{H}{H_c}\right)^2. \tag{4.27}$$

The line tension is therefore reduced to zero for:

$$\frac{H}{H_c} = \frac{1}{m} \left( \frac{8}{\pi} \ln \frac{m}{2q_0 r_c} \right)^{1/2}. \tag{4.28}$$

This calculation is a good indication that the double lines ($m = 2$) become unstable for $H \sim H_c/2$. For this value of $H$, $\theta$ is still small, and the different quantities can be approached in the way shown above. For $m = 1$, a more precise calculation is necessary. We must write indeed:

$$\frac{q_0}{q(H)} \sim \left(\frac{2}{\pi}\right)^2 \frac{1}{2} \left( \ln \frac{16}{1 - H^2/H_c^2} \right),$$

but the calculation in fact again leads to a dependence of $H/H_c$ on $m^{-1}$.

If the model of the double lines is well established by these experiments (and by the many observations of fan textures), that of single lines still remains a subject of discussion. Bouligand (1974) has shown, however, by using a topological analysis of kinks, and by comparing Cano wedges with rubbing perpendicular or parallel to the dihedron, that the *first line*, which is established in a zone of thickness less than the half-pitch is of the nematic type $|S| = \frac{1}{2}$, i.e. quasi-planar. It is a thin thread of the first kind. The thin threads which follow it (of the *second kind*) are decomposed into $\tau^- \lambda^+$ (not $\tau^+ \lambda^-$). This result seems to cope with the idea we have for respective energies of a $\tau^-$ and a $\tau^+$. It is clear that in the latter case the singularity is less spread, therefore the energy is greater.

# Chapter 5
# Smectics A

The term smectics refers to a very diversified collection of mesomorphic phases with properties between those of nematics and crystalline solids. Some can be classified satisfactorily as mesomorphic solids (i.e. smectics A which will be studied in this chapter) whereas others still arouse some controversy. However they are all lamellar phases.

## 5.1 SYMMETRIES OF SMECTIC PHASES

The term 'smectic' ($\sigma\mu\eta\gamma\mu\alpha$ — soap) was introduced by G. Friedel to designate a mesomorphic phase found in certain liquid crystals, soaps and lipid–water mixtures. This phase is characterized by the existence of a lamellar arrangement of elongated molecules, their axis of elongation being perpendicular to the lamella (Figure 5.1). G. Friedel only discovered smectics A which are characterized by the existence of liquid molecular layers. A large number of smectics phases has now been identified, and new ones are still in the process of being discovered. The study of the different structures which exist in an increasing number of compounds is an extremely active field of research.

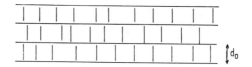

Figure 5.1. Molecular arrangement in a smectic A

### 5.1.1 Smectics A

The molecules are perpendicular to the layers and their centres of gravity are disordered within these layers. In the *basic* state of the crystal the layers are planar, but they easily slip over each other and the molecules diffuse readily inside the layers. Each layer can be considered as a two-dimensional liquid. The $z$ axis perpendicular to the layer is equivalent to the $-z$ axis. There are strong correlations of position between the layers and the liquid is optically uniaxial.

### 5.1.2 Smectics C

These are liquid layers with strong correlations of position between the layers as in smectics A, but the molecular axis is at an angle to the normal of the layers (Figure 5.2). Thus it has a less symmetrical structure than smectics A. If a material has both phases, it exists in a lower temperature domain than smectics A. The C phase has a plane of symmetry perpendicular to the layers (plane in Figure 5.2), a binary axis perpendicular to this plane and a centre of symmetry. It is also optically biaxial.

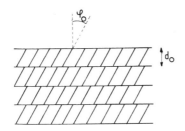

Figure 5.2. Smectic C

### 5.1.3 Chiral smectics C (C*)

If the molecule is active, the smectic C can be helicoidal, the helicoidal axis being perpendicular to the layers. There is no longer a plane of symmetry or a centre of symmetry, but there is a binary axis normal to the molecule and in the plane of the layers. This binary axis follows the molecule in its helicoidal movement. It may have an electric dipolar moment, which gives rise to a ferroelectric liquid (Meyer *et al.*, 1975) (Figure 5.3).

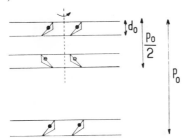

Figure 5.3. Smectic C*

### 5.1.4 Smectics B, E, F

In the structures, the molecules display a *two-dimensional order*. There is some confusion concerning the classification of these different phases, essentially because of controversies about their true nature. We adopt here the terminology proposed by Gray at the Garmish–Partenkirchen Conference on Liquid Crystals of One- and Two-Dimensional Order (January 1980).

*Smectic B.* The molecules are perpendicular to the layers and display a hexagonal order. They rotate freely about their axis. It appears at present that there is a long range correlation between the positions of these molecules from one layer to another. Therefore this phase is probably a three-dimensional crystal (Doucet, Levelut and Lambert, 1974; Moncton and Pindak, 1979).

*Smectic G* (formerly noted $B_c$ or H). The molecules are tilted with respect to the layers; the unit cell is monoclinic; a cross-section perpendicular to the molecules displays hexagonal packing (hence the name of pseudo-hexagonal structure for this monoclinic crystal). The molecules rotate freely about their axes. The three-dimensional order has been demonstrated (Doucet *et al.*, 1975).

*Smectic G\*.* This is the chiral modification of the Sm G phase.

*Smectic E.* The unit cell is orthorhombic; the molecules display in the layers the so-called 'herring-bone packing', which prevents them rotating about their axes. There is probably long-range correlation between layers with two-dimensional ordering.

*Smectic H* (formerly noted $E_c$ or G). This is the tilted modification of the Sm E phase.

*Smectic F.* This phase was first identified by Demus; only one compound was known to possess this order up to 1979. (Goodby and Gray, 1979). This order is remarkable in that it is intermediary between the Sm C order and the Sm G order: the layers display a pseudo-hexagonal packing but it seems that there are virtually no correlations between layers. This phase currently appears therefore as a possible candidate for the hexatic phase proposed by Halperin and Nelson (1978). Recent work by Benattar *et al.* (1980) seems to confirm this.

Other newly discovered phases, called Sm I and Sm J, bear some resemblance to the Sm F phase.

## 5.2 CHEMICAL SPECIES

*Thermotropic* types whose points of transition are well characterized will be distinguished from *lyotropic* types (soaps, lipid–water mixtures) whose points of transition depend on their water content. The principal compounds which have been or are being studied are given below, with the initials most commonly used by physicists and the chemical formula, which is of course essential.†

(a) DADB (diethyl 4.4′ azoxybenzoate) — perfect type of smectic material.

$$\text{Cr} \underset{114°C}{\longleftrightarrow} \text{Sm A} \underset{120°C}{\longleftrightarrow} \text{Liq.}$$

---

† The linear representations of the chemical formulae are given, but it is perhaps important to remember that these elongated molecules have a three-dimensional structure which determines their steric properties and influences the molecular dynamics. Some of the problems of molecular conformations in smectics are discussed at the beginning of Chapter 7.

Studied by G. Friedel.

$$C_2H_5-\overset{\overset{\displaystyle }{\|}}{\underset{\displaystyle O}{C}}-\langle\bigcirc\rangle-\overset{\overset{\displaystyle }{\downarrow}}{\underset{\displaystyle O}{N}}-\langle\bigcirc\rangle-O-\overset{\overset{\displaystyle }{\|}}{\underset{\displaystyle O}{C}}-C_2H_5 .$$

(b) CBOOA (p-cyanobenzylidene p′-octyloxyaniline)

$$Cr \underset{73°C}{\longleftrightarrow} Sm\ A \underset{83°C}{\longleftrightarrow} Nem \underset{109°C}{\longleftrightarrow} Liq$$

$$N{=}C-\langle\bigcirc\rangle-\overset{\overset{\displaystyle H}{|}}{C}{=}N-\langle\bigcirc\rangle-OC_8H_{17} .$$

(c) BBAA (butoxybenzylidene p · n-aniline acetophenol)

$$Cr \underset{84°C}{\longleftrightarrow} Sm\ A \underset{99°C}{\longleftrightarrow} Nem \underset{111°C}{\longleftrightarrow} Liq$$

$$H_9C_4-O-\langle\bigcirc\rangle-\underset{\underset{\displaystyle H}{|}}{C}{=}N-\langle\bigcirc\rangle-\underset{\underset{\displaystyle O}{|}}{C}-CH_3$$

(d) 8 CB

$$Cr \underset{21°C}{\longleftrightarrow} Sm\ A \underset{32.5°C}{\longleftrightarrow} Nem \underset{40°C}{\longleftrightarrow} Liq$$

$$N{\equiv}C-\langle\bigcirc\rangle-\langle\bigcirc\rangle-C_8H_{17}$$

(e) 8 COB (or M24)

$$Cr \underset{54.5°C}{\longleftrightarrow} Sm\ A \underset{66.6°C}{\longleftrightarrow} Nem \underset{72.3°C}{\longleftrightarrow} Liq$$

$$N{\equiv}C-\langle\bigcirc\rangle-\langle\bigcirc\rangle-O-C_8H_{17}$$

This compound is particularly stable.

(f) DOBAMBC (p-decyloxybenzylidene p′-amino 2-methyl-butyl-cinnamate)

$$Cr \xleftrightarrow{76°C} Sm\ C^* \xleftrightarrow{95°C} Sm\ A \xleftrightarrow{117°C} Liq$$
$$Sm\ G^*\ 63°C$$

The Sm C* and Sm G* phases are chiral and ferroelectric.

$$C_5^*H_{11}-C-O-CH=CH-\langle\bigcirc\rangle-N=CH-\langle\bigcirc\rangle-C_{10}H_{21}$$

C* designates asymmetric carbon.

(g) HOBHA (or 70.7) (heptyloxybenzylidene-heptylaniline)

$$Cr \xleftrightarrow[33°C]{} Sm\ G \xleftrightarrow[55°C]{} Sm\ B \xleftrightarrow[69°C]{} Sm\ C \xleftrightarrow[72°C]{} Sm\ A \xleftrightarrow[83.7°C]{} Nem \xleftrightarrow[84°C]{} Liq$$

$$C_7H_{15}-O-\langle\bigcirc\rangle-\underset{\underset{H}{|}}{C}=N-\langle\bigcirc\rangle-C_7H_{15}$$

(h) TBBA terephtalydene-*bis*-4-*n*-butylaniline

$$Cr \xleftrightarrow[113°C]{} Sm\ G \xleftrightarrow[144°C]{} Sm\ C \xleftrightarrow[172.5°C]{} Sm\ A \xleftrightarrow[200°C]{} Liq$$

$$C_4H_9-\langle\bigcirc\rangle-N=CH-\langle\bigcirc\rangle-CH=N-\langle\bigcirc\rangle-C_4H_9$$

Note that Sm B (more ordered than Sm C) is at a lower temperature.

(i) TBPA terephtalydene-*bis*-4-*n* pentylaniline

$$Cr \xleftrightarrow[]{76.5°C} Sm\ G \xleftrightarrow[]{144°C} Sm\ F \xleftrightarrow[]{153.5°C} Sm\ C \xleftrightarrow[]{182.5°C} Sm\ A \xleftrightarrow[]{213.5°C} N \xleftrightarrow[]{213.5°C} Liq$$
$$Sm\ H \quad 57°C$$

This compound, which differs from TBBA only by the aliphatic terminal chains, has a very similar polymorphism, except for the presence of the Sm F phase.

(j) Water–lecithin system (Luzzati and Tardieu, 1974). This is the type of lyotropic system which gives lamellar phases. The general formula of the molecule is:

$$R_1CO.OCH_2$$
$$R_2CO.OCH \qquad O$$
$$\overset{|}{CH_2} - \overset{\uparrow}{O}POCH_2CH^2N(CH_3)_3$$
$$\underset{O^-}{|}$$

$R_1$ and $R_2$ designate flexible, saturated or non-saturated paraffinic chains, whose essential chemical characteristic is that they are hydrophobic. The rest constitutes a hydrophilic polar head. The amphoteric properties of the molecule are the cause of the existence of mesomorphic phases in the presence of water (Figure 5.4).

Figure 5.4. Lecithin–water system

At room temperature, with water content lower than $\sim 30\%$, these systems form lamellar phases characterized by *fluid layers*. The paraffinic chains and polar heads are disordered: from the structural point of view it is a smectic A. Lyotropic systems give rise to a much richer variety of mesomorphic phases than liquid crystals.

A description of other lipidic systems can be found in Luzzati (1968).

(k) Soaps and detergents (Luzzati, 1968). This is another lyotropic system, forming mesomorphic phases in the presence of a polar solvent such as water.

## 5.3 ELASTICITY OF SMECTICS A

A phenomenological description of the elasticity of these materials is now presented. They have a liquid character along the plane of the layer and respond like solids to a force perpendicular to the layers. In other words, deformed states can be imagined in which the layers undergo great variations of curvature without the density of molecules per unit surface of the layer being fundamentally altered, while the thickness of the layers is only modified by quantities which are small in relation to the thickness.

As an introduction to effects resulting from this type of elasticity, two very different situations will first be considered. One is where the thickness of the layers is strictly invariable, but they have any curvature (this is the theoretical situation of the focal systems which will be studied later). The other is where their curvature is small and partly relaxed by the variations in thickness, which describes small distortions of the fundamental planar state. Furthermore it should be remembered that these are not the only deformations undergone by a lamellar medium (see section 5.8).

### 5.3.1 Introduction to the study of focal systems in lamellar media

Suppose that the thickness of the layers is invariable whatever the magnitude of the forces exerted on the layers. In every deformed state (Figure 5.5) these layers form a system of *parallel surfaces*. Their normals, according to an elementary theorem of geometry, are therefore common and form a congruence of straight lines depending on two parameters. (This situation is analogous to that of wave fronts in a medium of constant refractive index, the normals being analogous to light rays.)

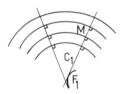

Figure 5.5. Parallel layers: the normals envelop two focal surfaces $F_1$ and $F_2$

The geometric aspects, which are fundamental to the study of these lamellar properties, will be discussed at a later point.† The study of these properties will begin with a look at the energy of such a deformation. A smectic A possesses, locally, the same symmetries as a nematic. We may then write first of all:

$$\rho f = \tfrac{1}{2}K_1(\text{div } \mathbf{n})^2 + \tfrac{1}{2}K_2(\mathbf{n} \cdot \text{rot } \mathbf{n})^2 + \tfrac{1}{2}K_3(\mathbf{n} \wedge \text{rot } \mathbf{n})^2.$$

It will be shown that the terms in $K_2$ and $K_3$ are zero in the geometry under consideration.

$\mathbf{n}$ represents the director, i.e. the normal to the layer. We can therefore write:

$$n_x = \lambda \frac{\partial \phi}{\partial x}; \qquad n_y = \lambda \frac{\partial \phi}{\partial y}; \qquad n_z = \lambda \frac{\partial \phi}{\partial z}, \tag{5.1}$$

† A remarkable exposé of surface properties will be found in the very qualitative book of Hilbert and Cohn-Vossen (1952). For a more analytical point of view refer to Julia (1954) or to the older book by Darboux (1888), recently republished (1954).

where $\phi(\mathbf{r}) = $ constant representing the set of surfaces, each layer corresponding to a value of the constant. $\lambda$ is a normalisation factor. From equation (5.1) there is no difficulty in extracting the relation:

$$\mathbf{n} \cdot \text{rot } \mathbf{n} = 0; \qquad (5.2)$$

which means that the system of directors does not possess twist. It is essential to note that relation (5.2) implies only that $\mathbf{n}$ can be written in the form (5.1), i.e. that there is a set of surfaces perpendicular to $\mathbf{n}$ but not that these surfaces are parallel. Conversely, if equation (5.2) is true, there is a set of surfaces perpendicular to $\mathbf{n}$.

To interpret the condition $\mathbf{n} \wedge \text{rot } \mathbf{n} = 0$ consider the lines of force of a field of directors $\mathbf{n}(x, y, z)$, without supposing now that a set of surfaces perpendicular to $\mathbf{n}$ exists. The significance of $\mathbf{n} \wedge \text{rot } \mathbf{n}$ is seen when placed in the local axes $\mathbf{n}$, $\mathbf{p}$ and $\mathbf{b}$; $\mathbf{p}$ and $\mathbf{b}$ are unit vectors of the principal normal and the binormal to the line of force. Now write:

$$\mathbf{n} = n_x \mathbf{n}_0 + n_y \mathbf{p}_0 + n_z \mathbf{b}_0$$

and apply Frenet's formula. We have: $d\mathbf{n}_0/ds = \mathbf{p}_0/R$, where $R$ is the radius of curvature of the lines of force.

An elementary calculation shows that:

$$\text{rot } \mathbf{n} = (\mathbf{n} \cdot \text{rot } \mathbf{n})\mathbf{n} + \frac{\mathbf{b}}{R}$$

$$\qquad (5.3)$$

$$\mathbf{n} \wedge \text{rot } \mathbf{n} = -\frac{\mathbf{p}}{R}.$$

The bend vector $\mathbf{B} = \mathbf{n} \wedge \text{rot } \mathbf{n}$ is therefore along the principal normal to the lines of force. If these are rectilinear, it is zero. It can be seen from equation (5.3) that this property does not necessarily imply that the medium is without twist $t(t = \mathbf{n} \cdot \text{rot } \mathbf{n})$, i.e. this property does not depend on whether the director has a set of perpendicular surfaces.

The quantity $\text{div } \mathbf{n}$ is linked to the mean curvature of the surface. We show that

$$\text{div } \mathbf{n} = -\left(\frac{1}{R_1} + \frac{1}{R_2}\right) \qquad (5.4)$$

if $\mathbf{n}$ is positively oriented along the increasing $z$, in a local reference frame.

In the vicinity of every regular point $M$ of a surface $\Sigma$, we can obviously write, choosing a $z$ axis perpendicular to $\Sigma$ at $M$:

$$z = \frac{1}{2}\left(\frac{x_1^2}{R_1} + \frac{x_2^2}{R_2}\right) + O(3) + \cdots . \qquad (5.5)$$

This means that a particular choice of the axes $x_1$ and $x_2$ in the tangent plane must be made in order to eliminate cross terms. These two particular directions are called principal directions in $M$. It can be recognized from equation (5.5) that $R_1$ and $R_2$ are the radii of curvature of the curves of the section of $\Sigma$ by the planes

$zx_1$ and $zx_2$. $\sigma_1 = 1/R_1$ and $\sigma_2 = 1/R_2$ are called the principal curvatures of $\Sigma$ in $M$. If $R_1$ and $R_2$ are of the same sign, the point $M$ is called elliptical. Every section of the surface $\Sigma$ in a plane parallel to the $x_1 x_2$ plane is an ellipse in the vicinity of $M$, and the two centres of curvature $C_1$ and $C_2$ of the principal sections are on the same side of the plane tangential to $M$ at the surface. It is easy to show that every section of $\Sigma$ in a plane passing through $z$ has a centre of curvature $C$ between $C_1$ and $C_2$. If $R_1$ and $R_2$ are of opposite signs, the point $M$ is called hyperbolic. Every section of $\Sigma$ in a plane parallel to the $x_1 x_2$ plane is, in the vicinity of $M$, a hyperbola. $C_1$ and $C_2$ are on either side of the $x_1 x_2$ plane and the centre of curvature $C$ is situated outside the segment $C_1 C_2$. The particular sections passing through the asymptotes of the hyperbola in $M$ have their centres of curvature at infinity (see Figure 5.6).

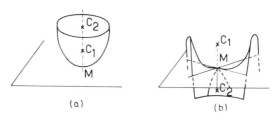

(a)   (b)

Figure 5.6. Regular points on a surface: (a) elliptical point, (b) hyperbolic point

The classical properties of the Darboux–Ribaucour moving trihedron can be applied in this case. Consider a line $\Gamma$ traced on an oriented surface $\Sigma$ ($\mathbf{n}$ is the normal to a point $M$). A direction of $\Gamma$ is chosen and $\mathbf{t}$ is the unit vector tangential to $\Gamma$ with this orientation. The Darboux–Ribaucour trihedron is completed by introducing a vector $\mathbf{N}$ in the plane tangential to $M$, such that $\mathbf{t}$, $\mathbf{N}$ and $\mathbf{n}$, form a direct trihedron. Let us coordinate this trihedron in relation to Frenet's trihedron $\mathbf{t}$, $\mathbf{p}$ and $\mathbf{b}$ from $\Gamma$ by the angle $\theta$ which permits $\mathbf{n}$ to be brought onto $\mathbf{p}$. $\mathbf{p}$ has a direction from $M$ towards the centre of curvature of $\Gamma$ (Figure 5.7). The formulae ruling the motion of the trihedron are written:

$$\frac{d\mathbf{t}}{ds} = \frac{1}{\rho_G}\mathbf{N} + \frac{1}{\rho_n}\mathbf{n},$$

$$\frac{d\mathbf{N}}{ds} = -\frac{1}{\rho_G}t + \frac{1}{\tau_G}\mathbf{n}, \qquad (5.6)$$

$$\frac{d\mathbf{n}}{ds} = -\frac{1}{\rho_n}\mathbf{t} - \frac{1}{\tau_G}\mathbf{N},$$

where

$$\frac{1}{\rho_G} = \frac{\sin\theta}{R} \text{ is the geodesic curvature;}$$

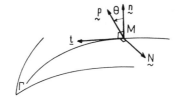

Figure 5.7. Darboux–Ribaucour trihedron

$$\frac{1}{\rho_n} = \frac{\cos\theta}{R} \text{ is the normal curvature; and}$$

$$\frac{1}{\tau_G} = \frac{1}{T} + \frac{d\theta}{ds} \text{ is the geodesic twist;}$$

$1/R$ and $1/T$ being, respectively, the curvatures and torsion of $\Gamma$.

It will be noted that the formulae (5.6) are written simply:

$$\frac{d\mathbf{v}}{ds} = \boldsymbol{\omega} \wedge \mathbf{v},$$

where $\boldsymbol{\omega} = (1/\tau_G,\ 1/-\rho_n,\ 1/\rho_G)$ is the instantaneous rotation vector of the Darboux–Ribaucour trihedron, and $\mathbf{v}$ any one of vectors $\mathbf{t}$, $\mathbf{N}$ and $\mathbf{n}$.

It is then easy to calculate div $\mathbf{n}$. To find the coordinate system in relation to a Darboux–Ribaucour trihedron at $M_0$

$$\mathbf{n} = n_x\mathbf{t}_0 + n_y\mathbf{N}_0 + n_z\mathbf{n}_0,$$

it is sufficient to note that $n_{x,x}$ and $n_{y,y}$ are the components of $d\mathbf{t}/ds$ in Darboux–Ribaucour trihedra deliberately chosen. By successively taking for $\Gamma$ the principal sections defined above (see equation (5.5)), we obtain:

$$\text{div } \mathbf{n} = -\left(\frac{1}{R_1} + \frac{1}{R_2}\right) = -(\sigma_1 + \sigma_2). \tag{5.7}$$

It appears therefore that div $\mathbf{n}$ is linked to the geometric properties of the surface and does not involve infinitesimally neighbouring surfaces. This geometric quantity is finally involved alone in the free energy.

$$\rho f = \tfrac{1}{2}K_1(\text{div } \mathbf{n})^2.$$

This is a remarkable property, which can be interpreted in the following manner (Rayleigh). Whatever the nature of the interactions inside a lamella which is of constant thickness and isotropic in all directions, the free energy of such a lamella is necessarily a positive defined symmetrical function of its two radii of curvature. Because of local symmetries, we can write:

$$\rho f = A(\sigma_1^2 + \sigma_2^2 + b\sigma_1\sigma_2), \tag{5.8}$$

118

where $A$ is positive and $b$ of modulus less than unity. Take the particular case of an elastic slab of thickness $h$, Young's modulus $E$ and Poisson's ratio $v$; it can be shown that:

$$\rho f_S = \frac{Eh^2}{3(1 - v^2)}[(\sigma_1 + \sigma_2)^2 - 2(1 - v)\sigma_1\sigma_2], \qquad (5.9)$$

where the subscript S means that this is the energy per unit surface of the layer.

Equation (5.9) does not include the distortion terms in the directions $x_1$ and $x_2$. It is applied to a slab of mean surface which is inextensible in two directions. On the other hand, it is the variations in dimension of the rest of the slab, along the directions perpendicular to the mean surface, which give the terms of free energy, which, summed up over the thickness, lead to equation (5.9). Thus, if the molecules of a layer are arranged to conserve a constant two-dimensional density, the mean surface where the aromatic parts, for example, are arranged, appears as inextensible while the interactions between aliphatic parts (through the thickness of the slab) are supposed elastic in the ordinary sense of solids and take the form of equation (5.9). This leads to the identification of $K_1 h/2$ and $Eh^3/[3(1 - v^2)]$. Experimentally, we have $K_1 = 10^{-6}$ dyne. This is compatible with the values $E = 10^8$ dynes/cm$^2$, $h = 50\,\mu$, which are not unreasonable.

It can be shown that the term $\sigma_1\sigma_2$ gives rise to a line integral when $\rho f_S$ is integrated on the surface of the layer. In fact we have:

$$\sigma_1\sigma_2 = \tfrac{1}{2}\text{div}\,(\mathbf{n}\,\text{div}\,\mathbf{n} + \mathbf{n}\wedge\text{rot}\,\mathbf{n}). \qquad (5.10)$$

This is obtained again by the Darboux–Ribaucour trihedron method. It is therefore not essential to take it into account in the expression of free energy.

It has been shown in this section that, in the first approximation, more precisely by supposing that the layers are of constant thickness, the free energy of a smectic A is reduced to a term $(\tfrac{1}{2})K_1(\text{div}\,\mathbf{n})^2$ per unit volume. It is convenient to note that this implies the existence of forces which are exerted normally to the layer. Suppose for example that the layers are stacked in a cylinder. We have: $\rho f = \tfrac{1}{2}(K_1/R^2)$; this indicates that there is a force $\mu$ (per unit surface) exerted perpendicular to the layer

$$\mu = -h\frac{dF}{dR} = h\frac{K_1}{R^3}.$$

This is equilibrated by stresses internal to the layer. It can be seen that, except for a very small $R$ (of the order of $h$), the behaviour in $R^{-3}$ makes certain that this force is weak and therefore can be treated as a perturbative term. A term $B\gamma^2$ is then introduced into the free energy, where $B$ is the elasticity modulus, and $\gamma$ the relative dilation of the layer:

$$\rho f = \tfrac{1}{2}K_1(\text{div}\,\mathbf{n})^2 + \tfrac{1}{2}B\gamma^2 \qquad (5.11)$$

### 5.3.2 Elasticity of a planar smectic

The form of the free energy per unit volume given by equation (5.11) becomes more applicable as a planar situation is approached. This case is to be studied here. Introduce the displacement $u(x, y, z)$ of the layer situated at level $z_0$ before every deformation at the point $x, y$ of this layer. The equation of this layer is now written:

$$\phi = z - u(x, y, z) = z_0. \tag{5.12}$$

$u$ is a quantity which varies slowly, and its derivatives, which express the deformations of the layer, are small. To second order, the direction cosines of $\mathbf{n}$ are written:

$$\mathbf{n} = \begin{cases} -\partial u/\partial x \,[1 + \partial u/\partial z], \\ -\partial u/\partial y \,[1 + \partial u/\partial z], \\ 1 - \tfrac{1}{2}(\nabla_\perp u)^2, \end{cases} \tag{5.13}$$

where

$$\nabla_\perp u = \begin{cases} \partial u/\partial x, \\ \partial u/\partial y, \\ 0. \end{cases} \tag{5.14}$$

Note that according to equation (5.13) $\mathbf{n} \cdot \mathrm{rot}\,\mathbf{n} = 0$. The layers are therefore preserved as such by the deformation. On the other hand $\mathbf{n} \wedge \mathrm{rot}\,\mathbf{n}$ is different from zero, and this will be taken into account by the introduction of a term in $K_3$. In fact the role of $K_3$ will be seen to be negligible. Let us also define

$$\gamma = 1 - \frac{d}{d_0} = 1 - \frac{1}{|\nabla\phi|} \sim \mathbf{n} \cdot \nabla\mathbf{u} \tag{5.15}$$

where $d_0$ is the thickness of the undeformed layer and $d$ is its thickness after deformation. To second order, $\gamma$ is written:

$$\gamma = \frac{\partial u}{\partial z} - \tfrac{1}{2}(\nabla_\perp u)^2.$$

For the moment, only those effects foreseen by the expansion of $\rho f$ to second order will be considered:

$$\rho f = \tfrac{1}{2}K_1 \left(\frac{\partial^2 u}{\partial x^2} + \frac{\partial^2 u}{\partial y^2}\right)^2 + \tfrac{1}{2}B \left(\frac{\partial u}{\partial z}\right)^2$$
$$+ \tfrac{1}{2}K_3 \left[\left(\frac{\partial^2 u}{\partial x \partial z}\right)^2 + \left(\frac{\partial^2 u}{\partial y \partial z}\right)^2\right]. \tag{5.16}$$

The equation governing the equilibrium is obtained by minimizing F. It reads:

$$\frac{K_3}{B}\frac{\partial^2}{\partial z^2}\Delta_\perp u + \lambda^2\,\Delta_\perp^2\,u = \frac{\partial^2 u}{\partial z^2}, \tag{5.17}$$

where

$$\Delta_{\perp} = \frac{\partial^2}{\partial x^2} + \frac{\partial^2}{\partial y^2}; \qquad \lambda = \left(\frac{K_1}{B}\right)^{1/2}$$

is a characteristic length. For $B \sim 4 \times 10^6$ dynes/cm$^2$ (i.e. quite elastic) and $K_1 \sim 10^{-6}$ dyne, we have $\lambda \sim 50\,\text{Å}$ (of the order of $d_0$). These values are of a reasonable order of magnitude in the middle of a domain of existence of a smectic phase. In the vicinity of a transition towards the nematic or isotropic phase, $B$ must become zero and $\lambda$ becomes infinite. The mechanical effects of these divergences have not yet been systematically studied. On the other hand, the nature of the nematic–smectic transition presents particular characteristics (de Gennes, 1972d, 1973a : possibility of a second order transition phenomenologically analogous to that between the normal state and the superconductive state, with divergences of $K_2$ and $K_3$ on the nematic side of the transition) which are very interesting and may be linked to certain observations of transitional textures. The elasticity presented here is too limited to understand these phenomena. This will be discussed in another context at the end of this chapter.

### 5.3.3 Effect of a localized perturbation (Durand, 1972)

Consider a semi-infinite medium bound by a surface displaying sinusoidal irregularities.

$$u = u_0 \cos qx. \tag{5.18}$$

We will suppose that this distortion has a small amplitude compared to its wavelength,

$$u_0 q \ll 1.$$

If the smectic layers take on this distortion (Figure 5.8) for $z = 0$, the solution of equation (5.17) which complies with this boundary condition is obviously:

$$u = u_0 \exp(-z/L)\cos qx, \tag{5.19}$$

with

$$Lq^2 = \left(\frac{1}{\lambda^2} + \frac{K_3}{K_1}q^2\right)^{1/2}. \tag{5.20}$$

where $q$ is assumed to be small (long wavelength). In this case the bend is negligible (this will practically always be the case) and we can write to a good approximation:

$$L = \frac{1}{\lambda q^2}. \tag{5.21}$$

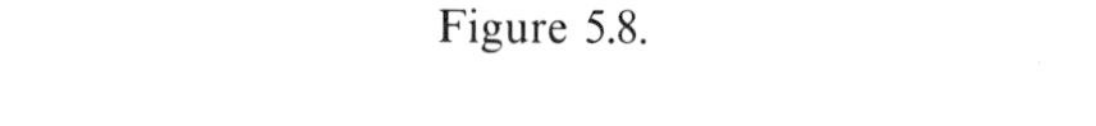

Figure 5.8.

It can be seen that the depth of penetration of the surface deformation is greater than the amplitude of this deformation, by a factor of $1/\lambda q$. Thus, for $q^{-1} = 10^{-4}$ cm, $\lambda = 5 \times 10^{-7}$ cm, we have $L = 200 \times 10^{-4}$ cm $(200\,\mu)$.

The total energy of the distortion is, per unit surface area.

$$\frac{u_0^2 q^2}{4} \frac{K_1}{\lambda}.$$

It is useful to compare this quantity with the contribution of surface tension $A_0$. Dimensionally $A_0 \sim K_1/d$, where $d$ is a molecular length. The differential surface element being $(\mathrm{d}x^2 + \mathrm{d}u^2)^{1/2}$, we have:

$$A_0 \int (\mathrm{d}x^2 + \mathrm{d}u^2)^{1/2} \sim \text{const} + \tfrac{1}{2} A_0 \int \left(\frac{\partial u}{\partial x}\right)^2 \mathrm{d}x,$$

i.e. per unit surface area:

$$\frac{u_0^2 q^2}{4} A_0.$$

The surface distortions contribute a term $K_1/\lambda$ to the surface tension, which is certainly little different from $A_0$:

$$A = A_0 + K_1/\lambda. \tag{5.22}$$

### 5.3.4 Deformation caused by an imposed dilation (Clark and Meyer, 1973)

Let us consider a perfect homeotropic sample contained between two glass slides which are perfectly flat. A displacement $\delta$ is applied to the upper slide leading to a dilation of the layers. Such a dilation only gives rise to an elastic energy $\tfrac{1}{2} B(\partial u/\partial z)^2$. Above a certain threshold $\delta_c$ this deformation becomes unstable because of an undulation of the layers (Figure 5.9) which allows the elastic stresses to relax at the expense of the forces of curvature. This threshold will be calculated here.

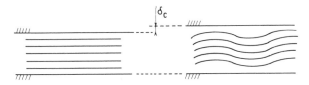

Figure 5.9. Deformation of layer caused by an imposed dilation

The displacement $u(x, z)$ can be expanded in a Fourier series. The first term is retained and it is written:

$$u(x, z) = \frac{\delta}{D} z + u_0 \cos q_1 x \cos q_3 z, \tag{5.23}$$

where $D$ is the thickness of the sample and $\delta$ the imposed dilation. $q_1$ is the wave number of the undulation, $q_3 = n(\pi/D)$ is the wave number of the deformation along $Oz$. $n$ is an integer. To demonstrate an instability, the free energy to third order in $u$ must be used, i.e. according to equations (5.13) and (5.14):

$$\rho f = \tfrac{1}{2} B \left[ \frac{\partial u}{\partial z} - \tfrac{1}{2} \left( \frac{\partial u}{\partial x} \right)^2 \right]^2 + \tfrac{1}{2} K_1 \left( \frac{\partial^2 u}{\partial x^2} - \frac{\partial^2 u}{\partial x^2} \frac{\partial u}{\partial z} - \frac{\partial^2 u}{\partial x \partial z} \frac{\partial u}{\partial x} \right)^2 .$$

The term in $K_3$ will not be introduced as we have seen that for quite a small $q_1$ (macroscopic distortion) its role is negligible. Furthermore, we shall see that the critical value of $q_1$ at an instability is small. Let us note also that this expression of $\rho f$, which contains the fourth-order terms, is only in fact valid up to third order. If $u$ is substituted into $\rho f$ and integrated over the thickness of the sample and over a unit length along $Ox$, it then becomes, by putting $\alpha = \delta/D$ and by restricting ourselves to the third order in $u$:

$$F = \frac{q_1}{2\pi} \int \rho f \, dz \, dx = \tfrac{1}{2} B \left[ \alpha^2 - \alpha u_0^2 \frac{q_1}{4} + \frac{u_0^2}{4} + \frac{9}{128} u_0^4 q_1^4 \right]$$

$$+ \tfrac{1}{2} K_1 \left( \frac{(\alpha - 1)^2}{4} q_1^4 u_0^2 + \tfrac{1}{8} u_0^4 q_0^4 q_3^2 \right).$$

For $u_0 \equiv 0$ this expression gives us the energy of pure dilation. There is an instability if, for a value of $\alpha$ greater than a threshold $\alpha_c$, $\partial F/\partial u_0$ is negative for $u_0 = 0$. This is written, by denoting $q_c$ as the value of $q_1$ at this threshold:

$$\alpha_c = (q_3/q_c)^2 + \lambda^2 q_c^2 (\alpha_c - 1)^2. \tag{5.24}$$

The minimum value of $\alpha_c$ as a function of $q_1$ must be sought. It is given by $\partial \alpha_c / \partial q_c = 0$, i.e.:

$$1 - \alpha_c = q_3/\lambda_c^2. \tag{5.25}$$

From these two equations we take:

$$\alpha_c = 2\lambda q_3 (1 + 2\lambda q_3)^{-1},$$

$$q_c^2 = \frac{q_3}{\lambda} (1 + 2\lambda q_3). \tag{5.26}$$

For $q_3 = \pi/D$, we have

$$\alpha_c = \frac{2\pi\lambda}{D} \left( 1 + \frac{2\pi\lambda}{D} \right)^{-1},$$

i.e.

$$\delta_c = 2\pi\lambda \left( 1 + \frac{2\pi\lambda}{D} \right)^{-1}.$$

The critical dilation is practically independent of the thickness of the sample ($\delta_c \sim 2\pi\lambda$), as long as we are not too close to the high temperature phase transition, where $\lambda \to \infty$. In this domain $\delta_c \to D$ and the present theory is insufficient.

This very remarkable phenomenon of instability actually allows $\lambda$ to be measured either by direct measurement of $\delta_c$ or by that of $q_c$ (diffusion of light). For CBOOA we obtain (Ribotta, 1974), by measurement of $\delta_c$, $\lambda = 17.5$ Å (78°C).

### 5.3.5 Action of electric or magnetic fields (Helfrich, 1970; Hurault, 1973)

Consider a homeotropic geometry (Figure 5.10) where the layers are maintained parallel to two slides containing the sample with a thickness $d$. If a magnetic field is applied in the horizontal direction, the molecules will lean towards one direction or the other, as the interaction with $\mathbf{H}$ does not depend on the sign of $H$. We have in fact:

$$\rho f = \tfrac{1}{2}K_1 (\operatorname{div}\mathbf{n})^2 + \tfrac{1}{2}B\gamma^2 - \tfrac{1}{2}\chi_a(\mathbf{n}\cdot\mathbf{H})^2, \tag{5.27}$$

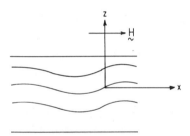

Figure 5.10. Helfrich–Hurault geometry

as in nematics. The distortion, which can be analysed in the form of $u = u_0(x)\sin k_z z (k_z = 2\pi/d)$ appears above a certain threshold, which is calculated now. $u_0(x)$ is expanded in a series and its first-order term retained. The director has components:

$$n_x = -\frac{\partial u}{\partial x} = \theta_0 \sin k_z z \sin kx,$$
$$n_y = 0, \tag{5.28}$$
$$n_z = 1,$$

where $\theta_0$ is a small angle. The mean free energy per unit thickness is:

$$\langle F \rangle = \tfrac{1}{8}\theta_0^2 B\left(\frac{k_z}{k}\right)^2 + K_1 k^2 - \frac{\theta_0^2}{8}\chi_a H^2. \tag{5.29}$$

124

The distortion is stable for $\langle F \rangle > 0$. It becomes unstable when the term between brackets is minimal, i.e.

$$k^2 = \frac{k_z}{\lambda} = \frac{\pi}{\lambda d},$$ (5.30)

and for a value of $H$ which makes $\langle F \rangle$ zero, i.e.:

$$\chi_a H_c^2 = 2\pi B \frac{\lambda}{d} = 2\pi \frac{K_1}{\lambda d},$$ (5.31)

It is interesting to compare this expression with that of Freedericks' instability, which corresponds to a similar geometry for nematics. In the latter case $H_c^2 \sim 1/d^2$, while here the dependence is on $1/d$. This explains that quite considerable dimensions of the sample would be needed to observe the effect foreseen by equation (5.31). In fact, for $K_1 = 10^{-6}$ dyne, $\lambda = 20$ Å, $d = 1$ mm, $\chi_a = 10^{-7}$, $H_c \sim 60$ KG would be necessary!

The effects of an electric field are phenomenologically analogous, making the substitution:

$$\chi_a(\mathbf{n} \cdot \mathbf{H})^2 \leftrightarrow \varepsilon_a(\mathbf{n} \cdot \mathbf{E})^2,$$ (5.32)

where $\varepsilon_a$ is the dielectric anisotropy. The necessary electric fields are of a reasonable order of magnitude, and recent experiments have shown how textures under an electric field can be observed (Hareng and Le Berre, 1975; Goscianski et al., 1975).

## 5.4 ELEMENTS OF COVARIANT ELASTICITY IN SMECTICS A
(Kléman and Parodi, 1975)

It has been indicated that smectic layers can undergo strong curvatures and still continue to form parallel systems. Locally, except at singular points, these layers can be assimilated into planes, and effects analogous to those which we have described in subsection 5.3.2 must be taken into account, i.e. small variations of thickness. In the next section, these systems of layers will be seen to have very well defined geometries. These represent Dupin cyclides in the majority of cases. For the moment, the problem of describing deformations in a smectic A as it undergoes both large curvatures and weak deformations of layers at the same time will be considered.

### 5.4.1 Covariant expression of free energy

On each layer a variable scalar $\phi(\mathbf{r})$ can be defined which is constant on the layer. For this variable to have a clear direction, it is taken as proportional to the numbering of the layer. If we start with a planar reference for which we have $\phi(z) = z_0$ and consider a curvature deformation of this planar situation, this leads to a function $\phi(r) = z_0$ which corresponds to the layer and has the same value as in the reference crystal. We therefore have:

$$\phi(r + dr) - \phi(r) = |\nabla\phi|\, dr = \frac{d_0}{d}\, dr,$$

where $d_0$ is the equilibrium thickness and $d$ the thickness of the layer in the perturbed medium. We have:

$$|\nabla\phi| = \frac{d_0}{d} = \nabla\phi \cdot \mathbf{n} \tag{5.33}$$

as $\nabla\phi$ is along the normal to the layer.

The variation in thickness of the layers being small, $|\nabla\phi|$ is little different from 1, and thus:

$$\varepsilon = 1 - \nabla\phi \cdot \mathbf{n}. \tag{5.34}$$

This quantity is little different from the relative dilation $\gamma$ introduced above. In fact:

$$\gamma = 1 - \frac{d}{d_0} = 1 - \frac{1}{1-\varepsilon} = \varepsilon + O(2) \tag{5.35}$$

and it is reasonable to write the free energy per unit volume in the form:

$$\rho f = \tfrac{1}{2}K_1(\operatorname{div}\mathbf{n})^2 + \tfrac{1}{2}B(1 - \nabla\phi \cdot \mathbf{n})^2 + \tfrac{1}{2}K_3(\mathbf{n} \wedge \operatorname{rot}\mathbf{n})^2, \tag{5.36}$$

which leads to equation (5.16) in the planar case. The introduction of the term $K_3$, as we have seen repeatedly, makes sense only near regions of very strong curvature of the lines of force of the director.

### 5.4.2 Equilibrium conditions: bulk terms

In order to minimize equation (5.36) it will be supposed that $\mathbf{n}$ (the director) and $\phi$ (which is known as the phase by analogy with wave systems) are independent variables. It is therefore necessary to introduce Lagrange conditions which express that in a virtual variation $(\delta\mathbf{n}, \delta\phi, \delta\mathbf{r})$, $\mathbf{n}$ and $\nabla\phi$ remain parallel. This is written:

$$P_{ij}\delta(\phi_{,j}) - (1 - \varepsilon)\delta n_i = 0, \tag{5.37}$$

where $P_{ij}$ is the projection operator on the layers,

$$P_{ij} = \delta_{ij} - n_i n_j. \tag{5.38}$$

Condition (5.39) expresses that $\mathbf{n}$ is a unit vector:

$$n_i \delta n_i = 0. \tag{5.39}$$

To minimize $F = \int \rho f \, dv$, taking these conditions into account, the following infinitesimal virtual transformations will be considered:

$$\mathbf{r} \to \mathbf{r}' = \mathbf{r} + \delta \mathbf{u}(\mathbf{r}),$$
$$\phi(\mathbf{r}) \to \phi'(\mathbf{r}') = \phi(\mathbf{r}) + \delta\phi(\mathbf{r}), \tag{5.40}$$
$$\mathbf{n}(\mathbf{r}) \to \mathbf{n}'(\mathbf{r}') = \mathbf{n}(\mathbf{r}) + \delta\mathbf{n}(\mathbf{r}).$$

Note the generality of this transformation: not only do we permit $\mathbf{n}$ and $\phi$ to be independent variables (in fact this generality is too great and no longer holds true in the presence of Lagrange conditions) but situations can be imagined where the fluid (which passes from $\mathbf{r}$ to $\mathbf{r}'$) and the $\phi$ layers are uncoupled. Helfrich (1969) called the phenomenon by which a molecule can leave its layer 'permeation'. This phenomenon certainly plays some role in the dynamics of smectics (de Gennes, 1975). Briefly, in equation (5.40), the first virtual variation envisaged corresponds to a *fluid displacement* while the two others (coupled under Lagrange conditions) correspond to a *displacement of the layers* or to a *rotation of the molecules*.

Let us write formally:

$$\rho\delta f = \rho \frac{\partial f}{\partial(\phi_{,i})} \delta(\phi_{,i}) + \rho \frac{\partial f}{\partial n_i} \delta n_i + \rho \frac{\partial f}{\partial n_{i,j}} \delta(n_{i,j}) + \rho \frac{\partial f}{\partial \rho} \delta\rho$$
$$+ \mu_i [P_{ij}\delta(\phi_{,j}) - (1-\varepsilon)\delta n_i] + \tilde{\lambda}_i n_i \delta n_i \tag{5.41}$$

where $\tilde{\lambda}_i$ and $\mu_i$ are Lagrange multipliers.

We also have, after (5.40):

$$\delta(\phi_{,i}) = \frac{\partial}{\partial x_i}\delta\phi - \phi_{,k}\delta u_{k,i},$$

$$\tag{5.42}$$

$$\delta(n_{i,j}) = \frac{\partial}{\partial x_j}\delta n_j - n_{i,k}\delta u_{k,j},$$

from which finally, after integration by parts, we have:

$$\delta F = \int_S \delta f_S \, dS + \int_V [\sigma_{ij}\delta u_{j,i} + g\delta\phi - (h_i - \lambda n_i)] \, dV, \tag{5.43}$$

where the $\delta f_S$ are surface terms, $\sigma_{ij}$ is a stress tensor, $\mathbf{h}$ the molecular field acting on the director, and $g$ the permeation force, a quantity which is conjugated with the displacement of the layers at constant fluidity. Let us introduce the notations:

$$\theta_i = P_{ij}\mu_j,$$
$$\lambda = \tilde{\lambda} - (1-\varepsilon)n_j\mu_j, \tag{5.44}$$

this yields:

$$\sigma_{ij} = -p\delta_{ij} - S_i\phi_{,j} - \frac{\partial(\rho f)}{\partial n_{k,i}}n_{k,j},$$

$$g = -\operatorname{div}\mathbf{S}, \tag{5.45}$$

$$h_i = \left(\frac{\partial(\rho f)}{\partial n_{i,j}}\right)_{,j} - \frac{\partial(\rho f)}{\partial n_i} + (1 - \varepsilon)\theta_i,$$

$$\delta f_S = \left(S_i \delta\phi + \frac{\partial(\rho f)}{\partial n_{i,j}} \delta n_j\right) v_i,$$

where $v$ is the exterior normal, and $p$ the pressure, defined as usual by:

$$p = -\frac{\partial(\rho f)}{\partial\left(\frac{1}{\rho}\right)};$$

we have used the conservation relation $\delta\rho = -\rho\delta u_{k,k}$.

Finally, the vector $S$ has components:

$$S_i = \frac{\partial(\rho f)}{\partial\phi_{,i}} + \theta_i. \tag{5.46}$$

$\delta F$ is zero at equilibrium. This reads:

$$\sigma_{ij,i} = 0,$$
$$g = 0, \tag{5.47}$$
$$P_{ij}h_j = 0.$$

The last one of these equations allows the elimination of the Lagrange multiplier $\mu$ from equation (5.44). This yields:

$$(1 - \varepsilon)\theta_i = -P_{ij}\left[\left(\frac{\partial(\rho f)}{\partial n_{j,k}}\right)_{,k} - \frac{\partial(\rho f)}{\partial n_j}\right]. \tag{5.48}$$

The second of equations (5.47) does not contain the pressure $p$. This is the equilibrium equation we are looking for. The first equation gives us $p$.

### 5.4.3 Equilibrium conditions: surface terms

The term $\delta f_S$ (equations (5.43) and (5.45)) includes several contributions.

(i) $$\delta f_{S\phi} = S_i \delta\phi v_i \tag{5.49}$$

represents the work done by the exterior medium on the surface $v_i$ when the layer is displaced by $\delta\phi$, the fluid being at rest. This is a permeation term, which itself can be decomposed (see equation (5.46)):

$$\frac{\partial(\rho f)}{\partial\phi_i} = -B\varepsilon n_i$$

Suppose that $n \wedge v = 0$ and $\varepsilon > 0$; the total energy diminishes if $\delta\phi > 0$, i.e. if new layers are created which relax the dilation $\varepsilon > 0$.

$\theta_i$ is a dominant term when the layers are normal to the surface. For the free energy given by equation (5.36), we have:

$$(1 - \varepsilon)\theta_i = -K_1 P_{ij} \nabla_j \operatorname{div} \mathbf{n}.$$

This term is therefore linked to the bend of the layers.

(ii)
$$\delta f_{Sn} = \frac{\partial(\rho f)}{\partial n_{j,i}} \delta n_j v_i. \qquad (5.50)$$

This corresponds to a surface torque. In fact, an infinitesimal rotation $\delta\omega$ can be introduced such that $\delta n = \delta\omega \wedge \mathbf{n}$. This yields:

$$\delta f_{Sn} = -\Gamma_l \delta\omega_l, \qquad (5.51)$$

with

$$\Gamma_l = t_{il} v_i,$$

$$t_{il} = \varepsilon_{ijk} \frac{\partial(\rho f)}{\partial n_{j,i}} n_k. \qquad (5.52)$$

(iii) The following stress contributions must be added to $\delta f_S$:

$$\delta f_\sigma = \sigma_{ij} v_i \delta u_j, \qquad (5.53)$$

which represents a surface force being exerted on the fluid, with constant layers.

## 5.5  FOCAL SYSTEMS IN SMECTICS A

Focal domains (section 5.5) and the dislocations (sections 5.6 and 5.7), which are the elementary defects of lamellar crystals, will be described in turn. The arrangement of a material in focal domains, according to various classical textures, was the subject of very interesting work at the beginning of this century (G. Friedel and Grandjean, 1910a, b). Furthermore, the observation of these focal domains led these authors to produce the hypothesis of the existence of a lamellar structure. It will be shown that the compact filling of space with these focal domains is not achievable. Imperfect focal domains have therefore to be considered and dislocations introduced. This discussion, which in the present state of our knowledge can only be incomplete, is in section 5.8.

### 5.5.1  Focal domains

Referring to subsection 5.3.1 (Figure 5.5) imagine a set of parallel surfaces whose common normals form a congruence of straight lines. These straight lines envelop a surface, called a focal surface, which in the most general case is composed of two sheets, $F_1$ and $F_2$, after the well-known theorems of surface geometry. The straight lines of the congruence are analogous to light rays, the parallel surface to wave fronts and the focals $F_1$ and $F_2$ to caustics. The surfaces $F_1$ and $F_2$ constitute elementary defects (of the wall type) in smectics. In fact surface defects are not observed, but lines are, which can be understood by admitting the viscous relaxation of the smectic (it is a liquid, after all!) leads to the disappearance of surface defects, and to their degeneracy into lines $F_1$ and $F_2$. The geometric problem of smectics is therefore the following: if a congruence of normals produces focal sheets which are degenerated into lines, then what is to be said of the surface normal to the congruence, i.e. the smectic layers?

This problem has a solution with very particular properties: the focal lines are necessarily focal conics, and the smectic layers take the form of Dupin cyclides (Figure 5.11). A particular case is where the ellipse is of zero eccentricity, the focal

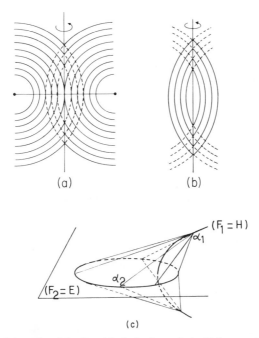

Figure 5.11. (a) Focal domain of the first kind: the hyperbola $H$ (here a straight line) and the ellipse $E$ (here a circle) constitute the singular loci of the director. $H$ is axis of revolution. The cyclides are here degenerated into tori centred on $E$; (b) focal domain of the second kind: same geometry as in (a) but here only $H$ is a singular locus; (c) bird's eye view of a focal domain of the first kind, in the general case

hyperbola being then the mid-perpendicular of the circle (Figure 5.11a). The Dupin cyclides are then tori. Note that (compare Figure 5.11a and b) two fundamental types of arrangement can be imagined, according to which the layers are between the focal domains (Figure 5.11a, focal conics of the first kind) or outside them (Figure 5.11b, focal conics of the second kind) (Bouligand, 1972a). Experiment seems to indicate the existence of focal conics of the first kind only. Remember that in the most general congruence, we have to distinguish two partitions of the set of straight lines of the congruence which can be imagined as follows: on a surface $\Sigma$ perpendicular to the congruence consider the two sets $C_1$ and $C_2$ of lines of curvature.† The straight lines of the congruence which rest on one of the $C_1$ lines touch the surface $F_1$ at the centres of curvature $\alpha_1$ of the lines of curvature $C_1$ of $\Sigma$. In the same way the normals of the congruence which rest on $C_2$ touch $F_2$ at the centres of curvature $\alpha_2$ of the line of curvature $C_2$. To say that $F_1$ and $F_2$ are degenerated into lines, is to say that $\alpha_1$ and $\alpha_2$ are fixed points when $C_1$ and $C_2$ respectively are traversed. From this it can be deduced that $C_1$ and $C_2$ are circles and the Dupin cyclides are surfaces whose lines of curvature are sets of circles. For the tori case, the sets of circles are obvious (Figure 5.11a). It should be noted that for one set of circles (in the plane of the diagram), the centre of curvature is the centre of the circle on $E$. For the other set of circles (centred on $H$), the centre of curvature is not in the plane of the circle.

Now consider a fixed point $\alpha_1$ on $F_1$ and the set of straight lines issuing from $\alpha_1$, which rest on different parallel circles $C_1$ traced on the successive Dupin cyclides. $\alpha_1$ is the centre of curvature for these different $C_1$. On the other hand, all the lines in the congruence issuing from $\alpha_1$ rest on $F_2$ (if $F_2$ were a true focal surface they would be tangential to it). Therefore $F_2$ is on the cones of revolution issuing from different points $\alpha_1$ situated on $F_1$. $F_2$ is therefore a planar curve and is a conic. The demonstration is similar for $F_1$. Finally it can be shown that $F_1$ and $F_2$ are two focal conics, the foci of one being at the apices of the other, situated in two rectangular planes.

Let us sum up the principal properties of focal domains.

(a) The smectic layers are all parallel and have common normals, parallel to the molecular directions in SmA phases.

(b) These normals rest on two focal conics (the focus of one is at the apex of the other; ellipse $E$, branch of hyperbola $F$) situated in two orthogonal planes.

(c) If (on $H$ for example) all the straight lines resting on $E$ are issued from a point $\alpha_1$, these describe a cone of revolution whose axis is along the tangent to $H$ at $\alpha_1$. This cone cuts the smectic layers along parallel cyclic sections. These cyclic sections are lines of curvature, which all have $\alpha_1$ as their centre of curvature.

(d) $\alpha_2$ plays an analogous role. We have two sets of lines of cyclic curves on the smectic layers which have the form of Dupin cyclides.

These properties are very well illustrated in Bouligand's article (1972a). The illustrations of Hilbert and Cohn-Vossen (1952) will also be referred to. Similarly the microphotographs of some smectic phases inserted in this monograph are indirect proof of the existence of focal conics.

The focal relation between two conics means that their orthogonal projections in every plane of space are composed of orthogonal conics. This property led Friedel and Grandjean to infer from observation the existence of focal conics, and therefore the existence of lamellar smectic phases (Figure 5.12).

Note a final property of focal domains. The eccentricities of $E$ and $H$ are the inverse of each other, $e_H e_E = 1$. Also the asymptotes of $H$ indicate the direction

---

† The lines of curvature are the envelopes of the tangents to the principal directions (see equation (5.5) and the section which follows).

Figure 12. Different projections of focal conics, after Friedel and Grandjean

normal to the layers at a long distance from the plane of the ellipse. This plane therefore *separates* smectic grains by making an angle $\alpha$ between them such that $\cos \alpha/2 = 1/e_H = e_E$.

The topological properties of focal domains are more easily accessible when the ellipse is degenerated into a circle and the hyperbola into a straight line. This is what we call a circular (or toroidal) focal domain (Figure 5.11a). The reader is advised to refer to this situation each time a non-trivial geometric property of focal domains is discussed.

*Circular focal domains.* A specific property of circular focal domains has to be kept in mind: since in that case $e = 0$, there is no disorientation from one side of the plane containing the singular circle to the other side. This property enables us to plunge a single circular domain in a perfect smectic phase without any difficulties of layers matching. The straight singular line has just to be perpendicular to the perfect phase; in the focal domain (here limited by a cylinder) the layers are folded into half tori (these parts of the tori where the Gaussian curvature $\sigma_1$, $\sigma_2$ is negative). And indeed isolated circular domains are often observed in homeotropic specimens; most frequently the singular circle is stuck to one of the glass slides, which means that we have here half circular domains.

*Parabolic focal domains* (Rosenblatt et al., 1977). When $e_H = e_E = 1$, we have the limiting case of parabolic focal domains, which were not studied by Friedel and Grandjean. The distortion of the layers extends to infinity but is extremely weak except in the region where the parabolae cross, i.e. in a volume of the order of $f^3$, where $f$ is the focal length common to the parabolae (excellent computed plots of the distortions can be found in the Rosenblatt et al. paper). Such domains can therefore exist isolated in an otherwise (quasi-) perfect smectic phase. Their energy is small compared to other types of domains (see below), and they have been invoked in the stabilization of the undulation instability (subsection 5.3.4) in the form of a square array (cf. also Asher and Pershan, 1979). They are probably numerous in planar geometries (layers perpendicular to the glass slides), where their parabolae have long been confused with elongated ellipses.

### 5.5.2  Principal textures made of focal domains

Reference should be made to the classical work of G. Friedel (1922) for the description of the different textures observed in smectics A. This text, remarkable for its rigorous workmanship, has not been surpassed. In short, the principal focal conic textures described by G. Friedel are as follows.

*Rod-like cell textures.*† These exist in materials possessing an isotropic liquid phase just above the smectic phase. When the isotropic phase is cooled, the transition is shown by the appearance of small focal domains independent of each other which stick together in *rods*. The smectic phase appears as a coalescence of these rods, without it being possible to distinguish true *phase relations* between adjacent focal domains. This is due essentially to their small size.

*Polygonal textures.* A particular case where the smectic is arranged in focal domains whose ellipses are stuck to glass slides, forming trellises. These arrangements are clearly visible, and prove the same laws as those observed for arrays of focal domains within the material. These trellises will be described later. (G. Friedel and Grandjean, 1910a, b; G. Friedel, 1922).

*Fan textures.* Complex arrays of domains with subvertical layers. These also exist in cholesterics.

'*Bubble raft*' *textures.* Here we do not adhere to G. Friedel's explanation of arrays of toric domains (Figure 5.11a) whose circles are stuck to the glass slide where they form a filling of tangential circles, of the type shown in Figure 5.13. The molecules would then be essentially tangential to the glass supporting the free droplet. In fact, this texture, although very important, must be classified with . . .

Figure 5.13. Bubble raft in a free droplet (after G. Friedel)

*Step textures.*‡ Grandjean terraces (Figure 5.14); characteristic of the homeotropic situation, i.e. where the anchoring conditions tend to align the director with the normal to the slide. These steps are very frequently decorated with focal domains, in the same way as the oily streaks described by G. Friedel. These textures will be studied in conjunction with translation dislocations, as this is their fundamental nature.

---

† *Translators' note:* In French, 'textures à bâtonnets'.

‡ *Translators' note:* In French, 'textures à gradins'.

Figure 5.14. Grandjean terraces (very simplified)

### 5.5.3 Laws of association of perfect focal domains

First, a *closed domain* should be distinguished from a *fragmented domain*.[†] In the former (Figure 5.11c) the domain is bound by two cones of revolution whose apices are on the hyperbola. The ellipse is complete. The maximum closed domain is obtained by taking the apices to infinity on the asymptotes. In fact it should be noted that the singularity of the cyclide at its intersection with the hyperbola is already strongly attenuated at a point on the hyperbola at a distance from the ellipse of the order of magnitude of the major axis of the ellipse. Therefore the two cylindrical zones of generatrices parallel to the asymptotes practically always form part of the domain if it is closed. The domain will be called fragmented if the ellipse is not complete. It is then bound by two sheets of cones of revolution whose apices are on the ellipse (Figure 5.15).

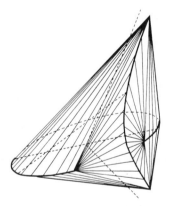

Figure 5.15. Fragmented domain (after G. Friedel)

The laws of association are of an essentially experimental nature. If certain aspects are clearly geometric (see Bouligand, 1972a), their explanation, which is still to be carried out in detail, can only rest on physical reasons. These are as follows:

(1) *Law of impenetrability of focal domains.* Two conics mutually interfere or are tangential to each other. More generally, two focal domains cannot penetrate each other; if they touch, they are tangential to each other along the generatrix of the cones which bound them.

[†] *Translators' note:* In French, 'domaine fragmenté'.

134

(2) *Law of corresponding cones.* When two conics are tangential to a point $M$, the cones of revolution, having an apex $M$ and resting on the focal lines of the two conics coincide. Then the two focal lines which are traced on this single cone intersect at two points $PP'$, and the four cones of revolution issuing from these points and passing through the two conics, cones which bound the focal domains, are tangential to each other along the two common generatrices $PM$, $P'M$ (from G. Friedel).

This law leads to a geometric construction which is very important in the case where the two tangential conics are in the same plane (Figure 5.16). The straight line of contact $Mm$ is, according to the law, the axis of the cone common to the two systems, which rests on the conics respectively confocal to each of the two tangential conics, i.e. passing through the foci $F_1$, $\phi_2$, $F_2$, $\phi_1$. Therefore the points $\phi_2 M F_2$ on one hand and $F_1 M \phi_1$ on the other are aligned. On the other hand, the point $A$ is a projection of the intersection of the focal conics (here the hyperbola).

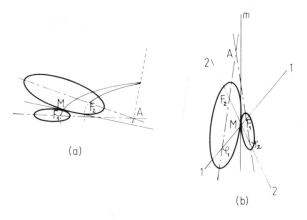

Figure 5.16. Law of corresponding cones: (a) overall view of the arrangement of conics; (b) view in the plane of the ellipses

(3) When two focal domains are tangential to each other along a generatrix, the corresponding conics touch each other in pairs, i.e. the pairs of conics either interrupt each other or are tangential to each other (G. Friedel, 1922).

It is clear that the three laws correspond to the tendency to fill space without defects other than focal conics, by assuring the continuity of the layers from one domain to the other, and by maintaining a constant thickness. They cannot therefore always be shown to be verified experimentally. Also, a repeated filling of space with focal domains according to the laws indicated requires the presence of focal domains of microscopic dimensions. At least in the experimental case (lecithin; Kléman *et al.*, 1977) where it was possible to make observations with electron microscope on freeze-etched specimens, it is certain that domains of microscopic dimensions do not exist. As regards divergences from the laws of association, these are evident in a number of cases. The case of *bubble rafts* will be analysed. All these departures require the presence of other types of defects, such as translation dislocations. Finally, it will be seen that, on certain conditions, it is possible to build domains in which the layers are still parallel but the singularities are no longer cofocal conics (subsection 5.8.2).

### 5.5.4 Trellis structures†

These structures appear to be composed of polygonal zones full of ellipses tangential to each other. These zones are stuck to the glass slides. In the case where the smectic is bound by the two slides, if one of the trellises is given (on one slide) the thickness of the slide and the other trellis can be completely determined. In particular, each edge of a polygon in a trellis can be associated to an adjoining edge in the other trellis, which is perpendicular to it in projection.

G. Friedel and Grandjean obtained these geometries by melting the solid phase or by adding impurities (colophane) to diethyl 4.4′ azoxydibenzoate (DADB). Trellises have recently been observed by Hareng and Le Berre (1975) by making a focal conic grow under an electric field. This experiment makes particularly evident the fact that the edges of the polygon are hyperbolas.

Inside a polygonal contour $P$ (Figure 5.17) the ellipses tangential to each other and to the contour obey the law of corresponding cones. In particular, their focal hyperbolas all intersect at a point $\Pi$, which is on the opposite slide. $P$ and $\Pi$ are conjugate. $\Pi$ is called the pole of $P$. Several edges of polygons originate from $\Pi$ in equal numbers to the edges of the polygon $P$, and are perpendicular to these edges.

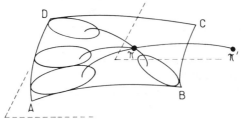

Figure 5.17. Relations of conjugation in trellises

Consider (Figure 5.17) the edge of the polygon $CB$. It is a focal line, obeying the law of corresponding cones, as do the ellipses which bound it on each side. Therefore $\Pi$ must be on the focal line conjugated to $CB$. $\Pi'$, the pole of P′ (the adjacent polygon) must also be on the same focal line. It should be noted that, as has been stated, $\Pi\Pi'$ and $CB$ are experimentally very elongated hyperbolas. This is not compatible with the laws of arrangement of focal conics. In fact, the difference from true focal conics is very little; they comply with the other laws of arrangement and also with the law of corresponding cones. It should finally be noted that the projections $\Pi\Pi'/CB$ are orthogonal.

† *Translators' note:* In French, 'structures en treillis'.

### 5.5.5 Energy of confocal domains

Let us differentiate the bulk energy and the line energy.

#### 5.5.5.1 *Bulk energy*

The cyclides being parallel, we must have rot $\mathbf{n} = 0$, i.e. $\mathbf{n} = \operatorname{grad} f$. The notations of section 5.4 are used again. We must have $\varepsilon = \varepsilon(f)$. Substitute this expression into the equilibrium equation div $\mathbf{S} = 0$. This yields,

$$F\left(\varepsilon, \frac{\mathrm{d}\varepsilon}{\mathrm{d}f}, O(f)\right) = 0, \tag{5.54}$$

where $O$ is a linear operator acting on $f$. $\varepsilon$ is a function of $f$ only (not of $\mathbf{r}$: $\varepsilon$ is constant on each layer). The preceding equation has a solution only if $f$ is an eigen function of $O(f)$. There is therefore only a discrete number of values of $f$ for which this condition is satisfied. The focal domain is therefore under a mechanical stress and the layers differ slightly from Dupin cyclides so that this constant can be relaxed. There is a term of elastic energy, which for dimensional arguments must be of the order of $(\frac{1}{2})B(\lambda^2/R^2)$ per unit volume, where $R$ is a typical dimension of the domain. The splay term $1/2K_1\,(\operatorname{div} n)^2$ is of the same order of magnitude.

There is an experimental result which seems to be fairly general. First, it should be noted that the Dupin cyclides which comprise focal domains are never complete; thus the tori of Figure 5.11a are always limited to their inner part. These parts of the tori have a *negative gaussian curvature* ($R_1$ and $R_2$, radii of the lines of curvature at a point of the torus, are of opposite signs), i.e. all the material points of the torus are hyperbolic points. This result is extended to the general case: on every Dupin cyclide belonging to a focal domain, the material points are generically hyperbolic. It is probable that this property is linked to the fact that, $R_1$ and $R_2$ being of opposite signs, this contributes to the reduction of div $\mathbf{n} = \pm(1/R_1 + 1/R_2)$, and that the layers have a tendency to approach the forms of minimum surfaces. A few cases invalidate this rule (convex forms of diethylene glycol laurate, Meunier and Billard, 1969; very hydrated lecithin, Kléman *et al.*, 1977).

The contribution of the splay term to the total energy of the focal domain has been calculated (Kléman, 1977). The method of calculation makes use of some elements of the theory of surfaces (Darboux), in particular the use of local axes at each point $M$ of each layer ($Mx_1$ and $Mx_2$ along the tangents to the lines of curvature, $Mz$ along the local normal) and of the fundamental quadratic form of the surface

$$\mathrm{d}s^2 = A^2\,\mathrm{d}u^2 + B^2\,\mathrm{d}v^2, \tag{5.55}$$

where $u$ and $v$ are coordinates along these axes. A complete discussion of these quantities is given in Chapter 7. The energy of curvature of the focal domain is then expressed as:

$$W = \int \tfrac{1}{2}K_1(\sigma_1 + \sigma_2)^2\,\mathrm{d}V = \tfrac{1}{2}K_1\int (\sigma_1 + \sigma_2)^2\,AB\,\mathrm{d}u\,\mathrm{d}v\,\mathrm{d}z. \tag{5.56}$$

Let us just write here the expressions of $A$ and $B$ in the cases of a generic focal domain and of a parabolic domain.

For a generic focal domain, defined by eccentricity $e$, major axis of the ellipse $a$, minor axis $b$, we have:

$$A = \pm b\sigma_2/(\sigma_1 - \sigma_2); \quad B = \pm b\sigma_1/(\sigma_1 - \sigma_2),$$
$$\sigma_1 = (c\cos v - z)^{-1}; \quad \sigma_2 = (a\cosh u - z)^{-1}, \tag{5.57}$$

where $c^2 = a^2 - b^2$ and $e = c/a$.

For a parabolic focal domain defined by a focal length $f$, we have:†

$$A = \pm \frac{2f\sigma_2}{\sigma_1 - \sigma_2}; \quad B = \pm \frac{2f\sigma_1}{\sigma_1 - \sigma_2}, \tag{5.58}$$

$$\sigma_1 = \left(-fv^2 - \frac{f}{2} - z\right)^{-1}; \quad \sigma_2 = \left(+fu^2 + \frac{f}{2} - z\right)^{-1}.$$

For small values of the eccentricity, $W$ reads:

$$W = 4\pi K_1(1 - e^2)K(e^2)a\ln\frac{a}{r_c}, \tag{5.59}$$

where $K(e^2)$ is the complete elliptic function of the first kind; $r_c$ is a core radius. It is easy to see that, for a fixed value of the major axis $a$, $W$ is a decreasing function of the eccentricity. Hence the situation in which the Dupin cyclides are degenerated into tori is not stable. However, the transformation from a torus to a general cyclide requires a change in the nature of the core of the singularities, which might well be prevented for topological reasons. In the case of a torus the singularity on the circle is a weak one: the different tori envelop cylindrically the circle but do not reach it. On the contrary, a general cyclide has singular conical points on the ellipse. Note incidentally that the topological barrier can be presumed to be higher in lyotropic systems than in thermotropic systems. It appears in conclusion that, independently of the topological barrier which can freeze a low-eccentricity focal domain, when the core energy is taken into account (and also maybe the elastic energy of variation in thickness of the layers), the total energy may reach a minimum for a value of $e$ different from zero.

It can also be shown that $W$ is a continuously decreasing function of $e$, for $e$ large, when $a$ is fixed.

The expression of $W$ in the case of a parabolic focal domain confined to a cylinder of radius $R$, whose axis is along the axis of the parabolae, is

$$W = 4\pi f K_1 \ln\frac{R^2}{4fr_c}. \tag{5.60}$$

---

† There is an error of calculation in Kléman (1977a) concerning parabolic focal domains.

138

It is not difficult to see that this quantity is in general small compared to the energy of a generic focal domain of the same extension. However, very little has been done concerning parabolic focal domains, and this is why they have, in fact, gone unnoticed in current observations of smectic domains.

### 5.5.5.2 Line energy

The only calculation which exists of this (de Gennes, 1972a) supposes a negligible curvature. The situation which is considered in Figure 5.18a is clearly reminiscent

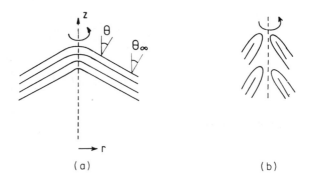

Figure 5.18. Core of a focal line; (a) non-dissociated; (b) dissociated

of a line of flare of the type described for cholesterics (section 4.4) and might well apply more to that case than to smectics. The free energy contains splay and elastic energy; in semi-polar coordinates it reads:

$$\tfrac{1}{2}K_1\left[\frac{1}{r}\frac{d}{dr}(r\theta)\right]^2 + \tfrac{1}{8}B(\theta_\infty^2 - \theta^2)^2, \tag{5.61}$$

the dilation $\gamma$ being indeed:

$$\gamma = \frac{d - d_\infty}{d_\infty} \sim (\theta_\infty^2 - \theta^2). \tag{5.62}$$

This expression of the energy is minimal for:

$$u'' + \frac{1}{\rho}u' - \frac{u}{\rho^2} + u(1 - u^2) = 0, \tag{5.63}$$

where

$$u = \theta/\theta_\infty \rho = \frac{r\theta_\infty}{2\lambda}.$$

This equation is identical in form to that of Ginsburg–Pitayevskii (1958) for the structure of a quantum of vortex in superfluid helium. According to the results obtained by these authors the solutions of these equations are:

(a) for regions $\rho \ll 1$ (inside a core radius $r = 2\lambda/\theta_0$), of the type $u = \text{const. } \rho$;
(b) for regions $\rho \gg 1$ of the type $u = 1 - \frac{1}{2}\rho^2$. The line energy is in this region:

$$W = \pi K_1 \theta_\infty^2 \ln \left(1.46 \frac{R}{r_F}\right). \tag{5.64}$$

This calculation, which indicates a line energy proportional to $K_1$, is valid only for low $\theta_\infty$. It supposes explicitly that the arrangement in parallel layers of *constant thickness* is broken in a zone of large radius $r_F$. It is possible that this condition could still be observed with particular arrangements displaying, for example, circular disclination lines (Figure 5.18b).

## 5.6 LINEAR THEORY OF DISLOCATIONS IN SMECTICS A

At a long distance from the dislocations, the induced deformation field can be dealt with by a linear theory. The analysis presented in this section therefore has great similarities to the classical theory of dislocations in solids. It will be noted, however, that this analysis strongly differentiates between edge dislocations and screw dislocations. In this section it will be shown that edge dislocations tend to have strong Burgers vectors, because of a peculiar core physics. The more detailed study of edge dislocations, in relation to experimental aspects which are appropriate to them (oily streaks, Grandjean terraces and tilt walls) is referred to in section 5.7. Screw dislocations, a class of objects which are particularly remarkable in smectics, are discussed in more detail in section 5.8.

We are dealing here with the linear theory of dislocations in a slightly distorted smectic A with quasi-planar layers, which obeys equation (5.17). The general linearized case of deformations due to dislocations in a Dupin cyclide has not yet been researched (see, however, subsection 5.6.7).

### 5.6.1 Screw dislocations

Equations (5.17) obviously possesses in semi-polar coordinates the solution analogous to that of a screw dislocation in a solid:

$$u = \frac{b}{2\pi}\theta \tag{5.65}$$

where $b$ is the Burgers vector and the continuity of the medium requires $b = nd_0$.

The line energy linked to equation (5.65) is zero (substitute (5.65) into (5.16)): thus the screw dislocations are objects whose energy is restricted to a core energy. It seems that screw dislocations are in fact extremely frequent: they have been observed with a density of the order of $10^8/\text{cm}^2$ in different lyotropic smectics A (Kléman *et al.*, 1977: electron microscopy of freeze-etched specimens) quenched

140

in liquid nitrogen, which then cleaves easily along the interfaces of the paraffinic chains. In these conditions screw dislocations piercing the cleavage surfaces determine the cleavage steps (Figure 5.19) which are grouped in rivers (Figure 5.20) according to a classical process, well known in solid crystals, and described by J. Friedel (1964, ch. 12). The height of the cleavage steps indicates that the Burgers vector is generally equal to $d$. This experiment indicates finally that the molecular arrangement at the core of the screw line is of type represented in Figure 5.21, the layers curving round the paraffinic chains and leaving the water on the outside. This model also shows that the core energy will be small when $b$ is small, which has been observed.

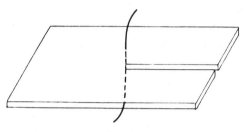

Figure 5.19. Cleavage steps; emergence of a screw dislocation through the cleavage plane

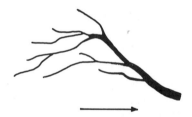

Figure 5.20. Cleavage steps producing screw dislocations and rivers. The arrow indicates the direction of propagation of cleavage

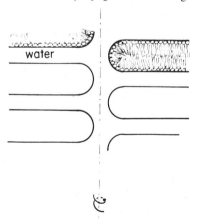

water

Figure 5.21. Arrangement of layers in the vicinity of a screw dislocation in a lyotropic smectic A

**5.6.2 Edge dislocations** (de Gennes, 1972b; Kléman and Williams, 1973, 1974)

Equation (5.17) is integrated with the following boundary conditions:

$$u(z = +0, x) = \begin{cases} 0(x < 0), \\ \\ \dfrac{b}{2}(x > 0). \end{cases} \qquad (5.66)$$

The symmetry $u(z) = u(-z)$ allows the calculation to be restricted to the region $z > 0$. Equations (5.66) is written again:

$$u(z = +0, x) = \frac{b}{4} + \frac{b}{4\pi} \int \frac{dq}{iq} e^{iqx} \qquad (5.67)$$

where we recognize in

$$\frac{1}{2\pi} \int \frac{dq}{iq} e^{iqx}$$

the unit step function, which is the integral of Dirac's function:

$$\delta(x) = \frac{1}{2\pi} \int dq \, e^{iqx}.$$

Equation (5.17) is solved in the form:

$$u(z, x) = \frac{b}{4} + \frac{b}{4\pi} \int \frac{dq}{iq} e^{iqx} g(z), \qquad (5.68)$$

which yields:

$$g(z) = \exp\left(-\frac{\lambda q^2}{\sqrt{1 + \lambda_3^2 q^2}} z\right). \qquad (5.69)$$

The term in $\lambda_3^2 = (K_3/B)$ should again be disregarded since it is only effective in the vicinity of the core ($q$ large). The angle of inclination of the layers is taken from equation (5.68):

$$\theta = \frac{\partial u}{\partial x} = \frac{1}{4\pi^{1/2}} \frac{d}{(z\lambda)^{1/2}} \exp\left(-x^2/4\, z\lambda\right). \qquad (5.70)$$

Note that $\theta$ decreases slowly along $z$ (over a length $x^2/4\lambda$: this is analogous, as an effect, to that which was analysed above in the study of the behaviour of distortions due to fluctuations in a layer). On the other hand, with constant $z$, the distortions quite quickly become zero. The parabolas $x^2 = \pm 4z\lambda$ therefore define on either side of the $z = 0$ axis regions of strong and weak distortions (Figure 5.22).

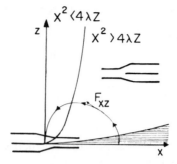

Figure 5.22. Distortions due to an edge dislocation; interaction of two dislocations of opposite sign (see text)

The energy of an edge dislocation line is calculated from expression (5.68). The energy density being divergent for $x = 0$ and $z = 0$, let us introduce a core radius $\xi$ which constitutes the lower limit to the integration over $x$. This yields:

$$W_1 = \int \rho f \, dx \, dz = \frac{K_1 b^2}{2\lambda\xi}. \tag{5.71}$$

The stages of calculation are as follows.

(a) Curvature term in equation (5.16):

$$\frac{\partial^2 u}{\partial x^2} = \frac{id}{4\pi} \int q \, dq \, e^{iqx} e^{-z\lambda q^2},$$

$$\tfrac{1}{2}K_1 (\text{div } \mathbf{n})^{1/2} = \tfrac{1}{2}K_1 \frac{\partial^2 u}{\partial x^2} \frac{\partial^2 u^*}{\partial x^2},$$

where $\partial^2 u^*/\partial x^2$ is the complex conjugate of $\partial^2 u/\partial x^2$.

Calculation of

$$\frac{\partial^2 u}{\partial x^2} \frac{\partial^2 u^*}{\partial x^2} \, dx \, dz:$$

integrate first in $x(-\infty, +\infty)$ using the relation:

$$\delta(q - q') = \frac{1}{2\pi} \int dx \exp[i(q - q')x].$$

The integral then takes the form:

$$\tfrac{1}{2}K_1 \int (\text{div } \mathbf{n})^2 \, dx \, dz = \frac{b^2 K_1}{8\pi} \int_0^\infty q^2 \, dq \exp(-2z\lambda q^2) \, dz.$$

The rest of the integration (in $z$) is trivial.

(b) Elasticity term $\gamma = \partial u/\partial z$: we proceed in the same way.

The summation of the two terms leads to:

$$W = \frac{Kd^2}{8\pi\lambda} \int_{-\infty}^{+\infty} dq.$$

It is now convenient to take the limits to the integration:

$$\left[ -\frac{2\pi}{\xi}, +\frac{2\pi}{\xi} \right].$$

The final result is deduced from this (equation (5.71)).

It is necessary to add a core energy to expression (5.71) of the line energy. It will be seen that experiment suggests the model of Figure 5.23, i.e. the splitting of the edge dislocation into two disclinations of opposite sign.

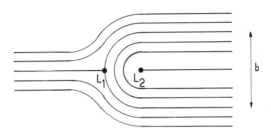

Figure 5.23. Decomposition of an edge dislocation into two disclinations $L_1$ and $L_2$ of opposite sign

In practice the core energy is that of the semi-cylindrical zone centred on $L_2$ and of radius $L_1L_2$. The exterior layers to the left (in Figure 5.23) are elastically distorted, but equation (5.71) takes this distortion into account. Suppose that between $L_1$ and $L_2$ the layers are of constant thickness: there is then only a curvature energy, which can be estimated as $(\pi K_1/2)\ln(b/2d_0)$ by integrating $\frac{1}{2}(K_1/r^2)$ between $r = d_0$ and $r = b/2$. The region in the vicinity of $L_2$ (of extension $d_0$), like that one in the vicinity of $L_1$, are regions where the elasticity which we have used up to now is invalid: these are core regions in the strict sense. Call their energy $\tau_c$. $\tau_c$ has little dependence on $b$. We therefore have finally, for the energy of an edge dislocation (model of Figure 5.23):

$$W = \frac{K_1 b^2}{2\lambda\xi} + \frac{\pi K_1}{2} \ln \frac{b}{2d_0} + \tau_c. \tag{5.72}$$

The different terms of equation (5.72) are now analysed. The first term is proportional to the square of the Burgers vector, as in the case of solids, but the dependence on the core radius $\xi$ is very original. According to the model in Figure 5.23, we have $\xi \sim b/2$. Therefore the first term is in fact proportional to $|b|$. Consider two dislocations of Burgers vector $b_1$ and $b_2$: the sum of the contributions: $W_1(b_1) + W_2(b_2) = W_1(b_1 + b_2)$. There is therefore an indifference towards splitting into smaller Burgers vectors $b_1, b_2$, for a dislocation of Burgers vector $b_1 + b_2$. This is very different from what is known of solids. However, the terms $W_2 = (\pi K_1/2)\ln(b/2d_0)$ and $W_3 = \tau_c$ lead to the stability of dislocations of large Burgers vector. This is obvious for the $W_3$ term. For the $W_2$ term, compare $W_2(2b)$ and $2W_2(b)$. It is clear that $W_2(2b) < 2W_2(b)$ for $b > 2d_0$. In fact, edge dislocations with strong Burgers vectors are observed, but with a difference: instead of being strictly rectilinear, as the model supposes, the dislocations present periodic variations of distortions along their extension. These will be analysed in due course.

### 5.6.3 Interaction between dislocations

It is known that interactions between dislocations in crystalline solids can be expressed by introducing a configurational force **F** which is exerted on the line and is expressed by Peach and Koehler's (1959) formula:

$$\mathbf{F} = (\mathbf{\sigma} \cdot \mathbf{b}) \wedge \mathbf{t}, \tag{5.73}$$

where $\sigma$ is the stress tensor (due to other defects or to applied stresses) calculated on the position of the line, **b** being the Burgers vector of the line and **t** the unit tangent to the oriented line. This formula can be applied with the necessary changes to smectics, since the definition of a stress field is known. This has been done in section 5.4 (equation (5.45)) where the covariant elasticity of smectics has been studied. In the case of quasi-planar structures, the expression of the stresses is written, when one restricts to linear terms in $u$:

$$\sigma_{13} = -K_1 \frac{\partial}{\partial x}\left(\frac{\partial^2 u}{\partial x^2} + \frac{\partial^2 u}{\partial y^2}\right),$$

$$\sigma_{23} = -K_1 \frac{\partial}{\partial y}\left(\frac{\partial^2 u}{\partial x^2} + \frac{\partial^2 u}{\partial y^2}\right), \tag{5.74}$$

$$\sigma_{33} = B\frac{\partial u}{\partial z}.$$

We have indeed:

$$\mathbf{n} = \left(-\frac{\partial u}{\partial x}, -\frac{\partial u}{\partial y}, +1\right)$$

to first order, and the pressure $p$ must be taken as zero in order to make the stresses zero at infinity. The other components of $\sigma_{ij}$ are zero.

It can be shown simply that the quantities in equations (5.74) are stresses without having to go through the 'heavy artillery' of section 5.4. Let us begin with expression (5.11) of the free energy density, which, in the case of planar distortions, reads, to second order in the derivatives of $u$:

$$F = \rho f = \tfrac{1}{2}K_1\left(\frac{\partial^2 u}{\partial x^2} + \frac{\partial^2 u}{\partial y^2}\right)^2 + \tfrac{1}{2}B\left(\frac{\partial u}{\partial z}\right)^2. \tag{5.75}$$

The free energy of a volume $V$ is $\int F \, dV$. Consider a virtual displacement $\delta u$. In this displacement, $\int F \, dV$ undergoes the variation:

$$\delta W = \delta \int F \, dV = \int K_1 \left[ \left( \frac{\partial^2 u}{\partial x^2} + \frac{\partial^2 u}{\partial y^2} \right) \right.$$
$$\left. \times \delta \left( \frac{\partial^2 u}{\partial x^2} + \frac{\partial^2 u}{\partial y^2} \right) + B \frac{\partial u}{\partial z} \delta \left( \frac{\partial u}{\partial z} \right) \right] dV. \quad (5.76)$$

It is interesting to make the quantities $\delta u$ appear on the right-hand side. This calls for integration by parts:

$$+\delta W = \int \left\{ K_1 \left( \frac{\partial^2}{\partial x^2} + \frac{\partial^2}{\partial y^2} \right)^2 u - B \frac{\partial^2 u}{\partial z^2} \right\} \delta u \, dV$$
$$+ \int \left\{ B \frac{\partial u}{\partial z} \, dS_3 - K_1 \frac{\partial}{\partial x} \left( \frac{\partial^2 u}{\partial x^2} + \frac{\partial^2 u}{\partial y^2} \right) dS_1, \quad (5.77) \right.$$
$$\left. - K_1 \frac{\partial}{\partial y} \left( \frac{\partial^2 u}{\partial x^2} + \frac{\partial^2 u}{\partial y^2} \right) dS_2 \right\} \delta u$$
$$- \int K_1 \left( \frac{\partial^2 u}{\partial x^2} + \frac{\partial^2 u}{\partial y^2} \right) \left[ \delta \left( \frac{\partial u}{\partial x} \right) dS_1 + \delta \left( \frac{\partial u}{\partial y} \right) dS_2 \right].$$

The volume integral is zero, because of the equilibrium equations. The surface terms have two parts: $\delta W_{S_1}$ (which is the factor of $\delta u$) is easily interpreted as the work done by the stresses in equations (5.74) in the displacement $\delta u$ (which is along $Oz$). The other surface term can be written:

$$\delta W_{S_2} = \int K_1 \left( \frac{\partial^2 u}{\partial x^2} + \frac{\partial^2 u}{\partial y^2} \right) (\delta n_x \, dS_1 + \delta n_y \, dS_2), \quad (5.78)$$

and is interpreted as the work done by the surface torque in an infinitesimal rotation $\omega$ such that $\delta \mathbf{n} = \omega \wedge \mathbf{n}$. Here we come back to the interpretations made in section 5.4 for the general case. It should be noted here that the stress tensor must not be symmetrical. Note finally that the bulk equilibrium equation can be written:

$$\sigma_{13,1} + \sigma_{23,2} + \sigma_{33,3} = 0. \quad (5.79)$$

Apply the Peach and Koehler formula (5.73) using the formulae (5.74) and $\mathbf{b} = (0, 0, b)$. This yields,

$$\mathbf{F} = \begin{cases} (\sigma_{23} t_3 - \sigma_{33} t_2)b, \\ (\sigma_{33} t_1 - \sigma_{13} t_3)b, \\ (\sigma_{13} t_2 - \sigma_{23} t_1)b. \end{cases} \quad (5.80)$$

### 5.6.3.1 Interaction between two screw dislocations $b_1$ and $b_2$

Suppose that the line $\mathbf{b}_1$ is situated along the $Oz$ axis of the coordinate system. We obviously have, since $(\partial^2 u/\partial x^2) + (\partial^2 u/\partial y^2) = 0$:

$$\sigma_{13} = \sigma_{23} = \sigma_{33} = 0.$$

There is no interaction. The result is trivial since we know that a screw dislocation is non-energetic.

### 5.6.3.2 Interaction between a screw dislocation and an edge dislocation

The force exerted by the screw on the edge is zero. It should be assumed that the force exerted by the edge on the screw is also zero, because of the principle of action and reaction. This is true for a total force, but not for a local force, which is non-zero.

### 5.6.3.3 Interaction between two edge dislocations

In Figure 5.22 we have shown the signs of the interactions between two parallel edge dislocations, one along $Oy$, the other of variable position $(x, z)$. The force lines $\mathbf{F}$ are also shown. It will be shown that $F_z$ is actually equal to zero on the parabolas $x^2 = 4\lambda z$. Two dislocations of opposite sign attract each other. Two dislocations of the same sign repel each other.

### 5.6.3.4 Exercise

It can be shown that two edge dislocations perpendicular to each other have no interaction. Discuss the validity of this result.

### 5.6.4 Interaction potential: influence of entropy terms

It is not difficult to show that the interaction energy of two parallel edge dislocations is given by:

$$W_I = \tfrac{1}{4} b_1 b_2 B \left( \frac{\lambda}{\pi |z_1 - z_2|} \right)^{1/2} \exp\left( -\frac{(x_1 - x_2)^2}{4\lambda |z_1 - z_2|} \right). \tag{5.81}$$

In this case an interaction potential exists, as the forces exerted by one dislocation upon the other obey a principle of action and reaction. We also have:

$$\mathbf{F}_i = -\operatorname{grad}_i W_I, \qquad i = 1,2. \tag{5.82}$$

In Figure 5.22 lines of equal potential have been drawn: they are perpendicular to the lines of force of $\mathbf{F}$. These potentials are high in the vicinity of $x_1 - x_2 = 0$, low for $z_1 - z_2 = 0$. It is interesting to compare them in this region to the entropy contribution which has been neglected up to now.

According to Nabarro (1967, ch. II) entropy contributes to the free energy of a dislocation in a solid crystal by a quantity $k_B T$ per atomic length of dislocation. Let us take as an analogy $k_B T/d_0$ per unit length of line in a smectic (Pershan, 1974). In the shaded region (Figure 5.22) the thermal effects dominate the elastic effects ($k_B T/d_0 > W_1$). Dislocations of the same sign as that shown at the origin, under the influence of the forces exerted on them by the dislocation at the origin (equation (5.72)) tend to gather in this region. The repulsive forces are then weaker than the forces of thermal origin, which have no privileged direction. Therefore under the action of these thermal forces the dislocations of the same sign on the line $z = 0$ can gather and form dislocations of very high Burgers vector, as indicated above.

### 5.6.5 Image forces

All the preceding considerations were applied to an infinite medium. In a finite medium the relaxation of surface stresses can be represented in the same way as dislocations in a solid in elastic theory, i.e. by introducing image dislocations whose stress fields added to those of existing dislocations comply with the boundary conditions. These conditions can be very different.

#### 5.6.5.1 Homeotropic geometry

The boundary is along a smectic layer.

(1) If the boundary is free, it is convenient to write that the stresses there are zero, i.e. $\sigma_{33} = B(\partial u/\partial z) = 0$.

It will then be shown that the image of an edge dislocation is an edge dislocation of opposite sign, situated in a position symmetrical to the first in relation to the boundary. There is therefore an *attraction* towards the surface. It has been proved experimentally that homeotropic free surfaces are zones of accumulation of edge dislocations. These dislocations also form particular textures (Grandjean terraces) which will be discussed below.

(2) If the molecules are anchored to the surface, which is fixed, we will write $u = 0$. It will then be shown that an edge dislocation is *repelled* by the surface.

#### 5.6.5.2 Planar geometry

The boundary is perpendicular to the smectic layers. This is only practically realizable with previously treated glass slides (deposit of a thin layer of SiO under oblique incidence, Urbach et al., 1974; Figure 5.24). The boundary condition is then $u = 0$. The image of an edge dislocation parallel to the surface is then an edge dislocation of opposite sign: there is *attraction* towards the surface.

148

Figure 5.24. Schematic representation of the effect of anchoring due to an evaporated layer under oblique incidence: the molecules of liquid crystal are arranged along the 'furrows'

### 5.6.6 General formulation of the linear elasticity of smectics

Equations (5.74) give the expression of stresses in a smectic slightly deformed with respect to the state where the layers are planar. Apply a point force of value equal to unity, directed along $Oz$, to a point $\mathbf{r}_0$ of the smectic. The deformation field created obeys the equation:

$$\sigma_{i3,i} + \delta(\mathbf{r} - \mathbf{r}_0) = 0. \qquad (5.83)$$

It is easy to show that in an infinite medium the solution of this equation can be expanded in a Fourier series in the form:

$$U(\mathbf{r}_0 \,|\, \mathbf{r}) = \frac{1}{(2\pi)^3} \int_{-\infty}^{+\infty} \frac{\exp\left[-i\mathbf{k}(\mathbf{r} - \mathbf{r}_0)\right]}{Bk_z^2 + Kq^4}\, d^3k, \qquad (5.84)$$

$$q^2 = k_x^2 + k_y^2.$$

$U(\mathbf{r} \,|\, \mathbf{r}_0)$ is a Green's function for the elastic problem of smectics A, and even in $\mathbf{r} - \mathbf{r}_0$. It can be shown that the fields of displacement of any dislocation line in an infinite medium can be simply expressed in function of this function.

If the medium contains both a unit point force (applied in $\mathbf{r}_0$) and a dislocation line $L$, the total displacement is the sum of the displacements due to each of these objects, and the same applies to the stress fields. There is no interaction between the point source and the line, as the stresses created by the point source constitute a field of *applied stresses*, while those created by the dislocation constitute a field of *internal stresses*. A well known theorem of dislocation theory has been used here (J. Friedel, 1964, ch. 2) which is only true because there is a linear superposition of displacements.

Suppose therefore that the line $L$ is created in the medium already under the applied unit force. As there is no interaction energy, the work, $u(\mathbf{r}) \times 1$ done by the unit point force when the dislocation introduces the field of supplementary displacements $u(\mathbf{r})$, is equal and opposite to the work done by the stresses due to the point force on the surface section. Let $\Sigma_{i3}(\mathbf{r}_0 \,|\, \mathbf{r})$ be these stresses. We therefore have (Kléman, 1974b):

$$u(\mathbf{r}_0) = -b_j \int \Sigma_{ij}(\mathbf{r}_0 \,|\, \mathbf{r})\, dS_i, \qquad (5.85)$$

which reads again, since $b_j = (0, 0, b)$,

$$u(\mathbf{r}) = -d \int (\Sigma_{13}(\mathbf{r}|\mathbf{r}') \, dy' \, dz' + \Sigma_{23}(\mathbf{r}|\mathbf{r}') \, dx' \, dz'$$
$$+ \Sigma_{33}(\mathbf{r}|\mathbf{r}') \, dx' \, dy'), \tag{5.86}$$

with:

$$\Sigma_{13}(\mathbf{r}|\mathbf{r}') = -K \frac{\partial}{\partial x'} \left( \frac{\partial^2}{\partial x'^2} + \frac{\partial^2}{\partial y'^2} \right) U(\mathbf{r}|\mathbf{r}'),$$

$$\Sigma_{23}(\mathbf{r}|\mathbf{r}') = -K \frac{\partial}{\partial y'} \left( \frac{\partial^2}{\partial x'^2} + \frac{\partial^2}{\partial y'^2} \right) U(\mathbf{r}|\mathbf{r}'), \tag{5.87}$$

$$\Sigma_{33}(\mathbf{r}|\mathbf{r}') = B \frac{\partial}{\partial z'} U(\mathbf{r}|\mathbf{r}'),$$

where $U(\mathbf{r}|\mathbf{r}')$ is the displacement field at the point $\mathbf{r}'$ due to the unit point force situated at $\mathbf{r}$. An obvious change of notation has taken place between equations (5.85) and (5.86).

The displacement fields of single lines can easily be deduced from equation (5.86). The calculations of the screw and edge dislocations will not be repeated. For the case of a circular planar loop of radius $R$ we find:

$$u(\mathbf{r}) = \varepsilon \frac{bR}{2} \int_0^\infty \exp(-\varepsilon\lambda q^2 z) J_0(q\rho) J_1(qR) \, dq, \tag{5.88}$$

where

$$\varepsilon = +1(z > 0); \qquad \varepsilon = -1(z < 0).$$

From the Green's function of a point force, the line energy and interaction energy between lines can again be simply expressed. Reference should be made to the article quoted for this demonstration. Only the results are given here:

(a) Self-energy of a line $L$:

$$W_s = -\frac{b^2}{2} K_1 B \oint_L \oint_{L'} \Delta \perp U(\mathbf{r}|\mathbf{r}')(dx \, dx' + dy \, dy'), \tag{5.89}$$

where $\Delta_\perp = \partial^2/\partial x^2 + \partial^2/\partial y^2$, and $L = L'$.

(b) Interaction energy of a line $L$ and a line $L'$:

$$W_i = -bb' K_1 B \oint_L \oint_{L'} \Delta \perp U(\mathbf{r}|\mathbf{r}')(dx \, dx' + dy \, dy'). \tag{5.90}$$

Starting from these expressions the results obtained above can be found without difficulty. In particular, it will be found that two perpendicular edge dislocations do not interact.

Note finally that equations (5.86), (5.89) and (5.90) are valid whether the medium is finite or not. The function $U(\mathbf{r}|\mathbf{r}')$ is then the displacement field due to the unit point force, with the boundary conditions of the situation under consideration.

### 5.6.7 Analogies between the elasticity of smectics and electromagnetism (Kléman, 1976b)

The linear theory of dislocations in smectics can be established on the basis of an analogy between lines of elastic currents and dislocations on the one hand, and magnetic induction and stresses on the other (Pershan, 1974). Such an analogy was developed for dislocations in solids (de Wit, 1960). There is a certain difficulty in this case as it requires the identification of the Burgers vector, a vectorial quantity, with the current $I$ which is a scalar quantity. But in the case of smectics, the Burgers vector is essentially scalar, since it has only one component, along the normal to the layers. Furthermore in a strongly deformed smectic (Dupin cyclide) the reasonable definition of the characteristic magnitude of a line is in terms of the phase $\phi$ of the layers. The Burgers number is then defined by:

$$n = \frac{1}{d_0} \oint d\phi = \oint \frac{\mathbf{n}(\mathbf{r})}{d} \cdot d\mathbf{l}, \qquad (5.91)$$

which is analogous to the current $I$, if we agree to identify the vector $\mathbf{m}(\mathbf{r}) = \mathbf{n}(\mathbf{r})/d$ with the magnetic field created by the line and the vector $\mathrm{rot}\,\mathbf{m}$ with the current density $4\pi/c\mathbf{J}(\mathbf{r})$, which obeys Maxwell's equation:

$$\mathrm{rot}\,\mathbf{H} = \frac{4\pi}{c}\,\mathbf{J}. \qquad (5.92)$$

$\mathrm{rot}\,\mathbf{m}$ must then be typically a defect density. In fact:

$$\int_S \mathrm{rot}\,\mathbf{m} \cdot d\mathbf{S} = \oint \mathbf{m} \cdot d\mathbf{l} = n \qquad (5.93)$$

is the total Burgers number of the dislocations piercing the surface $S$. Finally Peach and Koehler's force (equation (5.73)) is analogous to the Lorentz force, since the three components $\sigma_{11}, \sigma_{21}, \sigma_{31}$ of the stress tensor can be identified with the induction $\frac{1}{4}\pi\mathbf{B}$, the equation

$$\sigma_{i3,i} = 0$$

being analogous to Maxwell's equation

$$\mathrm{div}\,\mathbf{B} = 0. \qquad (5.94)$$

Pershan restricted this analogy to the case of a smectic with weakly deformed planar layers. There are no difficulties in extending it to the general case of covariant elasticity. It can then be shown that the generalization is to identify $\mathbf{S}$ and $-1/4\pi\mathbf{B}$ where (see equation (5.46)):

$$S_i = \frac{\partial(\rho f)}{\partial \phi_{,i}} + \frac{(n_i n_j - \delta_{ij})}{1 - \varepsilon}\left[\left(\frac{\partial(\rho f)}{\partial n_{k,k}}\right)_{,k} - \frac{\partial(\rho f)}{\partial n_j}\right] \tag{5.95}$$

$\mathbf{S}$ obeys the equation

$$\operatorname{div}\mathbf{S} = 0.$$

Note, moreover, that for the case of weakly deformed planar layers:

$$S_1 = K_1\frac{\partial}{\partial x}\left(\frac{\partial^2 u}{\partial x^2} + \frac{\partial^2 u}{\partial y^2}\right) = -\sigma_{13},$$

$$S_2 = -\sigma_{23},$$

$$S_3 = -\sigma_{33},$$

where the $\sigma_{ij}$ are given by equations (5.74).

In electromagnetism, the conjugate variables are $\mathbf{H}$ and $\mathbf{B}$. There is a free energy function such that at constant pressure and temperature

$$\delta F = -\frac{1}{4\pi}\int \mathbf{B}\cdot d\mathbf{H}. \tag{5.96}$$

In the case of smectics, if it is supposed that $\rho f$ only depends on the single variable $\phi$, we can write in the same way:

$$\delta F = \int S_i \delta(\phi_{,i})\,\mathrm{d}V + \int K_1\frac{P_{ij}}{1 - \varepsilon}(\operatorname{div}\mathbf{n})\,\delta(\phi_{,i})\,v_j\,\mathrm{d}S, \tag{5.97}$$

$v_j$ being the direction cosines of the exterior normal.

This expression is easily obtained from equation (5.43) by supposing that $\delta u_i = 0$, but it is important to be certain that it can be directly derived. As $\phi$ is the only independent variable here, it must be considered that the phase variations bring about correlative displacements of the fluid. $\delta\phi$ must not be given the same meaning here as in subsection 5.4.1.

A generalization of Peach and Koehler's formula is immediately derived from the identification of $+\mathbf{S}$ with $-\mathbf{B}/4\pi$. This is analogous to the Lorentz formula and yields:

$$\mathbf{F} = \mathbf{S}\wedge\operatorname{rot}\mathbf{m}, \tag{5.98}$$

i.e. the configurational force being exerted on an isolated dislocation of Burgers number $n$ and unit tangent $\mathbf{t}$ is:

$$\mathbf{F} = nd_0\mathbf{S}\wedge\mathbf{t}, \tag{5.99}$$

where $\mathbf{S}$ is the (vertical) stress field due to other dislocations or to forces and torques applied on the surface.

It is interesting to find this expression directly, without relying on the analogy, which is in any case not complete. Note that we can write $\delta F$ in the following way:

$$\delta F = -\int \text{div}\,\mathbf{S}\cdot\delta\phi\cdot dV + \int (S_j\delta\phi + K_1\,\text{div}\,\mathbf{n}\,\delta n_j)v_j\,dS \qquad (5.100)$$

where the surface integral of equation (5.97) is transformed by using equation (5.37). Using the fact that div $\mathbf{S} = \mathbf{0}$, only the surface integral remains. Suppose that a dislocation line $L$ of cut surface $\Sigma$ is situated in a field $\mathbf{S}$, div $\mathbf{n}$, and modify the cut surface $\Sigma \to \Sigma + d\Sigma$. This requires two operations:

(a) A translation of the layers $\Delta\phi = nd_0$ on $d\Sigma$ for the translation component of the dislocation. This translation can be made without rotation of $\mathbf{n}$. In the integral the two lips of $\Sigma$ are used for the surface of integration. The work necessary to create $d\Sigma$ is then, by the same argument as that developed by Peach and Koehler,

$$\delta\delta F_{\text{transl}} = \mathbf{F}\cdot d\mathbf{u}\,dl = nd_0(\mathbf{S} \wedge \mathbf{t})\cdot d\mathbf{u}\,dl \qquad (5.101)$$

where $d\Sigma := dl\cdot d\mathbf{u}$ is an infinitely small rectangle of dimensions $dl$ (along $L$) and $du$ (displacement of $L$ along a direction perpendicular to $\mathbf{t}$, unit tangent on $L$). The Lorentz force is in fact found.

(b) A rotation of the layers $\boldsymbol{\omega} = n'\pi\mathbf{k}$ which is the sum of a displacement $\boldsymbol{\omega} \wedge d\mathbf{u}$ and a local rotation of $\mathbf{n}$, for the rotation component of the dislocation. If $\boldsymbol{\omega} \wedge d\mathbf{u}$ is in the plane of the layers, the corresponding $\Delta\phi$ is zero ($\Delta\phi$ is in fact the projection of the displacement on the normal of the layers), and there is no work done by the Peach and Koehler force. There remains a configurational torque, which can be calculated in the same way as the Peach and Koehler force. Let us put $\delta\mathbf{n} = \delta\boldsymbol{\omega} \wedge \mathbf{n}$; this yields:

$$\delta F_{\text{couple}} = \int K_1\,\text{div}\,\mathbf{n}\varepsilon_{jkl}\,\delta\omega_k v_j\,dS,$$

from which, in a variation $d\Sigma = \mathbf{t} \wedge d\mathbf{u}\,dl$, for $\delta\omega_k = \omega_k$, we have:

$$\delta\delta F_{\text{couple}} = K_1\,\text{div}\,\mathbf{n}\varepsilon_{jkl}n_l\omega_k\varepsilon_{jpq}t_p\,du_q\,dl,$$

i.e. a configurational force per unit length of the line due to the torque:

$$\mathbf{F}_c = K_1\,\text{div}\,\mathbf{n}(\boldsymbol{\omega} \wedge \mathbf{n}) \wedge \mathbf{t}. \qquad (5.102)$$

An important difference must be noted between these formulae of configurational forces and those in the case of dislocations in a solid. In a strongly distorted smectic we are not dealing with a linear elasticity. $\mathbf{S}$ and div $\mathbf{n}$ are therefore not generally sums of independent terms each relative to a characterized defect, and in the formulae the action of the line tension cannot be separated out from that of the stresses due to other defects (see Friedel, 1964). Also the degrees of validity of formulae (5.101) and (5.102) are different. The Burgers number $nd_0$ may be small: it is possible to treat a translation dislocation like a linear object (of the small perturbation type) superimposed on a non-linear background. But $\omega$ is always large (equation (5.102) and this separation between imposed effects and perturbing effects is not possible here. Formula (5.102) will not therefore be used in this case.

In so far as translation dislocations are perturbing effects, equation (5.100) shows that, in order to introduce a translation line $L$, of Burgers number $nd_0$ into a deformed smectic, it is necessary to expend the energy:

$$W = nd_0 \int_\Sigma S_j v_j \, dS + \tfrac{1}{2} nd_0 \int \Delta S_j v_j \, dS, \qquad (5.103)$$

where $S_j$ is the field $S$ before the introduction of the line distortion and $\Delta S_j$ the supplementary field due to the presence of the line. The coefficient $\tfrac{1}{2}$ of the second integral is classical in linear calculations of self-energy. In the same way, the interaction energy between two lines $L$ and $L'$ $(n, n')$ is given by:

$$W_1 = nd_0 \int_\Sigma \Delta S_j' v_j \, dS = n'd_0 \int_\Sigma \Delta S_j v_j \, dS. \qquad (5.104)$$

The formulae (5.103) and (5.104) are valid when translation dislocations are introduced into a focal domain. It is easily verified that they do not depend on a choice of $\Sigma$, but only on $L$.

It may also be interesting to introduce a 'potential' vector $\mathbf{A}$ $(\mathbf{S} = \mathrm{rot}\,\mathbf{A})$. We then have:

$$W = nd_0 \oint_L \left( \mathbf{A} + \tfrac{1}{2}\Delta\mathbf{A} \right) dl,$$

$$W_1 = nd_0 \oint_L \Delta\mathbf{A}' \cdot \mathbf{dl} = n'd_0 \oint_{L'} \Delta\mathbf{A} \cdot \mathbf{dl'}.$$

Let us finish this analysis suggested by an electromagnetic analogy by returning to the linear problem of the quasi-planar layers. Pershan and Prost (1975) have made two suggestions in this connection.

(a) Smectics behave like linear media only if the stresses do not reach a critical value (analogous to the critical velocity of a superfluid or the elastic threshold of a solid crystalline medium).

(b) In elasticity the dislocations and impurities modify the macroscopic elastic properties, in the same way as fields produced by currents in a metal modify the lines of flux of applied fields. They have introduced the term diaelasticity by analogy with diamagnetism.

## 5.7 GRANDJEAN TERRACES, OILY STREAKS, INSTABILITIES OF EDGE DISLOCATIONS

In this section three types of observations are grouped together.

(1) *In contact with air* the smectic always takes on a homeotropic geometry. Suppose that the same occurs on a glass slide and deposit a drop of smectic on the glass. After the preceding discussion (on image forces) edge dislocations essentially group at the free surface: they are repelled by the glass and attracted by the

free surface. On the other hand, they tend to take the largest possible Burgers vectors. *Macroscopic* steps are then observed. These are Grandjean terraces (Figure 5.14).

These steps are frequently *decorated* at their edges by an *instability* which is repeated periodically along their length, at distances of the order of the thickness of a step. This instability is described by Friedel and Grandjean as a focal domain, or an element of a focal domain. The edge of the terraces therefore appears, according to these authors, as a linear assemblage of focal domains whose ellipses (or circles) are practically in the plane of the terraces and whose hyperbolae pierce this plane (see detailed analysis by Williams, 1976).

(2) *Placed between two glass slides in homeotropic geometry*, lyotropic smectics, observed between crossed nicols, show an array of very thick bright lines, whose size may attain the thickness of the sample. These lines, called 'oily streaks' by G. Friedel, are also found in cholesterics. An instability of the type observed in Grandjean terraces can be seen: the ellipses are clearly elongated perpendicularly to the axis of the line. In phase contrast, a fine longitudinal structure is also observed in the form of an array of striations parallel to the line. There is no doubt that these striations correspond to the splitting of the streak into numerous parallel dislocations of different signs.

It is interesting to note that these dislocations do not annihilate spontaneously, which confirms a model such as the one shown in Figure 5.23: the molecules must in fact pass from one layer to the other for annihilation to take place (*permeation*). It is clear that permeation is difficult in a lyotropic phase, where a hydrophobic parafinic chain cannot easily pass through the layer of water.

(3) *These observations have been confirmed by electron microscopy* (Colliex, Kléman and Veyssié, 1974); thin homeotropic layers ($\sim 2000$ Å) of lecithin have been observed spread out on a grid. The individual edge dislocations of small Burgers vector which are observed display a transverse instability analogous to that of macroscopic terraces.

### 5.7.1 Origin of instabilities in dislocations

The following qualitative explanation has been proposed for these instabilities. Let us begin with the model in Figure 5.23. The area between $L_1$ and $L_2$ has essentially curvature energy; on the other hand outside it (on the left of the diagram), the layers necessarily undergo variations in thickness. These can only be eliminated by the introduction of an abrupt discontinuity (of the wall type, see Figure 5.25).

In fact the solution chosen is an intermediate one. The energy of the wall is large. This can be deduced from dimensional arguments: the energy of the wall can be estimated at $K_1/d_0$ per unit surface, which would give $\sim (K_1/d_0)b$ per unit length of dislocation (wall of width $b$) which is of the same order of magnitude as $W_1$. There would then periodically be zones of the type in Figure 5.25 and zones where the layers are dilated. But this is again unstable regarding a transformation of wall *zones* into *focal lines* of the hyperbola type, while a focal line of the ellipse

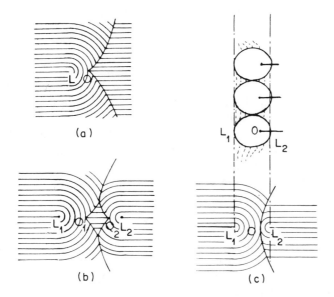

Figure 5.25. (a) matching of the layers along two parabolic half-walls in the vicinity of the core of an edge dislocation; (b) general case: walls linked to two edge dislocations; (c) formation of ellipses along an edge dislocation

type would be created on $L$ and these embryonic focal domains would then be enabled to develop.

The geometry of these instabilities can be precisely defined (Rault, 1975). If the layers are required to match without dilation along the wall (Figure 5.25) it is necessary for the wall to take the form of a parabolic arc of focus $L$, since we must have:

$$|MH - ML| = b \tag{5.105}$$

where $H$ is the foot of the perpendicular lowered on the line $OL$ and $b$ the Burgers vector of the dislocation. More generally (Figure 5.25b), it can be considered that a dislocation of Burgers vector $b$ is the difference between two dislocations of opposite sign $b_1 - b_2 = b$. If these two dislocations are on the same layer $L_1 L_2$ the same condition of non-dilation leads to the introduction of a wall in the form of a hyperbolic cylinder, such that:

$$|ML_2 - ML| = b. \tag{5.106}$$

The transformation into focal conics is then quite evident. The eccentricity of the hyperbolic cylinder is equal to

$$e_H = \frac{b_1 + b_2}{b_1 - b_2}.$$

The ellipse will then have the eccentricity $e_E = (1/e_H)$, with major axis

$$\alpha_E = \frac{b_1 + b_2}{2},$$

and minor axis $b_E = (b_1 b_2)^{1/2}$. A series of equal ellipses would be formed along the dislocation line and the edge dislocations would then join these ellipses in the non-transformed zones (Figure 5.25c). The universal transformation of the dislocation $b$ into a series of ellipses requires the energy of this series of ellipses to be less than that of the dislocation. According to Rault, the energy of the focal domain associated with an ellipse can be estimated as $\alpha K_1 b$ ($\alpha$ being a numerical constant), while the dislocation of Burgers vector $b$ occupying the same place has energy $K_1(b^2/\lambda)$ (see equation (5.72) and the discussion which follows). The transformation into focal conics is therefore favoured for:

$$b > b_c = \lambda \alpha. \tag{5.107}$$

There is therefore a critical Burgers vector above which transformation is favoured. Note that, if $\lambda$ increases with temperature ($\lambda$ becomes infinite in the vicinity of $T_c$ if the transition towards the highest temperature phase is a second order one), $b_c$ should increase in the same way. This conclusion should be regarded with some care, as $\alpha$ is a coefficient whose form is unknown.

### 5.7.2  Walls in a smectic A

We will be confined to the case of symmetrical walls. For a symmetrical wall, the following configuration can be expected a priori (Figure 5.26):

#### 5.7.2.1  *Discontinuity wall* (Figure 5.26a)

The layers change direction abruptly. According to our estimation of the preceding section, the energy of the wall would be, per unit surface:

$$\sigma_d \simeq \frac{K_1}{d_0} g(\theta_\infty), \tag{5.108}$$

where $g(\theta_\infty)$ is an increasing function of $\theta_\infty$ for small values of $\theta_\infty$, having probably a maximum (of the order of unity) towards $\pi/4$.

#### 5.7.2.2  *Curvature wall* (Figure 5.26b) (de Gennes).

The curves slowly change direction. If $\theta_\infty$ is quite a small angle (boundary with weak disorientation) the calculation of the energy may be made by using linear elasticity. We have in the zone of the wall curvature terms and dilation terms, which can be estimated thus: let $x$ be the direction perpendicular to the wall, $\theta(x)$ the variable angle of the layers with this direction. The curvature of the layers is $d \sin \theta \sim d\theta$. The local dilation $\gamma = d_0 - d$ is such that $\cos \theta_\infty = \cos \theta$. From which we have $\gamma \sim \frac{1}{2}(\theta_\infty^2 - \theta^2)$. Finally the free energy density is written:

$$F = \tfrac{1}{8} B(\theta_\infty^2 - \theta^2)^2 + \tfrac{1}{2} K_1 \left| \frac{d\theta}{dx} \right|^2, \tag{5.109}$$

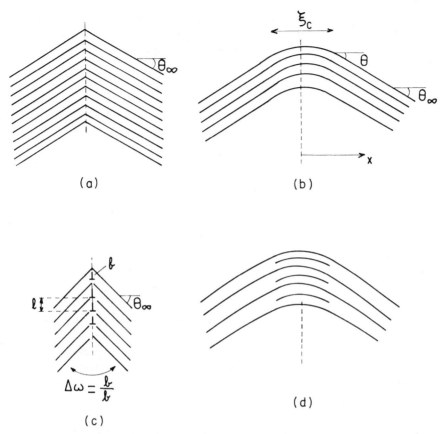

Figure 5.26. Symmetrical walls: (a) discontinuity walls; (b) curvature walls; (c) tilt walls of strong disorientation; (d) *double* tilt wall of weak disorientation: there are two series of parallel edge dislocations, shown here very schematically

which admits as equation of minimization:

$$2f(1 - f^2) + \xi_c^2 \frac{d^2f}{dx^2} = 0, \qquad (5.110)$$

where we have put:

$$\xi_c = \frac{2\lambda}{\theta_\infty} = \frac{2}{\theta_\infty}\left(\frac{K_1}{B}\right)^{1/2}; \qquad f = \theta/\theta_\infty.$$

The integration of equation (5.110) does not present any difficulties. First, we find for the first integral:

$$\xi_c \frac{df}{dx} = 1 - f^2, \qquad (5.111)$$

158

where the constant of integration has been chosen so that the boundary conditions are satisfied. (Equation (5.111) is obtained by making the change of variable $u(f) = (\mathrm{d}f/\mathrm{d}x)$.) From equation (5.111) we obtain:

$$f = \text{th}\,\frac{x}{\xi_c}. \tag{5.112}$$

The thickness of the wall, i.e. the zone where the rotation of the planes is felt, is therefore of the order of $\xi_c$. The calculation of the energy gives:

$$\sigma_c = \tfrac{2}{3}\frac{K_1}{\lambda}\theta_\infty^3. \tag{5.113}$$

### 5.7.2.3 *Dislocation wall* (Williams and Kléman, 1975; Williams, 1976)

As in crystalline solids, walls composed of dislocations can be constructed. The only Burgers vector which can be quantified being normal to the layers, it is Figure 5.26c (where $\theta_\infty$ is large) which is analogous to the tilt wall in Chapter 1. But this case is still at the hypothetical stage. Contra-distinctly we have found the case of Figure 5.26d (where $\theta_\infty$ is small). The experiment which led to the inference of the existence of this object consisted of shearing an initially planar sample (Figure 5.27a) where the anchoring of the molecules on the surface is strong. The formation on the surface of a series of dislocation lines, of large Burgers vector, should be noted (Figure 5.27b). This is visible under a polarizing microscope. The object obtained by experiment is equal to half of that in Figure 5.26d. It is interesting to note that this array of dislocations, whose presence relaxes the variations in thickness of the layer at the centre of the wall, is not stable. It gradually turns into an array of focal conics (Figure 5.25c) whose ellipses are in the plane of the boundary and major axes along the direction of shear. The creation of a boundary with the aid of focal conics is the purpose of the next section.

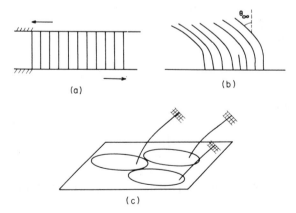

Figure 5.27. (a) Scheme of the shearing layer; (b) formation of dislocation line after shearing; (c) focal conics

### 5.7.2.4  Ellipses wall

We have seen above that the asymptotic direction of the hyperbolae is perpendicular to the smectic layers. The crystal, on each side of the plane of the ellipse, is therefore disoriented. If the plane is filled with quite a dense array of ellipses, all of the same eccentricity and the same axis, the corresponding hyperbolae are all parallel. Therefore, on each side of the plane, the crystal is disoriented by an angle equal to the angle of the asymptotes. A grain boundary has been formed.

The iterative filling of the plane with ellipses all of the same elongation and the same eccentricity is a problem which has still not been fully investigated. Such a filling must obey the law of corresponding cones, which implies (see Figure 5.16) that the contact between the two ellipses is not random. In fact, it can be seen that this contact is possible either along the minor axes, or along the major axes; but this is the first step of iteration. It is probable that the iteration stops for dimensions of ellipses which are still macroscopic, and it is therefore necessary to *terminate* the grain boundary with other defects of the structure. In lecithin, the observations already quoted from freeze-etched specimens give minimum sizes of the order of 1000 Å (Kléman *et al.*, 1977).

Let us therefore leave this problem aside, in spite of its practical importance, and deal with a similar problem from a theoretical point of view. This problem, which does not correspond to any physical reality, is that of the grain boundary of *no* disorientation (Bidaux *et al.*, 1973). This situation can be produced with toric focal domains, the ellipses being reduced to circles and the hyperbolae to straight lines perpendicular to the plane of iteration. The problem of the iteration of circles is known in mathematics as Apollonius's problem. In order to estimate the energy of the boundary, it is necessary to know:

(a) the number of circles of radius $R > \rho$, when the iteration reaches circles of radius $\rho$. Let $g(\rho)$ be this number;
(b) the total perimeter of these circles, $P(\rho)$; and
(c) the residual uncovered surface area $\Sigma(\rho)$.

We have to know the size of the biggest possible circles paving the plane to carry the iteration (let $L$ be their radius). The results are as follows:

$$g(\rho) = \text{const} \left(\frac{L}{\rho}\right)^n,$$
$$P(\rho) = \text{const}\, L^n \rho^{1-n}, \tag{5.114}$$
$$\Sigma(\rho) = \text{const}\, L^n \rho^{2-n}.$$

These quantities refer to a surface of the order of $L^2$. $n$ is an exponent and numerical calculations indicate it to be of the order of 1.32. For simplification it can be taken as equal to $\frac{4}{3}$. The existence of an exponent is very characteristic of a 'scaling' law, i.e. of a law relating to a problem where similar aspects are repeated in all dimensions. This is one of the original aspects of the problem in mind, which has led to its comparison with a large class of problems, in particular that of turbulence (Mandelbrot, 1975) and that of critical transitions (Toulouse and Pfeuty, 1975). We estimate the energy in the following manner (Bidaux *et al.*, 1973):

(a) An energy per unit line $xK_1$, where $x$ is a number, is associated with each ellipse. A circle therefore has energy $2\pi\rho xK_1$.

(b) An energy $yK_1\rho$, where $y$ is a number, is associated with each hyperbola. The dependence on $\rho$ comes from the fact that for a domain of a circle of radius $\rho$, the deformation of the layers in the vicinity of the hyperbola is attenuated at a distance $\rho$ from the plane of the boundary. The contribution of these two terms is therefore:

$$F_{\text{line}} = \text{const } K_1 P(\rho). \tag{5.115}$$

(c) The residual regions $\Sigma(\rho)$ are very strongly deformed over a distance $\rho$ from the plane of the boundary, this deformation being essentially elastic. We therefore have:

$$F_{\text{res}} = \text{const } B\rho\Sigma(\rho). \tag{5.116}$$

Let us minimize $F_{\text{line}} + F_{\text{res}}$ by taking the scaling laws (5.114) into account. The final iteration value is given by:

$$\rho^* = \text{const}\left(\frac{K_1}{B}\right)^{\frac{1}{2}} = \text{const } \lambda. \tag{5.117}$$

$L$ is not involved in this expression. It can therefore be seen that microscopic iteration values must be attained, and that all the plane is in fact covered with focal domains. The total energy is obtained by inserting $\rho^*$ into (5.115) and (5.116), i.e. for the surface $L^2$:

$$F_{\text{focal}} \sim \left(\frac{L}{\rho^*}\right)^n K_1\rho^*,$$

and finally:

$$\sigma_{\text{focal}} \sim K_1 L^{n-2}(\rho^*)^{1-n}. \tag{5.118}$$

Let us compare $\sigma_{\text{focal}}$ and $\sigma_c$. We have, by taking $\theta_\infty \sim 1$:

$$\frac{\sigma_{\text{foc}}}{\sigma_c} = \left(\frac{\rho^*}{L}\right)^{2-n}. \tag{5.119}$$

This expression is small, $L$ being always of macroscopic dimensions. The calculation therefore allows the supposition that the ellipse walls are more stable than curvature walls.

## 5.8 SCREW DISLOCATIONS. MINIMUM SURFACES. DEPARTURES FROM G. FRIEDEL'S RULES

Space cannot be paved in a regular manner with focal domains obeying the laws of association, unless they are reduced to the sizes of domains on the molecular scale. Experimentally, this is not true and it is also geometrically impossible (see subsection 5.7.2). Optical and electronic observations indicate the existence either of textures formed from very different elements of focal domains, but capable of being analysed as focal lines, or of departures from the laws of association of the domains. These two cases are still not clearly understood. First of all the concept of screw dislocation will be dealt with here in quite a new light, by relating it to the concept of minimum surfaces, then the observations in question will be described, leaving several points unanswered (subsection 5.8.1). Then the concept of focal relations will be introduced in a new light in harmony with some beautiful observations made by Williams (1975) (section 5.9).

### 5.8.1 Dislocations and minimal surfaces (Kléman, 1976d)

The confocal domains virtually produce the condition (see equation (5.11)) of conservation of thickness of the layers:

$$\gamma = 0, \tag{5.120}$$

and it is clear that the better this condition is obtained the larger is $B$ compared to $K_1/R^2$, where $R$ is a radius of curvature typical of the smectic. The condition $B \gg K_1/R^2$ reads:

$$R \gg \lambda. \tag{5.121}$$

The opposite inequality ($R \ll \lambda$) can be obtained, either in the vicinity of the core of the defect, where the radii of curvature are microscopic, or when the layer is observed in the vicinity of a smectic A $\leftrightarrow$ nematic second order transition. We know that $B$ tends towards zero in the nematic phase, therefore that $\lambda$ diverges. By referring to (5.11) the condition $R \ll \lambda$ leads us to consider *in opposition to the confocal structures*, the structures which satisfy:

$$\text{div}\,\mathbf{n} = 0, \tag{5.122}$$

i.e. the stacking of surfaces whose sum of curvatures $\sigma_1 + \sigma_2$ is zero at each point (minimal surfaces). It will be shown that *isolated* screw dislocations satisfy (5.122) and the equilibrium equation (5.47):

$$\text{div}\,\mathbf{S} = 0. \tag{5.123}$$

Let us use the phase variable $\phi(\mathbf{r})$ (see subsection 5.4.1) which is constant on each layer and equal to the level $z$ of the layer in the case of the perfect smectic. The normal to the layer has for direction cosines:

$$\mathbf{n} = \left\{ -\frac{\partial \phi}{\partial x}, -\frac{\partial \phi}{\partial y}, -\frac{\partial \phi}{\partial z} \right\} \frac{1}{|\nabla \phi|},$$

and equation (5.123), taking into account (5.122), reads:

$$\text{div}\,\mathbf{S} = \frac{B}{|\nabla \phi|} \sum_i \frac{\partial}{\partial x_i} \frac{\partial}{\partial x_i} |\nabla \phi| = 0 \tag{5.124}$$

In the same way, equation (5.122) is expressed with the inclusion of $\phi$,

$$\frac{1}{|\nabla\phi|}\left(\nabla^2\phi - \frac{1}{|\nabla\phi|}\sum_i\frac{\partial\phi}{\partial x_i}\frac{\partial}{\partial x_i}|\nabla\phi|\right) = 0. \qquad (5.125)$$

By comparing these last equations, we see that the problem involves a harmonic function ($\nabla^2\phi = 0$) obeying the condition (5.124). Let us consider an irrotational fluid in stationary flow, whose particles have the velocity:

$$\mathbf{v} = \nabla\phi.$$

The above conditions mean that this fluid is incompressible and that the stream lines are passed through at constant velocity. ((5.124) can be written $\mathbf{v}\cdot\nabla v = 0$, where $v = |\mathbf{v}|$). It can be shown that a particular case of a stream function which obeys these conditions corresponds to a folding of smectic layers in the shape of a right helicoid. i.e. an isolated screw dislocation:

$$\phi = -\frac{b\theta}{2\pi} + z. \qquad (5.126)$$

The free energy density is then written:

$$\rho f = \tfrac{1}{2}B\gamma^2 = \tfrac{1}{2}B\left\{\left(1 + \frac{b^2}{4\pi^2 r^2}\right)^{1/2} - 1\right\}^{1/2},$$

and is of the order of $Bb^4 r^{-4}$ for $r \gg d$. This is lower than the energy density of a screw dislocation in a solid. By integration we obtain:

$$W = \int_{r_c}^{R} 2\pi\rho fr\,\mathrm{d}r = \frac{Bb^4}{128\pi^3}\left(\frac{1}{r_c^2} - \frac{1}{R^2}\right), \qquad (5.127)$$

where R designates the external radius of the cylinder of integration (we can make $R \to \infty$, the line energy does not diverge) and $r_c$ is a core radius.

The behaviour in $b^4$ assures us that only the screw dislocations of very small Burgers vector ($b = d_0$ practically) will be stable, which differentiates them greatly from edge dislocations. Also the order of magnitude of equation (5.127) is 1000 times smaller than that of the edge dislocation of the same Burgers vector (see equation (5.66)).

The differences from linear theory developed in section 5.6 should be noted. We have shown here that equation (5.126) constitutes an *exact* solution of the covariant formalism, and we have calculated the energy to the fourth order, which is the first order not to vanish. Remember also that the focal domains ($\gamma = 0$) only constitute an approximate solution of the covariant formalism. The large number of screw dislocations observed can then be explained. But it must also be noted that these objects play a particularly important role in the nematic $\to$ smectic transition, where *arrays* of screw dislocations must appear.

Page 163

### 5.8.2 Arrays of screw dislocations: concept of minimal domain

The presence of a single screw dislocation transforms the infinite smectic into a minimal domain. Now suppose the presence of two parallel screw dislocations, at a distance $2L$ from each other (Figure 5.28). From the above demonstration the medium can no longer be of mean zero curvature. Essentially local curvature therefore appears in the cylinder $C$ of radius $L$ of generatrices parallel to the dislocations and centred on the median line of the latter. This energy is of the order of $K_1/L^2$ per unit volume, therefore of the order of $K_1$ per unit length of the dislocations. The total energy is of the order of:

$$E_t = 2W + K_1. \tag{5.128}$$

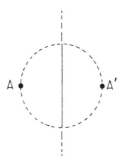

Figure 5.28. Parallel screw dislocations, $A$, $A'$. The existence of a wall element must be supposed along their median plane

It is clear that in the vicinity of $T_c$, $K_1$ is larger than $2W$. A configuration which leads to a minimum energy is the following: imagine that the two dislocations $A$ and $A'$ are separated by a wall situated equidistant and perpendicular to the plane containing $A$ and $A'$ and that on each side of the wall minimal solutions are conserved. The wall is effective over a distance of the order of $2L$, situated inside the cylinder $C$, outside of which the layers essentially again become perpendicular to the lines $A$ and $A'$ and therefore link up. Inside $C$ there is a real break in the layers, and they undergo on each side of the wall an abrupt rotation of the order of $\theta = b/L$, the axis of this rotation being along the normal common to $AA'$. This introduces an elastic energy which can be written

$$\frac{K_1}{\lambda} \left(\frac{b}{L}\right)^\alpha L,$$

where $K_1/\lambda$ is surface tension (zero at $T = T_c$, where $\lambda \to \infty$) and $\alpha$ is an exponent necessarily equal to, or greater than, unity. The total energy is written, after a simple calculation:

$$E_\alpha = \frac{Bb^4}{128\pi^3} \left(\frac{2}{r_c^2} - \frac{3}{L^2}\right) + \frac{K_1}{\lambda} b^\alpha L^{1-\alpha}. \tag{5.129}$$

164

To obtain this expression we have integrated $\rho f$ into each of the half-spaces containing the dislocations $A$ and $A'$.

It is clear that expression (5.129) leads to a still smaller energy than that given by (5.128) since $\alpha > 1$,† and $\lambda > b$. The medium therefore tends to divide into minimal domains, centred on screw dislocations. It is known that the condition $\lambda > b$ is easily obtained *near* $T_c$, since $b$ is small as seen above.

A wall composed of parallel screw dislocations all of the same sign constitutes a twist boundary. The preceding model can be applied to this case: a succession of minimal domains at distances of separation $L$ creates a twist of angle $\theta \sim b/2L$ between the two parts of the crystal separated by the wall (Figure 5.29). An essential but unresolved problem is to understand how such a wall, formed by transversal minimal domains, is deformed when the temperature is clearly lowered below $T_c$.

Figure 5.29. Twist wall formed by screw dislocations. The axis of twist is along the direction $BB$. It is supposed, in accordance with the model of minimal domains, that the dislocations are separated from each other by an element of wall

The model of minimal surfaces must be capable of being applied very much below $T_c$, in the core regions of defects, or at the intersection of focal domains. Thus the singular points observed by Cladis and White (1976) along the axis of smectic tubes, in which the layers are rolled almost cylindrically, could well be due to a periodic arrangement of the layers in the immediate vicinity of the axis, in catenoidal elements (Figure 5.30). Remember that the catenoid is the minimal surface of revolution obtained by rotation of a catenary around its base, and is also the only minimal surface of revolution. The layers parallel to a catenoid are not catenoids, which means that this geometric property of the system does not exist far from the core. It will be noted that a layer cannot be arranged along a single catenoid, but must undergo an abrupt rotation along one of the parallels and then follow another sector of catenoid. The period of this instability has been determined by linear calculations (de Gennes and Pincus, 1976) and is of the order of $2\pi\sqrt{\lambda R}$, where $R$ is the radius of the cylinder.

## 5.9 VARIATIONS TO PERFECT FOCAL DOMAINS

In this section we discuss some observations which put into evidence some large-scale singularities (by large scale we mean that their size is of the order of the sample thickness) which are not focal domains. Such objects have not received

† It can be shown (Kléman, 1976a) that $\alpha = 3$ or 4 according to whether the Van der Waals forces exerted between the molecules derive from non-retarded or retarded potentials.

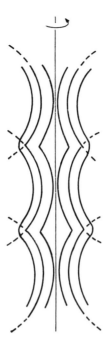

Figure 5.30. Rolling of the layers in the vicinity of an axis of a cylindrical structure. One of the layers takes the form of an array of catenoids joined together along the parallel of these surfaces of revolution. The neighbouring layers necessarily differ from catenoids

much attention up to now, and we just wish to formulate a few remarks; a starting point in all this discussion is that the system is able to arrange itself in geometries different from focal domains, but in which the parallelism of layers is practically observed. For example Léger (1973, 1976), Cladis and Torza (1975) and Huang, Pindak and Ho (1974) have observed that, in planar nematic specimens brought to the smectic phase just below $T_c$ ($T = 0.05°C$), there develops under an electric field a longitudinal structure very similar to Bouligand's (1973a) chevrons (see Figure 4.22). The layers would pile up in cones of revolution of equal size, favouring the equidistance of the layers. This explanation, although it ignores anchoring effects, is certainly close to reality. However, if the layers were everywhere equidistant there would not be much reason why the streaks observed transform to true conics when the temperature is lowered, which they do. Therefore it can also be imagined that the walls which border the chevrons are twist walls formed at the transition, separating regions of equal surfaces, close to minimal surfaces (or cones), in analogy with the discussion of section 5.8.

Other interesting objects, which exist even farther from the temperature transition, are the *helicoidal disclinations* which were first observed by Williams (1975) by abruptly cooling a nematic sample containing twist lines in the vicinity of which a twist zone has developed (Figure 5.31a). As the smectic refuses the twist of the director ($\mathbf{n} \cdot \text{rot } \mathbf{n} = 0$), this must relax in the form of twist walls attached to

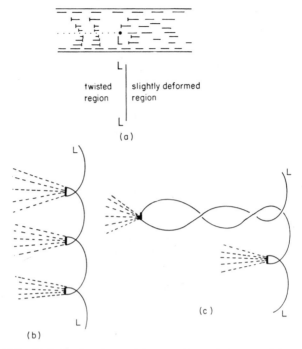

Figure 5.31. Helicoidal disclinations: (a) nematic sample containing a twist line, separating an almost perfect region from a region of twist: transverse view and view from above; (b) transformation of the twist line in a smectic (view from above): simple helix. The twisted region must contain a large number of walls of screw dislocations, attached to certain elements of the helix. These screw dislocations are indicated by dotted lines; (c) the formation of a double helix makes these screw dislocations disappear

the line. Experimentally the line itself takes an helicoidal shape as soon as the temperature is below the pretransitional zone (Figure 5.31b). A transformation of this helix to a double helix can eventually take place (see Figure 5.31c). This is a very stable geometry. The double helix is equivalent at a distance to a screw dislocation of huge Burgers vector. It has some remarkable features: two segments pertaining to the two different strands always intersect at right angles in projection (the angle of the strands with a plane perpendicular to the helicoidal axis is equal to $\pi/4$). Also, a small sphere of colloid, introduced inside the region delimited by the helices, moves along the axis without leaving this region (therefore in this zone the layers have a component parallel to the helicoidal axis), while the layers outside the helices are clearly perpendicular to it.

The geometry of the double helix is now understood (Frank and Kléman, 1977); it is a remarkable example of *large departure* from the focal domains and involves the introduction of the concept of *virtual focal surface*. However, before going on to a discussion of this new notion and the geometry of the double helix, let us discuss some specific features of *small departures* from focal domains.

### 5.9.1 Small departures from focal domains

They can be classified either as departures to the condition of constant thickness of each layer, (which means $\mathbf{n} \wedge \mathrm{rot}\,\mathbf{n} \neq 0$; see Figure 5.32) or as departures to the condition of absence of twist (which means $\mathbf{n} \cdot \mathrm{rot}\,\mathbf{n} \neq 0$, see Figure 5.33).

The first case $(\mathbf{n} \wedge \mathrm{rot}\,\mathbf{n} \neq 0)$ occurs frequently in cholesterics, where the variations in thickness are easy. In smectics, they could be relaxed by well located densities of edge dislocations. However, the whole situation is probably not much favoured energetically, although it would allow, because of the deformation of the focal lines, for a perfect stacking of the focal domains.

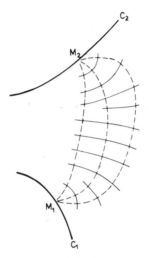

Figure 5.32. Variation in the thickness of the layers, presence of edge dislocations and focal relationship between $C_1$ and $C_2$ $(\mathbf{n} \wedge \mathrm{rot}\,\mathbf{n} \neq 0)$

In the second case $(\mathbf{n} \cdot \mathrm{rot}\,\mathbf{n} \neq 0)$, even if the molecules stay along straight lines, the layers fail to be well defined. This means that the medium contains a certain number of screw dislocations. In the simple case pictured Figure 5.33, the molecular alignments form a congruence of straight lines which lie upon two segments $\delta C_1$ and $\delta C_2$, which are assumed to be orthogonal. We soon see why. The whole region forms a small tetrahedron which contains a limited number of screw lines attached to the segments $\delta C_1$ and $\delta C_2$. Assume the distance

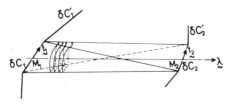

Figure 5.33. Tetrahedral focal domain built on two line arcs $\delta C_1$ and $\delta C_2$ in orthogonal perspective along $M_1M_2$ $(\mathbf{n} \cdot \mathrm{rot}\,n \neq 0)$

168

$M_1 M_2 = l$ to be much larger than the lengths $\delta C_1 = \delta C_2 = d$, then the density of dislocations attached to the segments is $|d^3/2l^3 b_0|$ per unit length, $b_0$ being the Burgers' vector of each of these dislocations.

Let us state here Darboux' theorem concerning congruences of straight lines: given two surfaces, $C_1$ and $C_2$, if there are tangents common to the two surfaces which form a congruence and such that the two surfaces $C_1$ and $C_2$ seem to cut at right angles when they are observed along the direction of any of these common tangents, then these tangents are normal to a family of parallel surfaces.

Therefore the situation in Figure 5.33 is the closest possible to one in which $\delta C_1$ and $\delta C_2$ are in perfect focal position. In fact we may consider $\delta C_1$ and $\delta C_2$ are small arcs on two focal conics, so that the content in screw dislocations is minimized. Therefore a possible model for the double helix observed by Williams could be a stacking of tetrahedra lying between the two strands. The energy would come essentially (apart from the usual curvature energy of the layers inside the tetrahedron) from the bad fitting between tetrahedra, since the self-energy of the screw dislocations is so small.

It is clear that, the smaller the tetrahedron, the smaller this energy. In the limit of zero volume tetrahedra, the sequence of imperfect 'focal domains' is reduced to a ribbon limited by two lines $C_1$ and $C_2$ such that $C_1$ and $C_2$ are pairwise connected by segments $M_1 M_2$ which put $C_1$ and $C_2$ in orthogonal perspective along $M_1 M_2$, according to Darboux' law (Figure 5.34). The problem is now to understand how this ribbon is connected with the smectic phase.

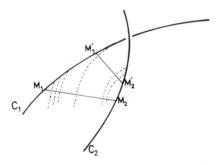

Figure 5.34. Intersection of smectic layers with a focal ribbon

The geometry of these focal ribbons possesses some peculiarities which have been studied in detail (Kléman, 1977, unpublished). In particular it can be shown that the double helix at an angle of 45° is such a focal ribbon. Let us study it in more detail.

### 5.9.2 Large departures from focal domains: virtual focal surface and the double helix (Frank and Kléman, 1977, unpublished)

The double helix focal ribbon, although it was introduced in the discussion of small departures from focal domains, leads in fact to the study of a completely different geometry. This is because the essential question now is how this focal ribbon matches with the external smectic phase. We first introduce for this purpose the notion of virtual focal surface.

Let us assume that in a given specimen boundary conditions are such that a layer must spouse a given surface $H(0)$ (see Figure 5.35). It is always possible to build a series of layers $H(1)$, $H(2)$, etc. parallel to the 'parent' surface $H(0)$. Such layers present singularities on the focal surface (presumably made of many sheets) $\Sigma_H$. In the example drawn here the layers on the left can cross the half-sheet $\Sigma_R$ corresponding to the normals of the right-hand side of the surfaces $H(n)$, and the layers on the right can cross the half-sheet $\Sigma_L$ originated from the left side. The multiple covering of the region inside $\Sigma_R$ and $\Sigma_L$ can be removed and the layers bound by a surface $\Sigma$ which is limited by the cusp edge $I$ and which is the locus of the centres of spheres which touch the parent surface $H(0)$ at two separate points ($\Sigma$ can be called the biequidistant focal surface). In this way the focal surface $\Sigma_H$ is physically reduced to a line; the biequidistant surface $\Sigma$ is a grain boundary which can eventually be redefined in terms of an array of dislocations.

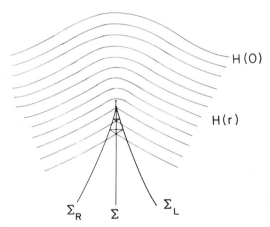

Figure 5.35. Virtual focal surface and focal line of $H(0)$

In the case of Williams' double helix, the parent surface is a ruled helicoid, generated by a straight line perpendicular to the helical axis and rotating with pitch $b$ about it. A right helicoid is a minimal surface (cf. Hilbert and Cohn-Vossen) and the splay energy $1/2K_1(\sigma_1 + \sigma_2)^2$ is vanishing on it. The surface $H(\lambda)$ parallel to $H(0)$ has a family of normals which envelope a focal surface (see Figure 5.36a and b) whose cuspidal edge is made of two helices, $C_1$ and $C_2$, around which the layers $H(\lambda)$ are folded as in a wedge disclination of strength $S = \frac{1}{2}$.

The family $H(\lambda)$ has an easy analytical description. Let $Z$ be the direction of the helical axis, $X$ and $Y$ two fixed axes perpendicular to $Z$. The equation of $H(0)$ reads:

$$X = r_0 \cos \theta_0; \qquad Y = r_0 \sin \theta_0; \qquad Z = b/2\pi \theta_0, \qquad (5.130)$$

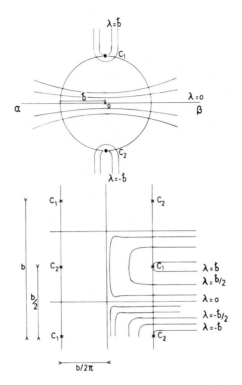

Figure 5.36. Geometry of the double helix: (a) cut perpendicular to the helical axis; $\flat = b/2\pi$ is the distance from the axis to the helical strands: (b) meridian cut

where $r_0$ and $\theta_0$ are cylindrical coordinates. Equation (5.130) describes only half of $H(0)$, i.e. the surface generated by the half-line $O\alpha$. Now consider the locus $H(\lambda)$ of the point at a distance $\lambda$ from $H(0)$; it reads:

$$X = r_0 \cos \theta + \frac{\lambda b}{2\pi} N_0^{-1} \sin \theta_0,$$

$$Y = r_0 \sin \theta_0 - \frac{\lambda b}{2\pi} N_0^{-1} \cos \theta_0, \tag{5.131}$$

$$Z = \frac{b}{2\pi} \theta_0 + \lambda N_0^{-1} r_0,$$

where $N_0^2 = r_0^2 + (b^2/4\pi^2)$. The energy of surface $H(\lambda)$ can be estimated using the method used above (subsection 5.5.5) by defining a local system of coordinates. There we choose $(r_0, \theta_0, \lambda)$.

It is an oblique system of coordinates and we get:

$$ds^2 = A^2 \, dr_0^2 + C^2 \, d\theta_0^2 + 2AC \cos \alpha \, dr_0 \, d\theta_0, \tag{5.132}$$

with  $A^2 = 1 + \lambda^2 b^2/4\pi^2 \cdot N_0^{-4}$,

$C^2 = N_0^2 A^2$,

$$tg\,\alpha = \pm \frac{(N_0^4 - \lambda^2 b^2/4\pi^2)}{(4\lambda b/\pi)N_0^2}.$$

The principal radii of curvature of $H(\lambda)$ can be obtained using standard formulae of the theory of surfaces (see Darboux):

$$R_1 = (\lambda b + 2\pi N_0^2)/b; \qquad R_2 = (\lambda b - 2\pi N_0^2)/b. \qquad (5.133)$$

The focal surface $\Sigma_H$ is obtained by saying that, on a given $H(\lambda)$, $R_1$ or $R_2$ are zero, which reads:

$$\lambda b = \pm 2\pi \left( \frac{r_0^2 + b^2}{4\pi^2} \right) \qquad (5.134)$$

The meridian section of $\Sigma_H$ is represented on Figure 5.37: there is a pattern (bold lines) which reproduces itself by a helicoidal translation of pitch $b$ forming a half-sheet of $\Sigma_H$. To this half-sheet must be added the half-sheet produced by the pattern (thin line) obtained by a symmetry about the line $Z = b/4$. This is not the total focal surface $\Sigma_H$ which also contains a sheet equal to the former and translated along $Z$ by a distance $b/2$.

It is a clear that these four half-sheets intersect an infinite number of times when one gets far from the helical axis. The inner regions of the sheets are covered by multiple layers.

The surfaces $H(-b/2\pi < \lambda < +b/2\pi)$ do not touch $\Sigma_H$; $H(b/2\pi)$ touches $C_1$ and $H(-b/2\pi)$ touches $C_2$; $\Sigma_H$ is met by any $H(\lambda)$ for $|\lambda| > b/2\pi$.

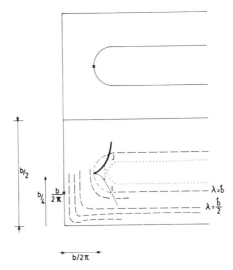

Figure 5.37. Focal surface and layers: (a) bold line: half-sheet; thin line: the other half-sheet pertaining to the same sheet; (b) dashed lines: layers $|\lambda| < b$, dotted line: a layer belonging to the set $|\lambda| > b$. These layers have two cusp points

The geometry of the contact of $H(\lambda)$ with $\Sigma_H$ can easily be studied on the meridian cut of $H(\lambda)$, which reads:

$$X_\lambda = \rho \cos \theta; \qquad X_\lambda = \rho \sin \theta; \qquad Z_\lambda = b\theta/2\pi + g_\lambda(\rho), \qquad (5.135)$$

where

$$\rho^2 = r_0^2 + \lambda^2 b^2 N_0^{-2} \quad \text{and} \quad g_\lambda(\rho) = \lambda r_0/N_0 + b/2\pi \, \text{arctg} \, (b/2\pi N_0 r_0).$$

The layers display a cusp point where they meet the focal sheets.

Of course the region $\rho > b/2\pi$, $Z \sim b/4$ cannot be covered by perfectly parallel layers, and elastic variations in thickness should take their source in this region, which can be relaxed all over the sample. The corresponding elastic energy is, however, probably small. An estimation of the energy of the double helix leads to a value of the order of $W \simeq 1.81 K_1 + 10^{-3} Bb^2 + W_c$ per unit length measured along the helical axes. The first term refers to the energy of the $S = \frac{1}{2}$ configurations around the helices corresponding to those $H(\lambda)$ which do not touch. The second term refers to those long-range deformations created by a screw line and discussed in section 5.8 and the third term to a core energy. One can easily convince oneself that the splitting in double helix is favoured only for large enough Burgers vector typically for $b \geqslant 25d$.

## 5.10  VARIATIONS TO THE RULES OF ASSOCIATION OF CONICS

This chapter is concluded by referring briefly to two results, which merit a fuller analysis.

(a) The hyperbolae of focal conics constitute a privileged position for the location of screw dislocations, where they assemble with *strong* Burgers vectors (C. Williams and Kléman, 1976). It would be important to analyse the influence of such a situation on the form of the associated ellipses.

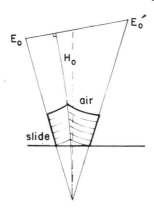

Figure 5.38. Geometry of focal domains in a free droplet. DADB has first order transition between the nematic phase and the smectic phase; this appears in the form of small focal domains which are arrayed in 'bubble rafts'. These domains intersect the substrate along cyclic sections of Dupin cyclides, which causes quasi-homeotropic anchoring

(b) Friedel's 'bubble rafts' are in fact made up of focal domains stuck to the supporting slab along *cyclic sections* of Dupin cyclides (see Steers *et al.*, 1974). This arrangement is incompatible with the orthodox laws of assocation but obviously minimizes the surface energy terms by allowing the smectic to be quasi-homeotropic. This is probably the only observation where the surface terms in a smectic play a simple and obvious role (Figure 5.38).

The subject of non-perfect (in the sense of breaking of Friedel's laws) association of focal conics is in fact a completely virgin field, both experimentally and theoretically.

# Chapter 6

# Rotation dislocations in continuous anisotropic media

The results in the preceding chapters did not exhaust all symmetries found in mesomorphic phases, for example the B and C phases of smectics, etc. and the hexagonal phases of lyotropes. In every one of these new phases, the characterization of the corresponding defects is not completely understood, but their physics is even less clear. This ignorance can be explained by the newness of these phases, and by the difficulty of the geometrical considerations necessary in this field of research. Meaningful and accurate results in this area are awaited and would be useful in the study of other physical systems where the geometry of field vectors or trihedrons have their role to play, e.g. the new phases of $^3$He or the description of velocity fields in turbulence. This chapter is concerned with a systematic study of defects in (fluid) continuous director fields and in (fluid) Cosserat media *independent of spatial symmetries* (nematics, cholesterics, etc.) which are related to them. A new definition of singularities is thus arrived at and the geometric frame which describes the distortions in these media will be examined, as well as the *contortion tensor* $K_{ij}$ (Kröner) until now familiar in elasticity within the linear limit where the non-commutability of rotations is not taken into account. The linear limit will be studied in Chapter 8 together with rotation dislocations in solids (where the presence of a strain tensor $e_{ij} \neq 0$ alters the circumstances).

## 6.1  RELATION BETWEEN DEFECTS AND CAUSTICS IN DIRECTIONAL MEDIA (Kléman, 1973a)

The importance of a geometrical description of directional media has already been stated. Thus, smectics A are studied as a system of layers perpendicular to the director, while the distortion of cholesterics is studied as that of a system of layers containing the director. Furthermore, in the latter case, there are no precise results confirming that the cholesteric planes continuously change shape between the perfect and imperfect cholesteric — whether their individuality is lost in this transformation or whether, more fundamentally, layers can still be defined in the deformed cholesteric phase which would not be the remnant of the perfect

174

cholesteric layers. Once this problem is solved, the solutions would be applicable to the nematic, since the cholesteric is never anything but a twisted nematic.

A geometrical method of describing directional media will be proposed here which recall Poincare's work on 'curves defined by differential equations'. configuration. This description is interesting in that it involves the idea of a defect in the most natural way. Moreover, from this point of view it is linked to methods used today to describe *qualitatively* solutions of ordinary differential equations, which recall Poincaré's work on 'curves defined by differential equations'. Rotation dislocations in a director field will be described as the caustics of a congruence of lines, and a hierarchical order of rotation and translation dislocations will be proposed. This theory will also allow the demonstration of the existence of a local orthonormal reference frame, linked, of course, to the defects of the directional media. The results obtained will extend quite naturally to Cosserat media, which will be defined as media in which an orthonormal reference frame exists at each point. Numerous physical cases can be represented by this geometry, particularly crystalline solids where the interest is in single distortions reducible to local rotations, or, a more recent example, the A phase of $^3$He (Mermin and Ho, 1976; Cross, 1975).

### 6.1.1 Congruence of lines

Let there be a distribution $\mathbf{n}(\mathbf{r})$ of the director; the $\mathbf{n}$ vectors clearly envelop an array of lines $L$ dependent on two parameters. These lines can then be described as the intersection of two surfaces:

$$f_1(\mathbf{r}, \alpha, \beta) = 0; \qquad f_2(\mathbf{r}, \alpha, \beta) = 0, \tag{6.1}$$

dependent on the parameters $\alpha$ and $\beta$.

In $L$, a sub-ensemble of lines of the congruence can always be chosen which form a surface $\Sigma$. This is possible in an infinite variety of ways, but it is advantageous to choose $\Sigma$ in such a way that the lines of the congruence forming $\Sigma$ *have an envelope in* $\Sigma$. This is expressed by a particular relation between $\alpha$ and $\beta$, which is obtained by writing that each line of $L$ has an envelope,

$$\frac{\partial f_1}{\partial \alpha} + \frac{\mathrm{d}\beta}{\mathrm{d}\alpha}\frac{\partial f_1}{\partial \beta} = 0; \qquad \frac{\partial f_2}{\partial \alpha} + \frac{\mathrm{d}\beta}{\mathrm{d}\alpha}\frac{\partial f_2}{\partial \beta} = 0. \tag{6.2}$$

The elimination of $\mathbf{r}$ between equations (6.1) and (6.2) leads to a first order differential equation

$$\phi\left(\alpha, \beta, \frac{\mathrm{d}\beta}{\mathrm{d}\alpha}\right) = 0. \tag{6.3}$$

The solutions of this equation lead to different partitions of $L$ into surfaces $\Sigma$, having the properties mentioned.

Let us clarify these properties with some examples. *Single unidirectional glide* in solid crystals is a trivial illustration. If the vector field $\mathbf{n}(\mathbf{r})$ is taken as the field formed by the unit vectors normal to the reticular glide planes (Figure 6.1), then

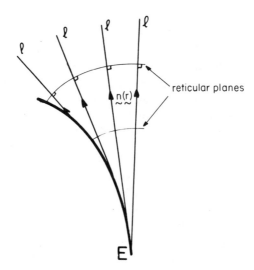

Figure 6.1    Single unidirectional glide in a solid crystal

in single glide, the ensemble of glide planes is transformed into parallel surfaces and the normals to these surfaces, which constitute $L$, envelop the evolute common to these surfaces. The planes containing the direction of glide and the lines $L$ form a partition $P$ of the congruence into surfaces $(\Sigma)$ responding to the required properties. There is no other similar partition. The singularity defined by this process is the evolute $(E)$.

A first example relating to liquid crystals is a wedge disclination $S = +1$ (Figure 6.2a) in a nematic. The planes perpendicular to the disclination constitute a partition into surfaces $\Sigma$. They contain the lines $L$, which in each $\Sigma$ plane envelops the intersection of the disclination with $\Sigma$. The envelope is reduced to a point on each $\Sigma$. Where $S = -1$ (see Figure 6.2b), the lines $L_1$, $L_2$, $L_3$, $L_4$ are some of the double lines of the congruence and constitute a singular part of the envelope in each $\Sigma$.

The case of focal conics is very instructive (Figure 6.2c): here there are two natural partitions of the congruence formed by the normals to the Dupin cyclides. The partition $P_E$ is composed of cones $\Sigma_E$ which rest on the hyperbola and whose apices are on the ellipse. On each $\Sigma_E$ the envelope is the apex of the cone, therefore a point on the ellipse. In the same way the partition $P_H$ is composed of cones $\Sigma_H$ which rest on the ellipse and whose apices are on the hyperbola.

More generally, there are as many partitions $P_i$ as there are *sheets* in the locus of the envelopes $E$ and the different $\Sigma$ solutions of equation (6.3). This locus is the focal surface (or caustic) of the congruence $L$. In liquid crystals, the focal surface is generally degenerate and the different sheets $E_i$ are reduced to (rotation) dislocation lines and even to singular points. In smectics, $L$ is a congruence of *normals*: there are necessarily two *sheets* (here degenerate into lines), and therefore two partitions.

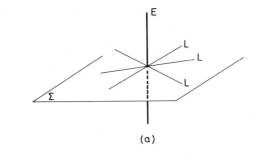

(a)

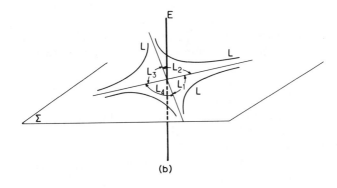

(b)

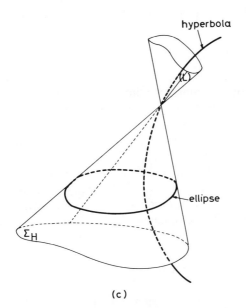

(c)

Figure 6.2 (a) Wedge disclinations $S = +1$ considered as an envelope of lines $L$; (b) *idem*: $S = -1$; (c) focal conics as envelopes of two different partitions of the congruence of normals $L$ to the Dupin cyclides

When the focal surface is not degenerate, the lines $L$ are tangential to all the sheets $(E_i)$. The lines $L$ must therefore be considered to meet all the dislocation lines or singular points. As this condition is not generally realized in the practical cases of liquid crystals, it is always necessary to divide the medium into *domains*.

Consider one such *domain* and suppose that the $E_i$ are degenerate into lines. Each $E_i$ is met by all the lines $L$: $E_i$ must therefore be either transverse to every surface $\Sigma$ of the partition $P_j$, or actually belong to it. If the $E_i$ belongs to a particular surface of the partition $P_j$, it must for the same reason belong to all the surfaces $\Sigma$ of this partition. Thus, we can state the following:

(a) If the $E_i$ is transverse to every surface of a given partition $P_j$, the $E_i$ appears as a wedge disclination for the ensemble of layers $\Sigma$ of $P_j$. The layers can, moreover, be considered as transforms from cholesteric planes. This indicates that the problem of the definition of the cholesteric planes in a directional medium does not have just one solution: there are as many solutions as there are partitions $P_j$.

(b) If the $E_i$ belongs to all the surfaces of a given partition $P_j$, the $E_i$ appears as a twist disclination for the ensemble of layers $\Sigma$ of $P_j$.

According to our definitions, there is only one partition which is transverse to a line $E_i$; this is precisely the partition $P_i$ whose layers contain points $E_i$ (envelopes of lines $I$) comprised in $\Sigma$. There is therefore, in general, only one way of defining a line $E_i$ as a wedge disclination. On the other hand, there are as many possibilities of defining $E_i$ as a twist disclination as there are partitions $P_j \neq P_i$ in the domain.

Figure 6.2c illustrates these considerations: the ellipse belongs to all the $\Sigma$ of the partition $P_H$ but is transverse to all the $\Sigma$ of the partition $P_E$.

### 6.1.2 Local trihedron; contortion tensor

Because of these properties, a local trirectangular reference frame can be defined at each point $\mathbf{r}$ of the directional medium which specifies both the position of the director $\mathbf{n}(\mathbf{r})$ and its relationship to the defects of the medium. Let us indeed attach to a given defect $E_i$ the partition $P_i$ which creates it and is transverse to it. At each point of the domain containing $E_i$ it is convenient to consider the trihedron as formed by the unit normal $\mathbf{N}$ to $\Sigma$ at $\mathbf{r}$, $\mathbf{n}(\mathbf{r})$, and $\mathbf{m} = \mathbf{N} \wedge \mathbf{n}$. This trihedron possesses both local and global properties.

The variation of this reference frame $\mathbf{R}(\mathbf{r})$ can be defined using a contortion tensor $K_{ij}(\mathbf{r})$ possessing nine components, in such a way that the infinitesimal rotation $\mathbf{d\omega}$ which links the orientation of $R(\mathbf{r})$ to $R(\mathbf{r} + \mathbf{dr})$ can be written:

$$d\omega_j = K_{ij}\,dx_i. \tag{6.4}$$

The variation $d\mathbf{p}$ of a vector $\mathbf{p}$ with constant components in the local reference frame reads:

$$d\mathbf{p} = d\mathbf{\omega} \wedge \mathbf{p}, \tag{6.5}$$

which gives:

$$p_{i,j} = \varepsilon_{ipq}K_{jp}p_q. \tag{6.6}$$

The gradients of **n** can also be simply expressed in the forms:

$$\text{div } \mathbf{n} = \varepsilon_{ipq} K_{ip} n_q$$
$$(\text{rot } \mathbf{n})_j = K_{ij} n_i - (K_{11} + K_{22} + K_{33}) n_j. \tag{6.7}$$

It should be noted that **N** is normal to the set of surfaces $\Sigma$. We therefore have $\mathbf{N} \cdot \text{rot } \mathbf{N} = 0$ which again reads:

$$K_{ij} N_i N_j = \Sigma K_{ii} = K. \tag{6.8}$$

**N** is obviously an axis of local helicity for the director **n**; the pitch is given by $q = 2\pi/p = K_{ij} N_i N_j = K$. But the local rotation axis $\Omega_j = K_{ij} N_i$ is not generally parallel to $N_j$.

The reference frame $R(\mathbf{r})$ must be defined in a unique manner: this property can be expressed by writing that the variation of $R$ between two points $M$ and $M'$ does not depend on the path followed between $M$ and $M'$. If the small successive rotations are commutable, this would mean that the formula (6.4) is a total differential. In fact we must write that

$$\mathbf{p}_{M'} = \mathbf{p}_M + \int_M^{M'} d\mathbf{p} \tag{6.9}$$

is independent of the path followed, $d\mathbf{p}$ being given by equation (6.5); this also reads (see Kléman, 1973):

$$\varepsilon_{lrs} p_k (K_{sk} K_{ri} + \varepsilon_{ijk} K_{sj,r}) = 0. \tag{6.10}$$

This expression must be true for any vector **p** having fixed components in $R(\mathbf{r})$. The coefficients of $p_k$ must be zero. If these coefficients are then multiplied by $\varepsilon_{ipk}$ this leads to the nine relations:

$$\theta_{ij} \equiv \varepsilon_{ipk} K_{kj,p} - \tfrac{1}{2} \varepsilon_{ipq} \varepsilon_{jmn} K_{pm} K_{qn} = 0. \tag{6.11}$$

It will be noted that the linear term of $\theta_{ij}$ is precisely the one which tends to zero if the formula (6.4) is required to be integrable. The non-linear term corresponds to the non-commutability of the small rotations: furthermore, apart from the sign, it is the minor $M_{ij}$ of the $(ij)$ term of the **K** matrix; we can write with the usual notations:

$$\boldsymbol{\theta} \equiv \text{rot } \mathbf{K} - \mathbf{M}. \tag{6.12}$$

The condition $\boldsymbol{\theta} = 0$ must be considered as a condition of *compatibility* of the contortion field $K_{ij}$, in the same way as the condition $\alpha_{ij} \equiv \varepsilon_{ikl} \beta_{lj,k}$ expresses, in a *solid*, that the distortions $\beta_{ij}$ are compatible, i.e. they are derived from a field of displacements ($\beta_{lj} = \partial u_j/\partial x_i$, cf. Chapter 2). Therefore, in the same way as the condition $\alpha_{ij} \neq 0$ expresses the fact that the solid contains translation dislocations, it must be noted here that the condition $\theta_{ij} \neq 0$ expresses the presence of rotation dislocations. This will be shown in the following section.

### 6.1.3 Wedge component of a disclination line

The trihedron $R(\mathbf{r})$ is not different from the Darboux–Ribaucour moving trihedron introduced in the previous chapter. On each surface $\Sigma$ of the partition $P$ under consideration, the lines $L$ define on $\Sigma$ the motion of $R$. In order to focus attention on the presence of one disclination, it is convenient to trace a Burgers circuit $\gamma$ on $\Sigma$, and to follow $R$ along $\gamma$ (Figure 6.3).

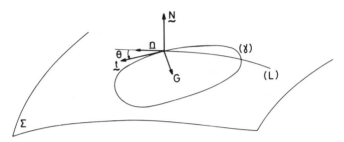

Figure 6.3. Burgers circuit on $\Sigma$, and motion of the Darboux–Ribaucour trihedron

Let us denote $\mathbf{t}$ the unit tangent to $\gamma$ and $\theta$ the angle between $\mathbf{t}$ and $\mathbf{n}$ at each point of $\gamma$. The reference frame $\mathbf{N}$, $\mathbf{t}$, $\mathbf{G} = \mathbf{N} \wedge \mathbf{t}$ defines a trihedron $R$ whose instantaneous motion is related to that of $R(\mathbf{r})$. Thus the instantaneous rotation of $R$ has components along $\mathbf{N}$:

$$\frac{1}{\rho_{\mathrm{G}}} = K_{ij}t_iN_j + \frac{\mathrm{d}\theta}{\mathrm{d}s}, \tag{6.13}$$

where $\rho_{\mathrm{G}}^{-1}$ is the geodesic curvature of $\gamma$ in $\mathbf{r}$.

Let us multiply both sides of equation (6.11) by $N_iN_j$ and integrate over the part of $\Sigma$ bound by $\gamma$. Note that the term rot $\mathbf{K}$ can be integrated by parts:

$$\iint \varepsilon_{ikl}K_{lj,k}N_iN_j\,\mathrm{d}\Sigma = 2\iint M_{ij}N_iN_j\,\mathrm{d}\Sigma + \oint K_{ij}t_iN_j\,\mathrm{d}s.$$

This yields:

$$\iint \theta_{ij}N_iN_j\,\mathrm{d}\Sigma = \iint M_{ij}N_iN_j\,\mathrm{d}\Sigma + \oint \rho_{\mathrm{G}}^{-1}\,\mathrm{d}s - \oint \mathrm{d}\theta. \tag{6.14}$$

This expression is interpreted geometrically in the following manner. Firstly it can be shown that

$$M_{ij}N_iN_j = \sigma_1\sigma_2, \tag{6.15}$$

i.e. that $M_{ij}N_iN_j$ is the gaussian curvature of $\Sigma$.

The derivation of equation (6.15) can make use of the identity (see Weatherburn, 1947):

$$\sigma_1\sigma_2 \equiv \tfrac{1}{2}\,\mathrm{div}\,(\mathbf{N}\,\mathrm{div}\,\mathbf{N} + \mathbf{N} \wedge \mathrm{rot}\,\mathbf{N}), \tag{6.16}$$

and of the fact that by using equations (6.7) and (6.11) it is possible to identify the right hand side of equation (6.16) with $M_{ij}N_iN_j$.

If the circuit $\gamma$ does not contain any singularity $L$, then $\oint d\theta = 2\pi$, and $\theta_{ij} \equiv 0$. Equation (6.14) is then none other than the well-known Gauss–Bonnet theorem:

$$\int\int \sigma_1\sigma_2 \, d\Sigma + \oint \rho_G^{-1} \, ds = 2\pi, \tag{6.17}$$

which, applied to a geodesic triangle, means that the sum of the angles of the triangle is less than $2\pi$ if $\sigma_1\sigma_2 < 0$, equal to $2\pi$ if $\sigma_1\sigma_2 = 0$ and greater than $2\pi$ if $\sigma_1\sigma_2 > 0$. Equation (6.17) is *independent* of the distribution of $\mathbf{n}$ on $\Sigma$. Equation (6.17) can therefore be substituted into equation (6.14) without any particular precautions.

Suppose that $\gamma$ encloses a singular point on $\mathbf{n}$, then by definition

$$\oint d\theta = 2\pi(1 - S),$$

where $S$ is the order of the singularity. By using equations (6.14) and (6.17) this becomes:

$$S = \frac{1}{2\pi} \int\int_{(\gamma)} \theta_{ij}N_iN_j \, d\Sigma. \tag{6.18}$$

where $\theta_{ij}$ is a density of rotation dislocations relative to the partition $P$. The integral (6.18) is relative to a surface element bound by $\gamma$ but situated on $\Sigma$. However, it should also be noted that $\theta_{ij}N_j$ is solenoidal

$$\theta_{ij,i}N_j + \theta_{ij}N_{j,i} = 0. \tag{6.19}$$

The integral $\int\int_\gamma \theta_{ij}N_j \, d\Sigma_i$ does not therefore depend on the surface which rests on $\gamma$. Also, this integral does not vary if $\gamma$ is displaced on a force tube of $\mathbf{N}$, so that $\gamma$ does not intersect the disclination line during this displacement. Formula (6.18) can therefore be applied finally to any circuit around the line in a region of *good* crystal (i.e. a region where $\theta_{ij} = 0$) expressed in the form

$$S = \int\int \Omega_i \, d\Sigma_i, \tag{6.20}$$

with

$$\Omega_i = \frac{1}{2\pi}\theta_{ij}N_j.$$

Condition (6.19) ($\Omega_{i,i} = 0$) expresses the fact that a disclination line can only be closed or go to infinity. This condition is analogous to that found for translation dislocations ($\alpha_{ij,i} = 0$).

## 6.2 COSSERAT MEDIA

The preceding study is peculiar in that one of the directions of the local trihedron is *integrable*: N possesses an ensemble of normal surfaces which is expressed as $\mathbf{N} \cdot \text{rot}\, \mathbf{N} = 0$.

The general case of a field of equal trihedrons is known as the Cosserat problem. Some of the geometric properties of such a system will be shown here, and it will be seen how singularities in them can be defined (Kléman, 1973a). This problem has recently acquired an immediate physical interest with the discovery of the A phase of $^3$He whose order parameter is actually a local trihedron (Anderson and Morel, 1961; Mermin and Ho, 1976). However, many other applications are possible.

### 6.2.1 General expression of the $K_{ij}$ and $M_{ij}$

Let there be a unit vector **n** with fixed components in the mobile reference frame $R(\mathbf{r})$ whose variation is defined by the $K_{ij}$.

It is clear that the rotation $d\theta_i$, undergone by **n** when $R$ changes, can be defined using only the coordinates of **n** and its gradients $n_{i,j}$. $d\theta$ is defined by

$$d\mathbf{n} = d\boldsymbol{\theta} \wedge \mathbf{n}. \tag{6.21}$$

It can be shown that

$$d\theta_i = \chi_{ji}\, dx_j, \tag{6.22}$$

with

$$\chi_{ji} = \varepsilon_{ipq} n_p n_{q,j}. \tag{6.23}$$

This tensor $\chi_{ji}$ cannot be the contortion tensor $K_{ij}$, as it does not act on the three axes of the trihedron in the same way.

However, one is allowed to add to $\chi_{ji}$ any term of the form $v_j n_i$, where the $v_j$ are the components of an arbitrary vector **v**. This brings us back, in fact, to adding a rotation along **n** to $d\theta_i$.

But in doing this, the most general form of the contortion tensor $K_{ij}$ has been obtained, when the trihedron $R$ changes under the single condition of 'conserving' **n**.† Let us then suppose the existence of three non-coplanar vectors $\mathbf{n}^1$, $\mathbf{n}^2$, $\mathbf{n}^3$ (chosen as orthogonal for greater ease) and their gradients $n_{i,j}^\alpha$ ($\alpha = 1, 2, 3$). The three tensorial equalities can be written (for $\alpha = 1$, 2 and 3):

$$K_{ij} = \varepsilon_{jpq} n_p^\alpha n_{q,i}^\alpha + v_i^\alpha n_j^\alpha \tag{6.24}$$

(here the repetition of the $\alpha$ does not imply a summation).

This time the $v_i^\alpha$ are no longer arbitrary. They must indeed obey:

$$K_{ij} n_j^\alpha = v_i^\alpha. \tag{6.25}$$

---

† There is said to be parallel transport of **n** (affine connection).

This equality can be written for any of the expressions of $K_{ij}$ from equation (6.24) and thus:

$$v_i^1 = n_q^3 n_{q,i}^2,\qquad(6.26)$$

the other $v_i^\alpha$ being obtained by permuting the exponents:

$$v_i^\alpha = \sum \tfrac{1}{2}\varepsilon_{\alpha\beta\gamma} n_q^\beta n_{q,i}^\gamma.$$

If it is supposed for the moment that $\theta_{ij} = 0$ (equation (6.11)), then the following relations, which derive from equation (6.11) and are valid in the non-singular regions of the medium, are obtained:

$$K_{ij}n_j = v_i,$$
$$M_{ij}n_j = \varepsilon_{ipq}v_{p,q} = -\mu_i,\qquad(6.27)$$
$$\det|K| + \boldsymbol{\mu}\cdot\mathbf{v} = 0,$$

these expressions being written for any vector $\mathbf{n}$ with constant components in the mobile reference frame. The notation $\boldsymbol{\mu} = \operatorname{rot}\mathbf{v}$ has been used; the last of the equalities in equation (6.27) indicates that the scalar product $\boldsymbol{\mu}\cdot\mathbf{v}$ is constant *whatever the value* of $\mathbf{n}$.

The first two equalities in equation (6.27) mean that $\mathbf{M}$ and $\mathbf{K}$ can be expressed as a sum of three bivectors:

$$K_{ij} = \sum_\alpha v_i^\alpha n_j^\alpha,$$

$$\qquad(6.28)$$

$$M_{ij} = -\sum_\alpha \mu_i^\alpha n_j^\alpha.$$

It will be verified that equation (6.28) leads to $\operatorname{rot}\mathbf{K} = +\mathbf{M}$.

The last equality in equation (6.27) which can be written for $\mathbf{n} = \sum_\beta a_\beta \mathbf{n}^\beta$, where the $a_\beta$ are any constant coefficients, leads to:

$$v_i^\alpha \mu_i^\beta = -A\delta_{\alpha\beta},\qquad(6.29)$$

while the classical relation $K_{lj}M_{pj} = A\delta_{lp}$ leads to:

$$\sum_\alpha v_p^\alpha \mu_p^\alpha = -A\delta_{lp},\qquad(6.30)$$

where $A = \det|K|$.

These relations, used with equation (6.27), give, where $A \neq 0$:

$$\boldsymbol{\mu}^\alpha = \tfrac{1}{2}\sum_{\beta,\gamma} \varepsilon^{\alpha\beta\gamma} \mathbf{v}^\gamma \wedge \mathbf{v}^\beta,\qquad(6.31a)$$

$$\boldsymbol{\mu}^\alpha \cdot \mathbf{v}^\beta = 0,\quad \text{if } \alpha \neq \beta.\qquad(6.31b)$$

Finally, let us note the expression of rot **n**:

$$(\text{rot }\mathbf{n})_j = K_{ij}n_i - Kn_j \qquad (K = K_{11} + K_{22} + K_{33}), \qquad (6.32)$$

where **n** is any constant vector, $\mathbf{n} = \sum a_\beta \mathbf{n}^\beta$, $\mathbf{v} = \sum a_\beta \mathbf{v}^\beta$.
Then obviously

$$\sum_\alpha \mathbf{n}^\alpha \cdot \text{rot }\mathbf{n}^\alpha = -2K. \qquad (6.33)$$

The integrability condition of **n** reads:

$$\mathbf{n} \cdot \text{rot }\mathbf{n} \equiv \mathbf{v} \cdot \mathbf{n} - K = 0, \quad \text{i.e. } K = \mathbf{v} \cdot \mathbf{n}. \qquad (6.34)$$

It is then seen that the three directions $\mathbf{n}^\alpha$ cannot be all integrable, unless $K \equiv 0$ (equation (6.33)). From equation (6.34) $\mathbf{n} \cdot \mathbf{v} = 0$ is obtained for every integrable direction.

### 6.2.2 Interpretation of the contortion tensor; density of translation dislocations

The process of deformation of solid crystals in which the lattice is curved without the distances between neighbouring atoms being changed is called glide. It can only occur in the presence of a non-zero density of dislocations, all essentially of the same sign. In *linear elasticity* this process can be characterized by vanishing strain $e_{ij}$ and non-vanishing lattice rotations $\omega_i$, i.e. by a density of dislocations

$$\alpha_{ij} = \delta_{ij}\omega_{k,k} - \omega_{i,j} \qquad (6.35)$$

The case of *single glide* is easily visualized: a family of parallel reticular planes is curved; the deformed reticular surfaces are parallel and the orthogonal trajectories are straight lines (see the analysis of polygonization in Chapter 2 and Figure 6.1.)

Nye (1953) was the first to study this geometry in detail and extended it to the case of large rotation, showing that a formula analogous to equation (6.35) still holds

$$\alpha_{ij} = \delta_{ij}K - K_{ji} \qquad (6.36)$$

where now $K_{ij}$ are the components of the tensor of contortion. We call *generalized glide* a situation which obeys equation (6.36). We refer the reader to Nye's (1953) article for the derivation of equation (6.36) or to Nabarro's (1967, ch. 1, p. 44 *et seq.*) book. Formula (6.36) is not approximate and can be obtained in another context besides dislocation theory, that of non-Riemanian geometry (Kondo, 1955; Bilby, 1960; Kröner, 1958). $\alpha_{ij}$ then appears as the 'torsion' of a certain space with euclidean metric, affine connection $\Gamma^k_{ij} = \varepsilon_{kil} K_{jl}$, and vanishing curvature $\theta_{ij}$. This is a very valuable point of view, capable of many generalizations, but one which this work will not dwell upon. Nabarro (*op. cit.*, ch. 8) gives an interesting analysis of this point of view.

The concept of glide is not imperatively limited to a *solid*. On the contrary, because of the condition of conservation of atomic density, it can be applied without any modification to an *incompressible anisotropic liquid*. The idea of a translation makes very clear sense, which can be appreciated in the following examples (which will be taken up again in subsection 6.2.3, in which the sense which can be attached to the $\theta_{ij}$ tensor will be discussed). For the moment, only regular points of the field will be considered, i.e. where $\theta_{ij} = 0$.

### 6.2.2.1 $\mathbf{N} \cdot \mathrm{rot}\,\mathbf{N} = 0$

This is the example which was discussed in detail in section 6.1. In the desire to be complete, the components of the $K_{ij}$, $v_i$ and $\mu_i$ in the *local reference frame* formed on the directions $\mathbf{n}$ (along the axis 1), $\mathbf{m}$ (along the axis 2) and $\mathbf{N}$ (along the axis 3) are written:

$$K_{11} = \frac{1}{\tau_G}; \qquad K_{21} = \frac{1}{\rho'_n}, \qquad K_{31},$$

$$K_{12} = -\frac{1}{\rho_n}; \qquad K_{22} = -\frac{1}{\tau_G}, \qquad K_{32}, \tag{6.37}$$

$$K_{13} = \frac{1}{\rho_G}; \qquad K_{23} = \frac{1}{\rho'_G}, \qquad K_{33} = K,$$

where $1/\tau_G$, $1/\rho_n$ and $1/\rho_G$ are respectively the geodesic torsion, the normal curvature, and the geodesic curvature of the envelope lines of $\mathbf{n} \cdot \tau'_G = -\tau_G$, $\rho'_n$ and $\rho'_G$ being the analogous quantities for the envelope lines of $\mathbf{m}$. $K_{31}$ and $K_{32}$ are the components of $\mathrm{rot}\,\mathbf{N}$ on the 1 and 2 axes.

Applying equations (6.28) we have:

$$v_i = K_{ij}N_j = \left\{ \frac{1}{\rho_G}, \frac{1}{\rho'_G}, K \right\}, \tag{6.38}$$

$$\mu_i = -M_{ij}N_j = \left\{ \begin{array}{c} -\dfrac{K_{32}}{\rho'_n} + \dfrac{K_{31}}{\tau_G} \\[2ex] \dfrac{K_{31}}{\rho_n} + \dfrac{K_{32}}{\tau_G} \\[2ex] \dfrac{1}{\tau_G^2} - \dfrac{1}{\rho_n\rho'_n} = \sigma_1\sigma_2 \end{array} \right\}. \tag{6.39}$$

### 6.2.2.2 $\mathrm{Rot}\,\mathbf{N} = 0$

The surfaces $\Sigma$ perpendicular to $\mathbf{N}$ are parallel to each other and the envelopes of $\mathbf{N}$ form a congruence of rectilinear normals. This is the case of single glide according to Nye. Thus, $K_{31} = K_{32} = 0$, from which

$$\boldsymbol{\mu} = (0, 0, \sigma_1\sigma_2). \tag{6.40}$$

186

Two partciular cases will be studied.

(a) The situation where **n** and **m** are also orthogonal to families of surfaces. This reads:

$$\mathbf{n} \cdot \mathrm{rot}\,\mathbf{n} = \mathbf{m} \cdot \mathrm{rot}\,\mathbf{m} = 0 \qquad (6.41)$$

and leads to

$$K = K_{11} = K_{22} = K_{33} = 0. \qquad (6.42)$$

The envelope lines of **n** and **m** on each surface $\Sigma$ are lines of curvature of $\Sigma$. We then have

$$\boldsymbol{\mu} \cdot \mathbf{v} = 0 \qquad (6.43)$$

for the three fields $\mathbf{v}^1$, $\mathbf{v}^2$ and $\mathbf{v}^3$, which are therefore also integrable. Finally, it should be noted that $A = 0$.

These properties are reciprocal. If $A$ and $K$ are zero, the fields $\mathbf{n}^z$ and $\mathbf{v}^\beta$ are all integrable, and the orthogonal sets of surfaces intersect two by two along their lines of curvature.†

(b) Let us suppose that the surfaces $\Sigma$ are the transforms of *reticular planes* $\Sigma_0$ where the directions $\mathbf{n}_0$ and $\mathbf{m}_0$ trace out a square lattice, and let us choose the transformed fields **n** and **m** in such a way that the transformation between $\Sigma_0$ and $\Sigma$, for each of these couples, is *conformal*;‡ this is in fact the *inverse problem* to that of geographic maps (Darboux). Its solution is as follows:

Let $u$ and $v$ be a rectangular coordinate system on $\Sigma$. Any line $u = $ const or $v = $ const forms a conformal representation of $\Sigma_0$ on $\Sigma$ if the linear element of $\Sigma$ can be written in the form:

$$ds^2 = \lambda^2(du^2 + dv^2), \qquad (6.44)$$

where $\lambda = \lambda(u, v)$. Then the element $dx(dy)$ corresponds to the infinitesimal elements $\lambda\,du$ ($\lambda\,dv$) on $\Sigma_0$.

These results are studied in detail in the section on smectics B (section 7.4) for which these considerations are important. It should be pointed out that there is always an infinite number of rectangular coordinate systems on any surface which can be represented by equations of the type (6.44). These coordinates define a conformal transformation with the plane and divide $\Sigma$ into infinitely small *squares*; they form a so-called isothermic coordinate system.

Is it possible that the system of lines of curvatures itself forms an isothermic coordinate system? This is only true for quite particular classes of surfaces (called isothermic surfaces) among which are minimum surfaces, which have a constant mean curvature, and Dupin cyclides. But only the latter form a system of parallel

---

† When three sets of surfaces intersect at right angles, the lines of intersection of two surfaces belonging to two different sets are lines of curvature common to these two surfaces (Dupin–Joachimsthal theorem).

‡ That is, the angle between any reticular directions on $\Sigma$ are the same as on $\Sigma_0$; but the lengths are not conserved. However, they are not far from being so (this is expressed quantitatively by equation (6.44)), which makes the consideration of the conformal transformation important also for solids.

surfaces. The interest behind this geometry must be stressed. According to our previous results, there exist in this case two sets of surfaces normal to the envelope lines of **n** and **m**. If on the contrary the conformal transformation envisaged does not make the lines of curvature an isothermic coordinate system then the envelope lines **n** and **m** bear densities of *screw dislocations* ($K_{11}, K_{22} \neq 0$) and the surfaces normal to the isothermic lines display helicoidal branchings.

### 6.2.2.3   $M_{ij} \equiv 0$

This particular case is interesting only for the discussion of the $\theta_{ij}$.

If all the $M_{ij}$ elements of a matrix $K_{ij}$ are zero, it is said that this matrix is of order *one*. It can then be shown (A. R. Curtis, quoted by Bilby and Smith, 1956) that $K_{ij}$ reads:

$$K_{ij} = \frac{\partial a}{\partial x_i} g_j(a), \tag{6.45}$$

where $a(\mathbf{r})$ is any scalar function. If a vector **N** is considered as constant in the mobile reference frame, according to equation (6.6) we have:

$$N_{j,k} = \varepsilon_{jpq} g_p(a) N_q \frac{\partial a}{\partial x_k},$$

i.e.

$$dN_j = \varepsilon_{jpq} g_p(a) N_q \, da. \tag{6.46}$$

It is deduced from this that the mobile trihedron is constant on the surfaces $a(\mathbf{r}) = $ constant.

### 6.2.3   Defects in Cosserat media. Rotation dislocations. Singular points

The density of translation dislocations $\alpha_{ij}$ measures, as Bilby (1960) has shown, the *local* Burgers vector. This is the Burgers vector resulting from an open circuit in a deformed crystal which corresponds to a closed circuit in a perfect crystal. The law of conservation which requires that the dislocations terminate on themselves then takes on a slightly different form from that of the theory of small deformations, there the local Burgers vector and the Burgers vector in the usual sense can be identified. We have:

$$\alpha_{ij,i} + \varepsilon_{klm} M_{lm} = 0 \tag{6.47}$$

Suppose now that the Cosserat field is not compatible everywhere ($\theta_{ij} \neq 0$). Using equation (6.11), equation (6.47) is altered to:

$$\alpha_{ij,i} + \varepsilon_{jlm} M_{lm} = -\varepsilon_{jlm} \theta_{lm}. \tag{6.48}$$

Only the antisymmetric part of $\theta_{ij}$ appears on the right-hand side. It should be stated that the translation dislocations, when they do not terminate on themselves, terminate on *twist* disclinations. The relation (6.48) is the continuous form of equation (2.14) established for lines of discrete rotation dislocations. In this sense, the $\alpha_{ij}$, for a nematic, represent these densities of translation dislocations spread out by viscous relaxation.

The symmetrical components of $\theta_{ij}$ can be worked out from equation (6.11) by bringing into it the expression for $\alpha_{ij}$ as a function of the $K_{ij}$.

### 6.2.3.1 *Relation between order of a rotation dislocation and circulation of a vector*

The physical sense of $\theta_{ij}$ in the case of a director system has already been studied. This is an example of a geometric case where one of the axes of the trihedron $R$ is integrable $(\mathbf{N} \cdot \text{rot} \, \mathbf{N} = 0)$. This condition is essential to the derivation of equations (6.13), (6.14) and (6.17). It is therefore necessary to look for a generalization of these expressions in the case where $\mathbf{N}$ is not integrable. Let there be therefore a circuit $\gamma$ bounding a cut surface $\Sigma$. The integral

$$S = \frac{1}{2\pi} \int\int_{\Sigma} \theta_{ij} N_j \, d\Sigma_i$$

depends only on $\gamma$, since $\theta_{ij}N_j$ is a solenoidal vector. This definition of $S$ can then again be adopted and is identical to the previous one where $\mathbf{N}$ is integrable. As the term rot $\mathbf{K}$ can be integrated by parts, as above, it becomes:

$$S = \frac{1}{2\pi} \int\int_{\Sigma} M_{ij} N_j \, d\Sigma_i + \frac{1}{2\pi} \int_{\gamma} K_{ij} N_j \, dx_i. \tag{6.49}$$

But the interpretation cannot be extended. In particular, $M_{ij}N_j \, d\Sigma_i$ is not equal to the element of curvature $\sigma_1 \sigma_2 \, d\Sigma$ of the surface $\Sigma$, and the Gauss–Bonnet theorem cannot be used. Thus, the quantity $S$ no longer has any reason to be an integer or a half-integer. *While formally analogous to the strength of a rotation dislocation, the quantity defined by equation* (6.49) *is in fact closer to a circulation.* Take, for example, the case of a field $M_{ij} = 0$; equation (6.49) appears as the circulation of an irrotational vector $v_i = K_{ij}N_j = \mathbf{g} \cdot \mathbf{N}(\partial a/\partial x_i)$. In the more general case where $M_{ij} \neq 0$, $S$ is the circulation of the irrotational part of a velocity. It is known, indeed, that a vector can always be split into a gradient part and a rotational part,

$$\mathbf{v} = \text{grad} \, \psi + \text{rot} \, \mathbf{a},$$

in a unique manner (see Sommerfeld, 1964). $M_{ij}N_j$ is the opposite of rot $\mathbf{v}$. The right hand side of equation (6.49) is therefore the circulation of grad $\psi$, i.e. the variation $\Delta\psi/2\pi$ on a circuit surrounding the singularities of $\mathbf{v}$.

These considerations can be applied to the well known case of a director field $\mathbf{n}$ in a nematic. Orthogonal fields $\mathbf{m}$ and $\mathbf{N}$ can be added to such a field to form trihedrons, in an infinite number of ways. Only a few of them restrict $S$ to integral values or half-integral values. These are the only ways which have a physical meaning. Moreover, the way in which these results have been introduced indicates that $S$ is the common strength of singularities which can be identified in **n** and **m** fields. This was discussed in section 6.1.

Suppose now that **n** and **m** both have a physical meaning (the applications to the A phase of $^3$He of the geometrical considerations which follow will be seen later). If the vector **m** ∧ **n** is not integrable, investigation of the *respective singularities* of rotation dislocations in the **n** and **m** fields introduces two integrable fields $N(n)$ and $N(m)$, according to the method proposed for nematics. It may then be supposed that the strengths $S(n)$ and $S(m)$ calculated from equations (6.14) or (6.18) are not going to be equal for the two fields **n** and **m**, the singularities of which are, moreover, not even necessarily located along the same lines. On the other hand, equation (6.49) means that whether these singularities are on field **n** or **m** or both, a quantity $S_c$ can be attributed to them (not quantized to half-integral values) which is the circulation of the vector $v_i = K_{ij}N_j = n_q m_{q,i}$ (see equation (6.26)) from which the rotational term has been substracted. This means that in general three topological invariants can be attributed to a line of singularities of the field **n, m**, two of them related to the wedge components of the director fields (or vector fields) **n** and **m**, the third being a circulation representative of the two fields combined. The lines of singularities under consideration are therefore both (rotation) dislocation lines of order $S(n)$ and $S(m)$ and vortex lines of strength $S_c$. Because of the solenoidal character of $\theta_{ij}N_j$, these three quantities are conserved along the line. $S_c$ is conserved in every small rotation about **n** ∧ **m**. Let $\delta\omega$ be this rotation. It can be shown that this leads to the modification of $K_{ij}$, $M_{ij}$, $\theta_{ij}$, in the following way:

$$K_{ij} \to K'_{ij} = K_{ij} + N_j \frac{\partial \delta\omega}{\partial x_i},$$

$$M_{ij} \to M'_{ij} = M_{ij} + \varepsilon_{ikl}\varepsilon_{jpq}N_q K_{kp} \frac{\partial \delta\omega}{\partial x_i}, \qquad (6.50)$$

$$\theta_{ij} \to \theta'_{ij} = \theta_{ij} + N_j\varepsilon_{ikl} \frac{\partial^2 \delta\omega}{\partial x_l \partial x_k} = \theta_{ij}.$$

$\theta_{ij}$ is not altered, therefore $S_c$ is not.

But if a local rotation of the system about any (given) axis in the plane **n, m** is carried out, then $S_c$ is altered, although $\theta_{ij}$ is not (to first order), for now **N** is altered, therefore $\theta_{ij}N_j$ is also. In these operations, $S(n)$ and $S(m)$ are not generally altered, as we are dealing with strengths of rotation dislocations.

In other words, for the circulation $S_c$ to be conserved, it must be identifiable with a strength of disclinations, i.e. the integrability condition $N \cdot \text{rot} N = 0$ must be conserved in any local deformation of the system.

### 6.2.3.2 *Defect lines in a system of trihedrons. Vortex lines and disgyrations in the A phase of $^3$He*

The formulae (6.14) or (6.18) are meaningful only if **N** is a vector with fixed components in the local reference frame. If the defect under discussion is to be obtained by the Volterra process, it is clear that $\mathbf{N}$ is the axis of rotation about which this process is put into practice. Consider firstly a perfect medium consisting of parallel trihedrons and a cut surface $C$ bound by a line $L$. Turn the trihedrons situated on each side of the cut surface through a relative angle $2\pi S\mathbf{N}$. It is clear that the $K_{ij}$ given by

$$K_{ij} = 2\pi SN_j\mu_i\delta(C) \qquad (6.51)$$

(where $\mu$ is the unit normal to $C$, $\delta(C)$ the Dirac distribution function) describe correctly the effect of this operation. $\theta_{ij}$ can now be derived:

$$\theta_{ij} = 2\pi S\delta(L)t_iN_j, \qquad (6.52)\dagger$$

where $t$ is the unit tangent to $L$ and $\delta(L)$ the Dirac distribution function peaked on $L$. (A summary of the formulae relating to the Dirac distribution on lines and surfaces will be found in section 7.2.2). It is obvious that by integrating $\theta_{ij}N_j$ over a surface intersected by $L$ and bound by a Burgers circuit, one obtains a disclination strength precisely equal to $S$.

In the end, it therefore appears that formulae (6.14) or (6.18) are only meaningful if $\mathbf{N}$ is the rotation vector. Also, as we have seen, $\mathbf{N}$ must be integrable.

The defects in a system whose order parameter is a trihedron can be classified both according to the direction $\mathbf{N}$ about which they are built, and according to their strength. $\mathbf{N}$ must obviously be an axis of symmetry of the system. If the strength is equal to $\pm 1$ or is a multiple of $\pm 1$, $\mathbf{N}$ can be any direction, every direction being an axis of symmetry of order 1. If $\mathbf{N}$ is an axis of symmetry of order 2, the fixed *vectors* of the local reference frame which are perpendicular to it are in fact directors. But the vector $\mathbf{v}$ associated with $\mathbf{N}$ by equation (6.27) $(K_{ij}N_j = v_i)$ is a true vector, the circulation of which is measured by equation (6.18). Finally, more complex cases exist if $\mathbf{N}$ is an axis of symmetry of a higher order.

Consider the example of the A phase of $^3$He, whose symmetries lead to the identification of an axis of the trihedron at the orbital angular moment of the Cooper pairs, $l$, and the two axes orthogonal to $l$, with the real and imaginary (vectorial) components $\Delta' + i\Delta''$ of the gap in anisotropy energy (Anderson and Brinkman, 1975). This combination constitutes a trihedron of 3 unequal axes, each axis being an axis of symmetry of order one. To complete the description of the order parameter, it is necessary to add a spin vector $\mathbf{V}$ which because of the weakness of the spin–orbit interactions, is practically uncoupled from $l$ over small distances, but is aligned with $l$ (along $l$ or $-l$) when the sample is considered over long distances, where the interactions again become predominant (see Anderson, 1973; Legett, 1975).

1. $\mathbf{V}$ is equivalent to $-\mathbf{V}$; this field can therefore display defects of integral and half-integral strength. De Gennes (1973b) gave these the name *spin disgyrations*. Disgyrations of strength $\pm\frac{1}{2}$ can only exist if $\mathbf{V}$ and $l$ are uncoupled. If they are not the defects of the $l$ field and the $\mathbf{V}$ field are the same, and are necessarily of strength

---

† More detailed calculations relating to the geometries of particular lines $L$ are in Kléman (1973a); Hall (1976) obtained the same results.

$\pm 1$, every fixed direction of the trihedron being a direction of symmetry of rotation of order 1.

2. Defects of the field **l** are also called disgyrations (of **l**) (de Gennes). The simplest which can be imagined are of the wedge type, $\Delta'$ or $\Delta''$ being along the axis of the line, and **l** describing a configuration of the type $S = \pm 1$. The axis of rotation being, for example, along $\Delta'$, $\Delta''$ also describes a wedge line of the same type as **l**. The surface defects are also probably of the disgyration type (Mermin, 1977), for the boundary conditions imposed on **l** restrict it to being perpendicular to the surfaces (Ambegaokar et al., 1974). (The normal to the surface is then along **l** or $-$**l**).

3. The singularities having axis of rotation **l** are of the *vortex line* type (de Gennes, 1973b; Mermin and Ho, 1976; Anderson and Brinkman, 1975), i.e. singularities of vector fields $\Delta$ or $\Delta'$. Their strength is a multiple of $\pm 1$. When discussing vortex lines, it is usual to understand by this singularities of a phase $\omega$ whose variations measure the rotations of $\Delta$ or $\Delta'$ about **l**. As the rotations which cause the passage from one trihedron $R$ to a neighbouring trihedron are not commutative, it is clear that the variations of $\omega$ are not defined in a unique way. There is therefore no quantization of $\Delta\omega/2\pi$ where $\Delta\omega$ would be an angle. On the other hand, we have seen that the circulation of $v_i = \Delta_k \Delta'_{k,i}$ (with the rotational term of **v** subtracted) is an invariant of the line, equal to $\Delta\psi/2\pi$ (by putting $\mathbf{v} = \text{grad}\,\psi + \text{rot}\,\mathbf{a}$). This invariant is equal to the strength of the line if $\mathbf{N} \cdot \text{rot}\,\mathbf{N} = 0$.

4. Defects of *two quanta* ($S = \pm 2$) can lose their core singularity like singularities in nematics for $S = \pm 1$.

These correspond to a rotation of $4\pi$ about a given axis of rotation. Hall (1976) proposed that the two quanta correspond respectively to rotations around rectangular axes. Thus in the experiment of the cylindrical tube discussed in Mermin and Ho (1976) where the boundary conditions restrict **l** to being perpendicular to the walls of the tube: along the perimeter of the tube, **l** turns through $2\pi$ about the axis of the tube, while $\Delta'$ and $\Delta''$ turn about **l** through an angle also equal to $2\pi$. When the axis is approached, **l** swings towards it, bringing $\Delta'$ and $\Delta''$ into positions which are the same whatever the vector radius along which the axis is approached might be. This is shown in Figure 6.4: only **l** and $\Delta'$ are shown, the non-singularity of the two distributions brings about that of the third. It is clear that this can be described either, as Hall proposes, as a defect with two rectangular quanta, or as a twist disclination. Also, a similar result on the spreading of the core is still true if the two quanta are coupled at any angle, the product of any two axes of rotation of $2\pi$ being a rotation of $4\pi$.

### 6.2.3.3 *Singular points in* $^3$He-A

Blaha (1976) and Toulouse and Kléman (1976) have shown that singular points in $^3$He-A are necessarily at the *termination* of singular lines. A singular line can therefore be imagined as being bound by two singular points (Blaha vortons). Blaha stresses the analogy of this configuration with Dirac's (1948) magnetic

192

monopoles (here vortons), linked by strings (here disclinations) (also see t'Hooft, 1974). This property can be proved by a very simple topological argument (see Chapter 10). But in this chapter we present this result by using the densities already introduced. The strength of a point singularity in a vector field $\mathbf{N}$ is obtained by the integral (6.53), summed over a closed surface which surrounds the singularity.

$$4\pi N = \int \varepsilon_{ijk}\varepsilon_{pqr}N_{q,j}N_{r,k}N_p \, dS_i. \tag{6.53}$$

It can be shown that this quantity represents the flux of the vector $\mathbf{N}$ (on this closed surface) transported to the origin through a sphere of unit radius centred

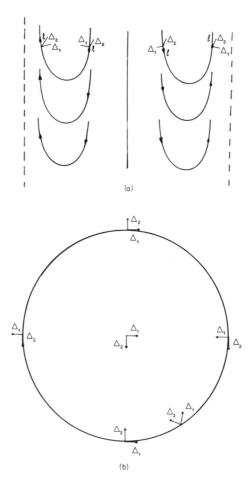

Figure 6.4. Scheme of a wedge disclination with non-singular core $S = 2$ in $^3$He. It is supposed that the trihedron, at a long distance from the line, has turned through $4\pi$ about the axis of rotation (Anderson and Toulouse). (a) Meridian section. (b) Section perpendicular to the line

at the origin. If $N$ possesses a point singularity, this sphere is covered an integral number of times; $N$ is therefore integral. On the other hand we have:

$$M_{ip}N_p = \varepsilon_{ijk}\varepsilon_{pqr}N_{q,j}N_{r,k}N_p, \qquad (6.54)$$

a relation which is obtained by noting that $M_{ip}N_p$ is not changed by making the trihedron turn about $N$ through any angle $\omega$. (This operation changes $K_{ij}$ into $K_{ij} + \partial\omega/\partial x_i\, N_j$). We therefore have:

$$N = \frac{1}{4\pi} \int M_{ij}N_j\, \mathrm{d}S_i, \qquad (6.55)$$

the interpretation of which in terms of flux is obvious if the surface of integration is perpendicular to $N$ throughout. This quantity is therefore equal to $\int (\mathrm{d}S/4\pi)/R_1 R_2$, i.e. the area of a unit sphere perpendicular to $N$ (transported to the origin) and divided by $4\pi$.

$M_{ij}N_j$ is rotational everywhere $\theta_{ij}$ vanishes. Therefore if equation (6.55) is not zero, this indicates that $\theta_{ij}$ is different from zero in at least one point on all the surface surrounding the singular point, therefore that the latter is linked to a disclination. It will be noted that singular points are linked to the rotational part of the vector $K_{ij}N_j$, whereas the singular lines are linked to the gradient part of the same vector.

# Chapter 7

# Lamellar phases

After the detailed study of smectics A, this chapter will be devoted to problems related to the other lamellar phases, in particular their elasticity, and the theory of defects in liquid lamellar phases (types Sm C and Sm C*) and ordered lamellar phases (Sm B). These problems are quite new and most of these phases have hardly been investigated experimentally, from the point of view of defects at least.

The polymorphism of lamellar phases is the first indication that they are a rich source of interest for researchers, although for a long time only physical chemists have studied this phenomenon (see Sackmann and Demus, 1966). Recent interest shown by physicists in lamellar phases, especially in the phase transitions observed in them (starting with de Gennes, 1973a, and McMillan, 1973a, b), has sustained and renewed the interest of physical chemists (Gray, 1973; Goodby and Gray, 1976, 1979; Jacques' group at the Collège de France; Strzelecki's group in Orsay and Gasparoux' group in Bordeaux). The study of the stability of smectic phases in relation to molecular conformations is an important subject, and it would have been difficult not to set aside at least one section (7.1) of the present work to this study. It will be seen, anyway, that macroscopic observation methods play an indispensible role in this type of research, and it is to be hoped that studies of miscibility and phase transitions will in future be oriented more clearly towards a deeper description of textures and defects.

The elasticity of liquid lamellar phases with inclined molecules requires quite a deep intrusion into the theory of surfaces in dealing with deformations of large curvature (subsection 7.2.1). It is undoubtedly the first physical case where the methods discussed in the preceding chapter can be appropriately applied. The free energy of a smectic C will be written as a quadratic form of the $K_{ij}$ components of the distortion tensor and the different stiffness coefficients allowed by symmetry will be linked to the quantities characterizing the trihedron representing the local order (normal curvatures, geodesic torsion and geodesic curvatures of certain lines described on the layers, etc.). The mobile trihedron method is particularly well adapted to this type of discussion. It will be seen, however, that the compatibility conditions of the $K_{ij}$ (see equation (6.11)) play a role in the research of the distribution of the layers and directors which is at least as important as Euler's equations which minimize the free energy: in this case these compatibility conditions are not reduced to the single equation $\mathbf{n} \cdot \mathrm{rot}\, \mathbf{n} = 0$

as found in the case of smectics A. It will be noted that the methods discussed in this section can be applied to any system whose order parameter can be represented by a trihedron (Cosserat media).

Subsection 7.2.2 gives a theoretical description of defects in smectics C, based on classification by the Volterra process. This description will be taken up again in Chapter 10, using different concepts (homotopy groups). Also some recent experimental results will be presented.

Chiralized liquid layer phases (Sm C*) have a particular physical interest: they are ferroelectric (Meyer, 1976). A discussion of their defects will also be given, supported by recent experimental results.

The geometry of smectics C leads to the redefinition of the concept of the director. As Frank has stressed (in a private communication) the director is defined as the mean value of an ensemble of fluctuating directions. This mean value can be obtained experimentally by various methods (optical methods, dielectric or diamagnetic susceptibility tensor, etc.), which give a unique response for smectics A when the director is normal to the layers, but the same does not apply to a biaxial medium such as smectics C. The angle $\psi_0$ introduced in Figure 7.1 is therefore somehow fictitious; moreover, it does not appear in the formalism and it is imagined essentially in order to indicate the *inclined* nature of the structural situation.

Ordered lamellar phases, (Sm B, Sm H, etc.) also open up a new field: the nature of interactions between layers, comparison with solid crystals and plastic crystals, the description of layer defects. Here again certain geometric properties found in the theory of surfaces will serve as our guide in the description of defects typical of a two-dimensional crystal (single layer or two layers) supposed to be strongly curved and in interaction with neighbouring layers. This discussion is given at a more theoretical level, just in order to introduce some ideas which might be useful in future experimental research.

Nothing will be said of recent studies on the role of defects in phase transformations in various lamellar systems (Huberman *et al.*, 1975; Helfrich, 1976), although these are interesting subjects which qualify well to be included here. Helfrich (1980) has extended defect models of phase transitions to numerous mesomorphic (lyotropic and thermotropic) phases and Halperin and Nelson (1978) have also discussed them in the frame of the Kosterlitz and Thouless (1972) two-dimensional transitions model.

Also, the descriptions by Harris (1970c, 1974, 1975) of surface topology in terms of defects are only mentioned here, despite their interest. Harris has shown that the genus and the orientability of a surface can be defined as resulting from a process of creation of disclinations (twist or wedge, according to the case) transverse to the surface.

This is also perhaps the place to mention that Scriven (1977) has stressed the interest of visualizing the bicontinuous equilibrium structures which occur in microemulsions of lyotropic phases (in the region of the phase diagram where direct micelles and inverse micelles are coexisting) as regularly ordered minimal surfaces. More generally, one can expect notions of surface geometry to be helpful in the study of the various transitions between lyotropic phases (lamellar, cylindrical, etc.).

## 7.1 CHARACTERISTICS OF SMECTIC PHASES. CONFORMATION PROBLEMS

Well before the systematic studies by X-ray diffraction (Levelut, 1976, Levelut and Lambert, 1971, Levelut et al., 1974), the astonishing polymorphism of thermotropic lamellar phases had been shown by optical observations both of miscibility relations between different smectic phases and of characteristic textures (Sackmann and Demus, 1966, 1973; see also for a recent example Billard and Urbach, 1972). Lyotropic phases, which possess just as rich a polymorphism as lamellar phases, have for a long time been characterized in the same way by their textures (Rosevear, 1954; Lachampt and Vila, 1969; Rogers and Winsor, 1969) before X-rays (Luzatti et al.) gave precise results.

### 7.1.1 Miscibility relations

Miscibility relations, which have been discussed by de Gennes (1975, ch. 7), are based on the following rule: two phases of different symmetries must be separated by at least one transition line. Reciprocally, if two phases are continuously miscible without crossing any transition line, they have the same symmetry. For example, introduce into a smectic A some molecules which, at the same temperature, are organized in a Sm C phase, in a concentration $x$; for low values of $x$, the A phase absorbs the C molecules without a change of symmetry. For a critical value $x_c(T)$ which depends on the temperature, a C phase appears. The transition can be either second order (the concentration of the C phase is higher than, or equal to $x_c$) or first order depending on the sign of the term in $\psi_0^4$ in the free energy expansion ($\psi_0$ is the tilt angle of the Sm C phase, and its coefficient $b$ depends on $x$ and $T$). In practice, two pure phases are placed on a slide, and the mixture is observed through a microscope. The method must both determine the differences in texture of the different phases and allow the nature of specific defects to be recognised. Frank (1977) has shown recently that de Gennes' analysis neglects the possibility of coexistence of two phases, even for a second order transition, when the transition line is a *spinodal* line.

The miscibility rule is only applied, according to de Gennes, if the molecules in solution have conformations quite similar to those of the solvent. The solute can then take the place of the molecule of the solvent as long as the interactions between the molecules remain negligible. The interaction of an isolated molecule of solute with the molecules of the solvent is of the same order of magnitude and the same nature as those of a molecule of solvent with its neighbours. They are therefore typically short range interactions. We have here the liquid analogue of the crystal solid containing *coherent* foreign inclusions (see J. Friedel, 1964).

When the inclusion is *incoherent* we have on the other hand the analogue of a solid solution. A typical case is that of the solution of active molecules in a racemic mixture: although the molecules are similar, the *local* change of symmetry induces long-range effects which distort the sample, even for weak concentrations of foreign molecules, and create twist. The introduction of molecules which are very different in size or form does not necessarily lead to the formation of precipitates, as would be the case if the comparison with solids were pursued. Liquid *diffusion*, which tends to *make the densities uniform*, is the essential phenomenon in this case (unless the phases are separated by gravity, but this effect is not considered here). The privileged sites for the formation of accumulations would perhaps be defects. But what probably determines the nature of the dispersion of foreign molecules is now going to be the selective affinity of *different parts* of these molecules with different parts of the host molecules. Applying the Latin adage, *similia similibus solvantur*, we may rely upon the fact that the paraffinic parts attract each other, while there is repulsion between the aromatic and paraffinic parts, in spite of the increase in entropy which a total mixture would represent. Such studies of the influence of the molecular conformation would be worth pushing forward. Remember that for the same reason the mixture of *incompatible* polymers leads to ordered solutions at long range (Skoulios, 1967), and the lipid–water mixture gives lamellar phases. Thus the concept of segregation is replaced by that of long range order, essentially because we are dealing with a liquid medium!

### 7.1.2 Stability of pure smectic phases

Although the discussion above relates to *mixtures* of different constituents, it has a close relationship with ideas recently proposed by Jacques and his collaborators concerning the stability of smectic phases of a given compound. For these authors, a given organic molecule leads to a *nematic phase* if the aromatic (or fatty) parts are indifferent to the presence in their immediate vicinity of fatty (or aromatic) parts of another molecule of the same species. This explains why the displacement of the aromatic centre along the axis of the molecules in a single *isometric* series (thus

$$\text{R}-\langle\bigcirc\rangle-\text{C}=\text{C}-\langle\bigcirc\rangle-\text{OR}'$$

with $R + R' = 10$ carbons, constitutes an isometric series) would have little influence on the nematic phase (existence and temperature range). On the other hand, the stability of different smectic phases, characterized by compatibility phenomena (fusion of fatty chains and dipolar interactions between similar radicals), would be very sensitive to this displacement. Some results from the work of Jacques' group, and those of Gray (see Goodby and Gray, 1976) which are relevant to *homologous* series should be mentioned here.

(a) Conditions seem favourable to Sm C phases when the aromatic part is near to the centre of gravity of the molecule (Jacques). This result, which amplifies the role of permanent dipolar moments perpendicular to the axis in the rigid part of the molecule (as McMillan, 1973a, foresaw) is not sustained by Gray, who sees the length of the chain as the predominant factor. Jacques' point of view attributes the liquidity of the layers to the fusion of the chains, while the central parts remain strongly coupled. But the fusion of the chains is favoured if the chains are long.

(b) The Sm A phase would obey analogous laws; in fact, *liquid* layer phases would be favoured by long chains at the expense of *ordered* layer phases.

(c) The *parity* of the number $n$ of methylene groups along a chain could also play a role (Gray); $n$ odd: stability of the B and G phases, diminishing of the stability zone of A. N.M.R. studies (Deloche and Charvolin, 1975) show the *intramolecular* dynamic effects of global rotation of the molecule, and the fluctuation of collective orientation, which are linked to a similar parity property, and there is perhaps a relationship between these two results.

The chiral smectic phases are obtained either by mixing an active molecule into an achiral compound, or with a smectogenic molecule carrying an asymmetric carbon. (Keller *et al.*, 1976). In this latter case, search for ferroelectric compounds has led to the inclusion in the molecule of transverse dipolar moments in substitutional position (same authors). Compared to the non-substituted achiral material, a greatly diminished temperature range is noted, together with a disappearance of the ordered layer phases when the substitute is lateral. In these compounds we again find that the long molecules (but also the forked molecules) favour liquid layer phases: Keller *et al.* indicate the existence of a critical length below which the C (or C*) phase does not exist.

Apart from these studies of polymorphism and stability of liquid crystals, which put emphasis on chemical properties, numerous works exist which rely upon statistical physics considerations, trying especially to understand the nature of phase transitions (Sm A–nematic, Sm C–nematic, Sm A–Sm C transitions, etc.). The theories are either mean field theories (of the classical Maier–Saupe type for the order parameter of nematic phases, or of the Landau type), or more elaborate statistical theories taking into account short range order effects or fluctuations near the transitions. It is outside the scope of this book to deal with such theories which, being statistical in nature, do not always have to use a clear physical model of the interaction between molecules. But some recent results might influence our viewpoint on defects. This is why we give now a short account of results from the mean-field theory. More refined results (introducing the fluctuations of the order parameter and using renormalization group methods) are entirely outside our field of interest at the moment, since an already considerable amount of experimental work has not yet been able to bring decisive answers (CBOOA is in this respect a very instructive example).

The order parameter of the nematic phase is, according to the classical Maier–Saupe theory (see, for instance, Saupe, 1964):

$$S = \tfrac{1}{2}\langle 3\cos^2\theta - 1 \rangle,$$

where $\langle \ \rangle$ denote a thermal average. $\theta$ is the angle of the director with respect to the mean axis of orientation. McMillan (1971) introduces a second order parameter to describe the layering of the smectic A phase:

$$\sigma = \langle \cos\frac{2\pi z}{d_0} \cdot (\tfrac{3}{2}\cos^2\theta - 1) \rangle,$$

where $d_0$ is the layer thickness. The model predicts essentially that an increase in length $d_0$ of the molecules decreases the nematic–isotropic temperature $T_{NI}$; that it does not change very much the smectic–nematic transition $T_{NA}$, and that the transition Sm A–Nem is first order when the nematic ordering $S$ is far from unity at $T_{NA}$ (i.e. when $T_{NA}/T_{NI}$ is not very much smaller than unity) and second order when nematic ordering is achieved at $T_{NA}$. The coupling between the nematic order parameter $S$ and the layering (here introduced via cos $(2\pi z/d_0)$) is extremely strong from the start in this model, and more refined models closer to experimental results (measurement of order parameter by analysis of X-ray diffraction peaks) have been developed since, by McMillan and others. Let us indicate the essential points of the developments.

The 'smectic' order parameter is nothing else but the density $\rho(z)$, whose various terms $\rho_k \cos(2\pi k z/d_0)$ in a Fourier series expansion describe various harmonics of the layering. Kobayashi (1970) was first to introduce $\rho_1$ as the fundamental order parameter and soon after de Gennes (1972d) and McMillan (1972) recognized the analogy with superfluids. Meyer and Lubensky (1976), discuss in depth a free energy density with two order parameters $\rho_1$ and $\rho_2$ coupled together, without a nematic order parameter, in a mean-field approximation. They show that the coupling increases the tendency towards a first order nematic–smectic A transition; this is a phenomenological result, which might be amended or even changed if more attention were given to the physical origin of a non-vanishing $\rho_2$. The natural situation in which this occurs is probably when the constitutive molecules are polar (like CBOOA) and like to couple by pairs. Then the layer thickness is larger than the molecular length. The transition in CBOOA is probably second order (McMillan, 1973a), contrarily to the prediction of Meyer and Lubensky, although a slight first order character seems to have been observed by Cladis (1973); it has been argued that polar coupling between molecules might minimize polarity effects. Chu and McMillan (1975) explain differently the second order character: by the proximity of a tricritical point.

It is noticeable that, for some years (and probably because nothing seems settled, either on the theoretical or the experimental side), the effort is no longer on the nature of the transition, but on building correct free energy expansions including pertinent order parameters which could explain a number of new transitions which have been discovered, without emphasis on their nature. Let us quote the discovery of the re-entrant nematic (a nematic phase appearing at a lower temperature than the smectic one; see Cladis, 1975; Cladis et al., 1978; Liébert and Daniels, 1977), and of the Sm A–Sm A transition (Sigaud et al., 1979). Prost (1980) has attempted to build a general phenomenological model for 'bilayer smectics' in which $\rho_1$ is replaced by a vectorial order parameter figuring the polarization $\mathbf{P}$ of the molecules, which is peaked at a value of the wave vector $k_1$ not necessarily equal to half the value of $k_2$ (for $\rho_2$). In this way he claims that he explains the various transitions indicated above and predicts some others, in particular the so-called commensurate–incommensurate transitions between a monomolecular Sm A phase and a bimolecular Sm A phase, analogous to a paraelectric–ferroelectric phase. More recently an intermediary phase Sm $\tilde{A}$ has been discovered in which exists a transverse undulation in the layers (Hardouin *et al.*, 1980). The mixture of the two compounds:

$$C_5H_{11} - \bigcirc - OOC - \bigcirc - OOC - \bigcirc - CN$$

$$C_5H_{11} - \bigcirc - CH{=}CH - \bigcirc - OOC - \bigcirc - CN$$

possess the three phases, Sm $A_2$, Sm $\tilde{A}$ and Sm $A_1$, in this order with increasing temperature.

It is most probable that this newly enriched polymorphism of lamellar phases will give rise to active research in the following years. Let us indicate that similar 'commensurate–incommensurate' transitions have already been seen in ordered lamellar phases (Brownsey and Leadbetter, 1980).

### 7.1.3 Characteristic textures

The Sm C phase possesses focal conic textures, which, when obtained from the focal conic structure of the A phase, appear *broken*. However, they can be obtained without defects when, for example, there is a direct Sm C → N transition. The *schlieren texture* or 'plage à noyaux' (texture with nuclei) is also characteristic and is obtained by cooling a homeotropic phase: the projection **t** of the optical axis **v** on the smectic layers has all the properties of a nematic director, but the visible defects are integral nuclei, because the rotation axis of these defects is of order one. This point will be discussed below.

Some ordered smectic phases possess a texture with nuclei which indicates that, like the Sm C phase, the molecules are inclined on the layer (furthermore we distinguish Sm B (models perpendicular to the layer) and Sm G (tilted molecules). Very good focal domains are also obtained, with less streaked fans than those of the Sm A phase, which the Sm B phase comes from. If the material possesses the series Sm A ↔ Sm C ↔ Sm B, it is noted that the broken conics of the Sm C phase become good once again, isomorphic with the texture of the Sm A phase when the temperature is lowered in the Sm B phase. Such a sequence has been followed here with the compound HOBHA (see section 5.2) with or without application of a magnetic field, and the regularity of the conics is striking, which is surprising since the layer is ordered. Coming from the solid crystalline phase, the classic texture of the Sm B is the mosaic structure, characteristic of what may be expected in a system of ordered layers.

The chiral phases possess the same textures as the achiral phases: the pitch decorates them in a characteristic manner.

In a general way, our knowledge of textures and defects in lamellar phases is still very poor, except for the Sm C phases. An extremely useful monograph on textures in liquid crystals (beautifully illustrated) has been written by Demus and Richter (1978).

## 7.2  Sm C

### 7.2.1  Elasticity

#### 7.2.1.1  *Quasi-planar elasticity*

The elasticity of smectics C has been established within the limit where the layers are weakly distorted with respect to the planar situation (Orsay Liquid Crystal Group, 1971; Helfrich, 1974). In this case, it is convenient to take as variables, in the $xyz$ axes such that $Oz$ is along the non-perturbed normal to the layers (Figure 7.1):

(a) the angle $\Omega_z$ of the projection $\mathbf{t}$ of the optical axis $\mathbf{v}$ with the $Ox$ axis;

(b) the vertical displacement $u$ of the layers, whose derivatives $\Omega_x = \partial u/\partial y$ and $\Omega_y = -\partial u/\partial x$ represent small angles of rotation.

It is supposed that the angle $\Psi_0$ of $\mathbf{v}$ with the layer is constant.

The energy $\rho f$ must obviously be a function of the derivatives of $\Omega$, since a rotation of the whole must not change $\rho f$. On the other hand, because of the symmetries of the Sm C, it must be invariant by the simultaneous transformation $x \to -x$, $y \to -y$, $z \to -z$ (existence of a centre of symmetry) and by the transformation $y \to -y$, the vertical $xOz$ being a plane of symmetry. This leads to the formula:

$$\rho f = \tfrac{1}{2}B_1(\Omega_{z,x})^2 + \tfrac{1}{2}B_2(\Omega_{z,y})^2 + \tfrac{1}{2}B_3(\Omega_{z,z})^2$$
$$+ B_{13}\Omega_{z,x}\Omega_{z,z}$$
$$+ \tfrac{1}{2}A(\Omega_{x,x})^2 + \tfrac{1}{2}\{A_{12}(\Omega_{y,x})^2 + A_{21}(\Omega_{x,y})^2\} \qquad (7.1)$$
$$+ C_1\Omega_{x,x}\Omega_{z,x} + C_2\Omega_{x,y}\Omega_{z,y}$$
$$+ \tfrac{1}{2}\bar{B}\gamma^2$$

(a)    (b)

Figure 7.1. (a) Sm C. Lamellar elasticity, definition of coordinates. The axes $Ox$, $y$, $z$ are fixed in the laboratory. (b) Sm C. Covariant elasticity; definition of coordinates (mobile trihedron $\mathbf{t}$, $\mathbf{N}$, $\mathbf{n}$; these axes are used as curvilinear coordinate axes)

where $\gamma = \partial u/\partial z$ describes the variations in thickness of the layers, which, in linear elasticity, are small. For the same reason, we have supposed that the derivatives $\Omega_{x,z}$ and $\Omega_{y,z}$ are negligible, as the condition

$$\Omega_{i,z} = 0, \qquad i = 1, 2, \tag{7.2}$$

coupled to $\Omega_{x,x} + \Omega_{y,y} = 0$ signifies:

$$\operatorname{rot} \mathbf{n} = 0. \tag{7.3}$$

The effect of the variations in thickness is therefore included in the single term $\frac{1}{2}\bar{B}\gamma^2$.

In the centre of symmetry operation, the components of $\mathbf{\Omega}$ do not change. In the operation $y \to -y$, the components $\Omega_x$ and $\Omega_z$ change sign simultaneously ($\mathbf{\Omega}$ is a pseudo-vector). But the operator $\partial/\partial x_i$ changes sign if $x_i$ changes sign. These rules only allow the existence of the following cross-terms:

$$\Omega_{x,x}\Omega_{z,x} \qquad (= -\Omega_{y,y}\Omega_{z,x}),$$
$$\Omega_{x,y}\Omega_{z,y},$$
$$\Omega_{y,x}\Omega_{z,y}$$
$$\Omega_{z,z}\Omega_{z,x},$$
$$\Omega_{x,y}\Omega_{y,x}.$$

The term $\Omega_{y,x}\Omega_{z,y}$ can again be written:

$$\Omega_{y,x}\Omega_{z,y} = (\Omega_z\Omega_{y,x})_{,y} - (\Omega_z\Omega_{y,y})_{,x} + \Omega_{z,x}\Omega_{y,y};$$

it is therefore the sum of two gradients (which have no contribution to the minimization of free energy) and of a term analogous to $\Omega_{x,x}\Omega_{z,x}$. An analogous decomposition can be made for $\Omega_{x,y}\Omega_{y,x}$.

### 7.2.1.2 *Covariant elasticity; discussion of compatibility conditions*

Rapini (1972b) gave an expression of the covariant free energy, as a function of the gradients of $\mathbf{t}$, $\mathbf{N}$ and $\mathbf{n}$ (see Figure 7.1b) and has drawn some interesting consequences as far as the magnetic instabilities of smectics C are concerned. A covariant expression different from that proposed by Rapini will be given here, noting that at each point of the smectic the natural reference frame $(\mathbf{t}, \mathbf{N}, \mathbf{n})$ plays the role of the $R(\mathbf{r})$ introduced in the preceding chapter. It is therefore tempting to introduce the distortion tensor $K_{ij}$. We can then write:†

$$\rho f = \tfrac{1}{2}C_{ijkl}K_{ij}K_{kl} + \tfrac{1}{2}\bar{B}\gamma^2, \tag{7.4}$$

choosing the phenomenological coefficients $C_{ijkl}$ so that the symmetry properties of the Sm C are locally satisfied. This study is much simpler if, to write the free energy at a point $\mathbf{r}$ and in its vicinity, the reference frame $R$ is chosen as the *fixed*

---

† We have not specifically used $\mathbf{t}$, $\mathbf{N}$ and $\mathbf{n}$ here since we are using the components of $K_{ij}$ in the reference frame $R$.

reference frame. The different $K_{ij}$ then have simple geometric interpretations which are obtained by noting that the instantaneous rotation $d\omega/ds$ of a mobile reference frame being displaced on a line $L$ of the layer, along the direction $\mathbf{t}$, has components in the axes $\mathbf{t}$, $\mathbf{N}$ and $\mathbf{n}$:

$$\frac{d\omega}{ds} = (\tau_G^{-1}, -\rho_n^{-1}, \rho_G^{-1}), \tag{7.5}$$

where $\tau_G^{-1}$, $\rho_n^{-1}$ and $\rho_G^{-1}$, are respectively the geodesic twist, the normal curvature and the geodesic curvature of the curve $L$ (see Chapter 5).

We also have by definition:

$$d\omega_j = K_{ij}\,dx_i. \tag{7.6}$$

From equation (7.6) we deduce the components of $K_{ij}$ in the reference frame $R$:

$$
\begin{aligned}
K_{11} &= \tau_G^{-1}; & K_{21} &= \rho_n'^{-1} = -\rho_n^{-1} + \sigma_1 + \sigma_2, \\
K_{12} &= -\rho_n^{-1}; & K_{22} &= \tau_G'^{-1} = -\tau_G^{-1}, \\
K_{13} &= \rho_G^{-1}; & K_{23} &= \rho_G'^{-1},
\end{aligned} \tag{7.7}
$$

where the primes relate to the lines $L'$ enveloping $\mathbf{N}$. The relations between $\rho_n$, $\rho_n'$ on the one hand, and $\tau_G$, $\tau_G'$, on the other, originate from the known expressions of $\rho_n$ and $\tau_G$ in the reference frame of the lines of curvature:

$$
\begin{aligned}
\rho_n^{-1} &= \sigma_1 \cos^2 \omega + \sigma_2 \sin^2 \omega, \\
\tau_G^{-1} &= (\sigma_2 - \sigma_1) \sin \omega \cos \omega,
\end{aligned} \tag{7.8}
$$

where $\omega$ is the angle of the line $L$ with the line of curvature $\xi_1$, $\sigma_1$ the curvature of $\xi_1$ and $\sigma_2$ the curvature of $\xi_2$.

The quantities $K_{3i}$ are interpreted in the same way by considering the line $\Lambda$ enveloping $\mathbf{n}$ on surfaces normal to the layers. $K_{33}$ is therefore the geodesic twist of $\Lambda$. We also have

$$\operatorname{rot} \mathbf{n} = (K_{31}, K_{32}, 0). \tag{7.9}$$

As it has been agreed not to involve the variation of thickness of the layers except by the term $\frac{1}{2}\overline{B}\gamma^z$, the components $K_{31}$ and $K_{32}$ will not appear in $\rho f$.

The symmetry properties of the $C_{ijkl}$ are easily deduced from the results obtained for the planar case. In fact, within the limit where $\sigma_1 \sigma_2 \to 0$, we have $K_{ij} = \Omega_{j,i}$. We therefore have exactly the same form as in the planar case by replacing $\Omega_{j,i}$ by $K_{ij}$, except that it is no longer possible to eliminate terms such as $K_{12}K_{23}$ and $K_{12}K_{21}$ without a deeper discussion. This is based on the study of the conditions of compatibility $\theta_{ij} \equiv 0$ (see equation (6.11)) of the $K_{ij}$.

*Conditions of compatibility.* In the nine components $\theta_{ij}$, the $\theta_{3i}$ components are related to the compatibility of the $K_{ij}$ in a movement of the reference frame $R$ on a given layer. (It is understood from now on that the components $\theta_{ij}$, $K_{ij}$, $M_{ij}$ of the tensors $\boldsymbol{\theta}$, $\mathbf{K}$, $\mathbf{M}$, etc. are written in the local reference frame.)

In fact the problem of the compatibility of $R$ on a *surface* has long been solved under the Codazzi–Mainardi conditions (see Darboux). The fundamental quadratic form of the surface is written:

$$ds^2 = A^2 \, du^2 + B^2 \, dv^2, \tag{7.10}$$

where $A \, du = ds_u$ and $B \, dv = ds_v$ designate the infinitesimal displacements over the surface along the coordinate lines $L$ enveloping $\mathbf{t}$, and $L'$ enveloping $\mathbf{N}$. $A$ and $B$ are functions of $u$ and $v$, the coordinates in a layer, and of the chosen layer $(w)$. It is not generally possible to write $ds^2 = ds_u^2 + ds_v^2$ giving $ds_u$ and $ds_v$ the meaning of total differentials.

The Codazzi–Mainardi conditions then read:

$$\frac{\partial(AK_{11})}{\partial v} - \frac{\partial(BK_{21})}{\partial u} - AB(K_{12}K_{23} - K_{13}K_{22}) = 0,$$

$$\frac{\partial(AK_{12})}{\partial v} - \frac{\partial(BK_{22})}{\partial u} - AB(K_{13}K_{21} - K_{11}K_{23}) = 0, \tag{7.11}$$

$$\frac{\partial(AK_{13})}{\partial v} - \frac{\partial(BK_{23})}{\partial u} - AB(K_{11}K_{22} - K_{12}K_{21}) = 0.$$

This form of the compatibility equations is the most convenient for our problem. It is interesting to compare these equations with equations (6.11) which are the components of $\theta_{ij}$ on *fixed axes*. The projection of equation (6.11) on the axes of the mobile reference frame $R$ leads to the following expressions:

$$\theta_{ij} = \varepsilon_{ikl}\frac{\partial K_{ij}}{\partial s_k} + M_{ij} + K_{ki}K_{kj} - KK_{ij} = 0, \tag{7.12}$$

where all the quantities are related to the mobile axes. The Codazzi–Mainardi equations correspond to the three components $\theta_{3j} \equiv 0$. To show this, it is sufficient to use the relations (given by Darboux) which link the coefficients of the fundamental quadratic form to the geodesic curvatures of the coordinated lines:

$$\rho_G^{-1} = K_{13} = -\frac{1}{AB}\frac{\partial A}{\partial v} = -\frac{1}{A}\frac{\partial A}{\partial s_v},$$

$$\tag{7.13}$$

$$\rho_G'^{-1} = K_{23} = +\frac{1}{AB}\frac{\partial B}{\partial u} = \frac{1}{B}\frac{\partial B}{\partial s_u}.$$

Let us substitute these expressions into equation (7.11); they then read:

$$\theta_{31} = \frac{\partial K_{21}}{\partial s_u} - \frac{\partial K_{11}}{\partial s_v} + (K_{12} + K_{21})K_{23} + (K_{11} - K_{22})K_{13} = 0,$$

$$\theta_{32} = \frac{\partial K_{22}}{\partial s_u} - \frac{\partial K_{12}}{\partial s_v} + (K_{12} + K_{21})K_{13} - (K_{11} - K_{22})K_{23} = 0, \tag{7.14}$$

$$\theta_{33} = \frac{\partial K_{23}}{\partial s_u} - \frac{\partial K_{13}}{\partial s_v} + K_{11}K_{22} - K_{12}K_{21} + K_{13}^2 + K_{23}^2 = 0.$$

It is important to note that the expressions of $\theta_{1j}$ and $\theta_{2j}$ are not obtained by carrying out a circular permutation on equations (7.11); the reason is that the curves enveloping the vectors $\mathbf{t}$, $\mathbf{n}$ and $\mathbf{N}$, do not form a tridimensional lattice and that this geometric particularity forbids the definition of a *unique set of parameters* ($u$, $v$ and $w$). Consider, in fact, a normal $A$ to the layers, and two successive layers $\Sigma_1$ and $\Sigma_2$ pierced by this normal at $A_1$ and $A_2$. Let us establish an arbitrary set of parameters for $A_1$ using the three numbers ($u = 0, v = 0, w = 0$) and for $A_2$ using ($u = 0, v = 0, w = h$). This choice of origins determines entirely the set of parameters in $\Sigma_1$ and $\Sigma_2$. Let us displace $A$ along $L_1$ (line of coordinates enveloping $\mathbf{t}$ in $\Sigma_1$) to a new position $A'$ ($u = u_1, v = 0, w = 0$). In this new position the normal $A'$ does not pierce $\Sigma_2$ at ($u = u_1, v = 0, w = h$) as the lines $L_1$ and $L_2$ do not both rest on $A$ and $A'$. It can be seen therefore that it is not possible to define a set of parameters ($u, v, w$) which assigns two given values ($u$ and $v$ for example) to every envelope line (here of $\mathbf{n}$) independent of the position on the line. On the other hand equations (7.12) are independent of the set of parameters and have a local value. The considerations we have just made indicate the nature of the complexity of the geometric description of a system of trihedrons not arranged in a lattice.

*Use of the compatibility conditions in the study of cross terms of the free energy.* To deal with the existence of cross terms in $\rho f$, expressions (7.11) are more simple than equation (7.14); the total energy being written $\int \rho f \, d\Sigma \, dn$, we can write $d\Sigma = AB \, du \, dv$; it can then be seen that the integration over a surface of a cross term such as $K_{12}K_{21}$ is analogous to the integration of the sum of a cross term such as $K_{11}K_{21} = -\tau_G^{-1}$, and of two gradients. The gradients do not contribute to the minimization of energy, and the term $K_{11}K_{22}$ already appears in $\frac{1}{2}A\tau_G^{-2}$ (see equation (7.1)). In the same way, by integration over the surface, the term $K_{12}K_{23}$ will be analogous to a cross term $K_{13}K_{22}$ which already appears in $C_1\tau_G^{-1}\rho_G^{-1}$.

The compatibility condition which we have not used involves cross terms which are not invariant for the transformation $\mathbf{N} \to -\mathbf{N}$.

In the same way, it can be verified that the compatibility equations relating to $\theta_{1i}$ and $\theta_{2i}$ do not involve any cross terms used in equation (7.1). They cannot lead to a reduction in the number of terms but they must be obeyed by the tensor $K_{ij}$ in the same way as the second equation (7.11).

*Density of covariant energy*

$$\rho f = \tfrac{1}{2}B_1\rho_G^{-2} + \tfrac{1}{2}B_2\rho_G'^{-2} + \tfrac{1}{2}B_3K_{33}^2 + B_{13}\rho_G^{-1}K_{33}$$

$$+ \frac{A}{2}(\sigma_1 - \sigma_2)^2 \sin^2 \omega \cos^2 \omega$$

$$+ \tfrac{1}{2}\{A_{12}(\sigma_1 \cos^2 \omega + \sigma_2 \sin^2 \omega)^2$$

$$+ A_{21}(\sigma_1 \sin^2 \omega + \sigma_2 \cos^2 \omega)^2\}$$

$$+ C_1\tau_G^{-1}\rho_G^{-1} + C_2\rho_n'^{-1}\rho_G'^{-1} + \tfrac{1}{2}\bar{B}\gamma^2.$$

$$(7.15)$$

Several possible solutions of this equation will be considered, highlighting the very important role played by the compatibility equations.

*Strictly parallel layers.* This hypothesis is written:

$$\gamma = 0; \qquad K_{31} = K_{32} = 0. \tag{7.16}$$

As we postulate the existence of layers, the compatibility equations $\theta_{31} = \theta_{32} = \theta_{33} = 0$ are automatically verified for every set of trihedra of which two axes (here $t$ and $N$) are on the layers. These are the Codazzi–Mainardi equations for this set of surfaces.

The other compatibility equations are all verified when we take:

$$K_{33} = \frac{\partial \omega}{\partial n}. \tag{7.17}$$

This condition could indeed have been guessed from the beginning. The compatibility equations do not therefore impose any constraints, either in the choice of the lines $t$ and $N$ on the different layers or in the choice of layers.

In particular it is reasonable to suppose, as for smectics A, that the layers fold into Dupin cyclides, if the lengths

$$\left(\frac{A_i}{B}\right)^{1/2}; \qquad \left(\frac{B_i}{B}\right)^{1/2}; \qquad \left(\frac{C_i}{B}\right)^{1/2}$$

(the $A_i$, $B_i$ and $C_i$ are the stiffness coefficients of equation (7.15)) are small compared to the radii of curvature of the layers, and all of the same order of magnitude. The distribution of $\omega$ on each layer is then obtained as the solution of minimizing differential equation (7.15), where certain terms would first have been replaced by quantities characteristic of Dupin cyclides. Remember that in the reference frame of lines of curvature the quadratic form of a cyclide is written (Kléman, 1976a):

$$ds^2 = A_1^2 \, d\alpha_1^2 + A_2^2 \, d\alpha_2^2, \tag{7.18}$$

$$A_1 = \frac{b\sigma_2}{\sigma_1 - \sigma_2}; \qquad A_2 = \frac{b\sigma_1}{\sigma_1 - \sigma_2},$$

$$\sigma_1^{-1} = c \cos \alpha_2 - n, \tag{7.19}$$

$$\sigma_2^{-1} = a \cosh \alpha_1 - n,$$

where $n$ is a length measured along the normal to the layers and $a$, $b$ and $c$, are the characteristic dimensions of the focal ellipse and hyperbola ($c^2 = a^2 - b^2$). $\alpha_1$ and $\alpha_2$ are the parameters of the conics. The geodesic curvature of the lines of curvature are given by (Darboux):

$$\rho_1^{-1} = -\frac{1}{A_1 A_2} \frac{\partial A_1}{\partial \alpha_2} = \frac{1}{b} \frac{\sigma_{1,\alpha_2}}{\sigma_1}$$

$$\tag{7.20}$$

$$\rho_2^{-1} = \frac{1}{A_1 A_2} \frac{\partial A_2}{\partial \alpha_1} = \frac{1}{b} \frac{\sigma_{2,\alpha_1}}{\sigma_2},$$

and the geodesic curvatures of a line making an angle $\omega$ with the direction $\alpha_1$

$$\rho^{-1} = \frac{\partial \omega}{\partial s} + \rho_1^{-1} \cos \omega + \rho_2^{-1} \sin \omega$$

$$= \cos \omega \left( \frac{\partial \omega}{\partial \alpha_1} + \rho_1^{-1} \right) + \sin \omega \left( \frac{\partial \omega}{\partial \alpha_2} + \rho_2^{-1} \right). \tag{7.21}$$

Substituting these expressions into $\rho f$, and expressing the free energy in the form of the integral

$$W = \int \rho f\, AB\, d\alpha_1\, d\alpha_2\, dn, \tag{7.22}$$

the search for the differential equation obeyed by $\omega$ leads to the minimization of $W$ with respect to $\omega(\alpha_1, \alpha_2, n)$. It is evident that this problem does not give rise to any difficulty.

If one of the coefficients $A_i$, $B_i$ or $C_i$ is much greater than the others, then Dupin cyclides are no longer necessarily favoured. Such situations can give rise to *broken textures* as we shall explain below.

*t along asymptotic directions.* This possibility, which was drawn to our attention by a remark of R. Thom, occurs in the case where the coefficient $A_{12}$† is much greater than the others $A_i$, $B_i$, $C_i$. This requires

$$K_{12} = \sigma_1 \cos^2 \omega + \sigma_2 \sin^2 \omega = 0. \tag{7.23}$$

We will continue to suppose that the effects of compressibility perpendicular to the layers are always sufficiently great to make the layers parallel ($K_{31} = K_{32} = 0$). But if the distribution of **t** is imposed, these layers are no longer necessarily Dupin cyclides. In fact our problem is the opposite of the previous one: to find the form of the layers which minimizes the free energy $\int \rho f\, dV$, when the distribution of **t** obeys a constraint of the type in equation (7.23).

The basis of the solution consists of seeking differential equations which are obeyed by the coefficients of the fundamental quadratic form

$$ds^2 = A^2(u, v, n)\, du^2 + B^2(u, v, n)\, dv^2 \tag{7.24}$$

on each of the layers $n$. It is therefore necessary to express $\rho f$ (see equation (7.15)) as a function of $A$, $B$ and their derivatives. We have first of all, using the Darboux formulae:

$$\rho_G^{-1} = -\frac{1}{AB} \frac{\partial A}{\partial v}; \qquad \rho_G'^{-1} = \frac{1}{AB} \frac{\partial B}{\partial u},$$

$$\sigma_1 \sigma_2 = -\frac{1}{AB} \left[ \frac{\partial}{\partial u} \left( \frac{1}{A} \frac{\partial B}{\partial u} \right) + \frac{\partial}{\partial v} \left( \frac{1}{B} \frac{\partial A}{\partial v} \right) \right]. \tag{7.25}$$

---

† According to Johnson and Saupe (1977) and to Allet *et al.* (1978), $A_{12}$ is greater than $A_{21}$ in certain current Sm C materials. Whether it is 'much' greater is not known.

On the other hand, equation (7.23), with the compatibility conditions $\theta_{11} = \theta_{22} = 0$ added to it, leads to:

$$\tan^2 \omega = -\frac{\sigma_1}{\sigma_2}; \qquad K_{33} = -\tfrac{1}{2}\tau_G^{-1} = \pm\tfrac{1}{2}(-\sigma_1\sigma_2)^{1/2}. \qquad (7.26)$$

Only the terms $C_2$ and $A_{21}$ are not used in these relations; the quantity $K_{21} = \sigma_1 + \sigma_2$ appears and is not directly linked to the coefficients of the quadratic form, but appears in the compatibility equations, either directly (in $\theta_{12} = 0$ and $\theta_{31} = 0$) or in the form of its derivatives. It is therefore sufficient to express $K_{21}$ beginning with any one of the two conditions, $\theta_{12} = 0$ and $\theta_{31} = 0$, and substituting into equation (7.15). The energy density then appears as a function of $A$ and $B$ and their derivatives. It is then sufficient to minimize equation (7.22) with respect to $A$ and $B$.

Thus a set of surfaces is obtained with a negative gaussian curvature ($\sigma_1\sigma_2 < 0$) which must necessarily give rise to broken textures: for on one hand it is not obvious that *thick* subsets of parallel surfaces can be extracted from it, on the other hand these subsets do not consist of cyclides because their singular caustics will be surfaces and not lines. It is therefore the competition between these energies of singular surfaces and the volume energy (7.22) which will determine the size of these textures.

### 7.2.2 Defects in smectics C

#### 7.2.2.1 Topology of defects

The topology of defects can be tackled by methods taken from the Volterra process (Chapter 2): some recent studies (Bouligand and Kléman, 1979; Williams, unpublished) lead to the following classification.

(a) Translation dislocations of the system of layers; let us call them $d$ lines.

(b) Rotation dislocations multiple of $2\pi$ about the direction of rotation $\mathbf{n}$. These dislocations affect the fields $\mathbf{t}$ and $\mathbf{N}$. The most obvious defects which they can display are those analogous to Friedel's nuclei (wedge lines), but the possibility of core relaxation by escape into the third dimension requires the introduction of defects multiple of $4\pi$. Twist lines and mixed loops may also be found. These lines will be denoted by the letter $m$; their order is defined as for the lines in nematics, from which we get the notations $m(+2\pi)$, $m(-2\pi)$ etc.

(c) Rotation dislocations multiple of $\pi$ about the direction of rotation $\mathbf{N}$. These dislocations affect both the $\mathbf{n}$ field (i.e. the lamellar structure) and the $\mathbf{t}$ field (i.e. the director field). These lines will be denoted $l(-\pi)$, $l(\pi)$ etc. (see Figure 7.2a).

(d) Rotation dislocations multiple of $2\pi$ about any given direction of the trihedron $\mathbf{n}$, $\mathbf{t}$, $\mathbf{N}$. These objects, which affect the whole trihedron and will be denoted as $t(\pm 2\pi)$, $t(\pm 4\pi)$ etc., contain as particular cases some of the above objects. Here we have the analogue of what was described in the preceding chapter regarding $^3$He. Remember that the core of the multiple lines of $4\pi$ can

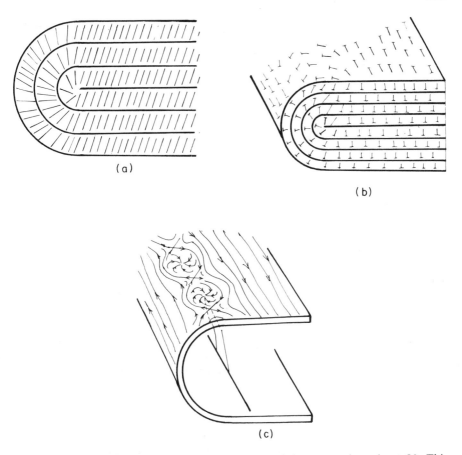

Figure 7.2. (a) Perfect disclinations $l(\pi)$ constructed by a rotation about **N**. This disclination affects both the **t** and **n** fields. (b) Imperfect disclination $l(\pi)$: dissociation of the imperfection in a 'Néel wall'. (c) Imperfect disclination $l(\pi)$: the Néel wall' is in turn broken down into circular and cross 'Block lines', i.e. into nuclei $m(+2\pi)$ and $m(-2\pi)$. (From Bouligand and Kléman, 1979)

relax by a sort of escape into the third dimension (see Figure 6.4). This is the case here for the lines $t(\pm 4\pi)$ of which the lines $m(\pm 4\pi)$ and $l(\pm 4\pi)$ are particular cases, obviously less energetic.

(e) Focal conics: these objects in Sm C (as in the case of Sm A) are peculiar rotation dislocations of the lamellar system, as they are not obtained by the classic Volterra process, and this is why they were introduced for Sm A by their traditional definition, after G. Friedel. A general definition of focal conics will be given in Chapter 10, which attaches them effectively to $l(\pm \pi)$ dislocations. Sm A and Sm C are very different from the point of view of $l$ type defects: in Sm A all the directions of the layer are axes of symmetry, while in Sm C only the direction **N** plays this role. On the other hand, in Sm A, the **t** field is not a physical one. This difference gives rise to a large number of new problems which can be experimentally investigated by observing the Sm A–Sm C transition in a polarizing microscope (see below).

This classification of the different curvilinear defects of Sm C will be taken up again in Chapter 10 in a new light.

Note that the *l* lines affecting the lamellar structure and the **t** field must take on a definite form. Let $\tau$ be the tangent unit vector to such a line. By analogy with the nematic case, a density of translation dislocations must be adjoined to the lines and terminate on the line,

$$\frac{d\mathbf{b}}{ds} = \tau \wedge \mathbf{N}. \tag{7.27}$$

This condition necessarily limits the form of *l*, since the translations are quantified in a smectic. One therefore expects predominance of wedge lines.

However, this observation must be understood in a particular manner. Note in fact that this formula can be applied to the field of the lamellar structure as well as to the **t** field. Therefore these fields must be separated in the following discussion, and a Sm C must be considered as a Sm A on which inclined molecules are superimposed.

Figure 7.2b represents precisely the case of an imperfect $l(\pi)$ where the lamellar structure is that of a wedge dislocation of the system of layers, while the **t** field possesses a twist singularity. It is therefore convenient to apply equation (7.27) to this field following the example of the nematic case.

Figure 7.2b represents a *Néel wall*, which can be broken down, according to a process well-known in magnetism (see Chapter 9), into a succession of 'Bloch lines' which are alternatively circular and cross-lines (i.e. $m(+2\pi)$ and $m(-2\pi)$ nuclei; see Figure 7.2c terminating on the line *l*. This scheme was first proposed by Bouligand.

Situations can be imagined where the line has a twist character with respect to the lamellar structure; this can be brought about according to equation (7.27) by the addition of quantified translation dislocations. C. Williams (1976) used one such model to describe focal lines in a smectic A. But the field **t** does not then necessarily have a twist character. Figure 7.3 represents a line $l(2\pi)$ which clearly illustrates this property. Discussion of focal lines in Sm C is to be found in the paper by Bourdon *et al.* (1982).

Note finally that the **t** field, if it behaves like a director (rotations about **N**) has defects typical of a vector for rotations about **n**.

A detailed discussion regarding the possibility of coupling between $l(\pi)$ and $l(-\pi)$ lines leading to translation dislocations will not be given here.

### 7.2.2.2 *Observations*

Various textures and defects are described in Taylor *et al.* (1970), Lefevre *et al.* (1971), Scheffer *et al.* (1972), Meyer and Pershan (1973) and Demus (1975).

The emphasis on individual defects is more stressed in recent work. Lagerwall *et al.* (1978) have been the first to observe dislocations of unit Burgers vector in a

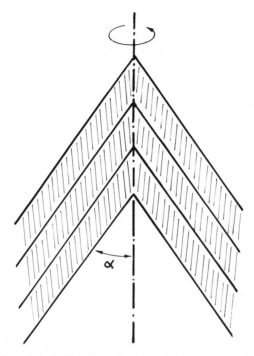

Figure 7.3. Imperfect $(2\pi)$ disclination: the imperfection relaxes through emission of translation dislocations. The object is then a focal line

lamellar phase, by taking advantage of the characteristics of the second order Sm A–Sm C transition. The transition temperature is higher in compressed layers, smaller in depressed ones. This gives rise to a strong optical contrast near the transition. The geometry is that of a wedge with homeotropic anchoring conditions. Bourdon (1980), in the same geometry, has been able to follow the evolution of dislocations down to large temperature variations ($T_c - T \sim 10°$C): they move through the forest of screw dislocations, to which they can eventually be pinned and gather in dislocations of large Burgers vector, taking a characteristic sinusoidal shape. Bartolino and Durand (1978) have observed in homeotropic samples with parallel slides the appearance of dislocations relaxing the variations in thickness of the layers due to change in temperature.

Allet *et al.* (1978) have observed free droplets of Sm C phase whose layers are strongly anchored perpendicularly to the substrate. They interpreted the stripes they observed in terms of imperfect disclination lines of opposite signs ($l(\pi)$ and $l(-\pi)$) of the type of Figure 7.2b.

Perez *et al.* (1978) have observed how focal conics transform at the transition. A disclination line appears whose role is to relax topologically the elements of symmetry which are lost from Sm A to Sm C. If the orientation of the Sm C phase is uniform at a distance from the focal domain, the line is a $m(-2\pi)$ line linking two points on the ellipse and passing through the physical focus. Other configurations may exist for other boundary conditions (Bouligand and Kléman, 1979). A complete experimental investigation of these configurations in Sm C, but also in chiral Sm C, can be found in Bourdon *et al.* (1982).

## 7.3 CHIRAL LAMELLAR PHASES AND EFFECTS OF MOLECULAR POLARITY

### 7.3.1 Effects of polarity in non-chiral media (induced polarity)

As indicated in subsection 3.2.2 in connection with nematics, non-chiral molecules can present a coupling, known as a flexoelectric coupling, between distortions of splay or bend and the dipolar electric moment $\mathbf{P}$ on each molecule. In the perfect ordered state, this dipolar moment is averaged at zero, because the molecules can take all the rotation positions about their axis. This can be done either statically on a volume of material small enough for it to include a differential element of the continuous theory, or dynamically by rapid rotation of each molecule about its axis. But, as illustrated in Figure 3.2, the local asymmetry of the molecule linked to the existence of a dipolar moment parallel $(\mathbf{P}_{\parallel})$ or perpendicular $(\mathbf{P}_{\perp})$ to the axis of the director, can lead to macroscopic effects which demonstrate these dipolar moments when the medium is no longer in the perfect ordered state (Meyer, 1969).

After Meyer (1976), let us introduce a density of free energy in the form:

$$\rho f = \tfrac{1}{2}K_1^0 S^2 + \tfrac{1}{2}K_2^0 t^2 + \tfrac{1}{2}K_3^0 \mathbf{B}^2 + \tfrac{1}{2}\alpha_1 \mathbf{P}_{\parallel}^2 + \tfrac{1}{2}\alpha_3 \mathbf{P}_{\perp}^2$$

$$-\frac{1}{8\pi}(\varepsilon_1^0 E_{\parallel}^2 + \varepsilon_3^0 E_{\perp}^2) - e_1^0 \mathbf{E}_{\parallel} \cdot \mathbf{S} - e_3^0 \mathbf{E}_{\perp} \cdot \mathbf{B} \qquad (7.28)$$

$$- a_1 \mathbf{P}_{\parallel} \cdot \mathbf{S} - a_3 \mathbf{P}_{\perp} \cdot \mathbf{B} - u_1 \mathbf{P}_{\parallel} \cdot \mathbf{E}_{\parallel} - u_3 \mathbf{P}_{\perp} \cdot \mathbf{E}_{\perp}.$$

In this equation, $\mathbf{S} = \mathbf{n} \cdot \operatorname{div} \mathbf{n}$, $t = \mathbf{n} \cdot \operatorname{rot} \mathbf{n}$ and $\mathbf{B} = \mathbf{n} \wedge \operatorname{rot} \mathbf{n}$ designate quantities linked to the three fundamental deformations of splay, twist and bend. $\mathbf{E}_{\parallel}$ and $\mathbf{E}_{\perp}$ are the components of the electric field. $\varepsilon_1^0$ and $\varepsilon_3^0$ are the usual dielectric constants. The different coupling terms are written to second order. The contribution of the space charges $\operatorname{div} \mathbf{P}$ will be supposed negligible. Its introduction would mean that non-local effects would have to be considered which furthermore are probably screened by the free ions always present in the medium.

Suppose that the distortions $\mathbf{S}$, $t$ and $\mathbf{B}$ are given and minimized in relation to $\mathbf{P}_{\parallel}$ and $\mathbf{P}_{\perp}$. This yields:

$$\mathbf{P}_{\parallel} = \alpha_1^{-1}(a_1 \mathbf{S} + u_1 \mathbf{E}_{\parallel}),$$
$$\mathbf{P}_{\perp} = \alpha_3^{-1}(a_3 B + u_3 \mathbf{E}_{\perp}). \qquad (7.29)$$

These values, substituted into equation (7.28) lead to the expression:

$$\rho f = \tfrac{1}{2}K_1 S^2 + \tfrac{1}{2}K_2 t^2 + \tfrac{1}{2}K_3 \mathbf{B}^2 - \frac{1}{8\pi}(\varepsilon_1 \mathbf{E}_\parallel^2 + \varepsilon_3 \mathbf{E}_\perp^2)$$

$$- e_1 \mathbf{E}_\parallel \cdot \mathbf{S} - e_3 \mathbf{E}_\perp \cdot \mathbf{B}, \tag{7.30}$$

where we have put

$$K_i = K_i^0 - a_i^2/\alpha_i; \qquad \varepsilon_i = \varepsilon_i^0 + 4\pi u_i^2/a_i,$$

$$e_i = e_i^0 + u_i a_i/\alpha_i. \tag{7.31}$$

From equation (7.30) one can extract the electric displacement vector $\mathbf{D} = -4\pi(\partial(\rho f)/\partial \mathbf{E})$ (see Landau and Lifshitz, 1962, ch. 10):

$$\mathbf{D} = 4\pi(e_1 \mathbf{S} + e_3 \mathbf{B}) + \varepsilon_1 \mathbf{E}_\parallel + \varepsilon_3 \mathbf{E}_\perp. \tag{7.32}$$

This is indeed equation (3.7), by interpreting $\mathbf{P}_i = e_1 \mathbf{S} + e_3 \mathbf{B}$ as a polarization induced by the field.

It will be noted that the electric field $\mathbf{E}$ acts on the medium by quadratic effects (dielectric terms) and linear effects (terms of induced polarization). This means that they can be separated: in a periodic geometry, the dielectric effects must be of a period twice as small as the effects of induced polarization, which have the period of the geometry. Prost and Pershan (1976) have tried to show induced polarization in this way in a Sm A. Clearer evidence seems to come from the more recent article by Marcerou and Prost (1978).

### 7.3.2 Effects of polarity in chiral media (spontaneous polarity)

There is a subtle relationship, observed by Meyer and, in a roundabout way, foreseen by Frank (1958), between the properties of chirality of a molecule and the fact that polarization may or may not be spontaneous.

Note first that, in a medium of chiral molecules, the fundamental state is helicoidal. This is translated into the expression of the free energy by a term $\tfrac{1}{2}K_2(t - t_0)^2$ replacing the term $\tfrac{1}{2}K_2 t^2$. It is possible to fill the space without recovery or vacancies and without director singularities, with a spontaneous constant twist $t = t_0$. A cholesteric phase can therefore be homogeneous.

The same does not apply to a phase where a spontaneous splay $S^2 = S_0^2$ or a spontaneous bend $\mathbf{B} = \mathbf{B}_0$ is imposed, the other distortions being zero. In both cases, there would necessarily be defects (Frank). On the other hand, as Meyer noted, it is possible to construct a homogeneous state of constant twist $t = t_0$ and bend $\mathbf{B} = \mathbf{B}_0$. This is the state of a conical cholesteric

$$n_x = \sin\phi\cos t_0 z: \qquad n_y = \sin\phi\sin t_0 z: \qquad n_z = \cos\phi, \tag{7.33}$$

or the state of a chiral smectic C (Sm C*). Although in this case a coupling between the curvature of the layers and the distortion of the director imposes on $\rho f$ a more complex form than that given in equation (7.28), there is hardly any doubt that the following argument, which links a spontaneous value of **B** to the existence of a spontaneous polarization, applies here.

Let us modify equation (7.28) by introducing a spontaneous polarization $\mathbf{P}^0 = \mathbf{P}^0_\parallel + \mathbf{P}^0_\perp$, which means that the components of **P** should be replaced everywhere they appear by those of $\mathbf{P} - \mathbf{P}^0$. Minimizing **P** yields:

$$\mathbf{P}_\parallel = \mathbf{P}^0_\parallel + \alpha_1^{-1}(a_1 S + u_1 \mathbf{E}_\parallel),$$
$$\mathbf{P}_\perp = \mathbf{P}^0_\perp + \alpha_3^{-1}(a_3 \mathbf{B} + \tilde{u}\mathbf{E}_\perp),$$
(7.34)

which replace equations (7.29). It can be seen that solutions of spontaneous twist and bend exist for $\mathbf{P} = \mathbf{E} = 0$. On the other hand, there is spontaneous induced polarization

$$\mathbf{P}^0_i = (e_1 S_0 + e_3 \mathbf{B}_0) = -\frac{\alpha_1}{a_1} e_1 \mathbf{P}^0_\parallel - \frac{\alpha_1}{a_3} e_3 \mathbf{P}^0_\perp.$$
(7.35)

The only simple case where a constant value of the induced polarization corresponds to a homogeneous spontaneous distortion is where the spontaneous bend is accompanied by a spontaneous twist, i.e. the conical cholesteric or the chiral smectic C. A spontaneous polarization $\mathbf{P}^0$ perpendicular to the molecule therefore arises.

According to Meyer, the argument can be reciprocated in the following form: for a molecule in the Sm C phase to display a spontaneous polarization, it is sufficient that the molecules be chiral and polar. Polarity alone is not sufficient. If the molecule is not chiral the dipolar moment is in the plane of symmetry of the Sm C and tends to zero on average by compensation. On the other hand, if it is chiral, it does not possess a plane of symmetry but a binary axis perpendicular to the molecule. As there is obviously a component of the moment along this axis, there also exists a local spontaneous polarization component along this axis.

### 7.3.3  Defects in Sm C* phases

These will first be classified by the Volterra process, using the local symmetries. Then we shall describe the rare experiments which have been performed up to now.

#### 7.3.3.1  *Rotation dislocations multiple of $2\pi$*

These can have any axis of rotation. Here, the arguments of section 7.2.2.1 are referred to. If this axis is along **t**, the projection of the optical axis on the layers, the dislocation is a singularity of **n** (normal to the layer) and of $\mathbf{P}^0$, the spontaneous polarization. If the axis is along $\mathbf{P}^0$, the dislocation affects **n** and **t** etc. In a system of physical trihedrons, **n**, **t** and $\mathbf{P}^0$, there can be no disappearance of the core singularity by any sort of 'escape' of the trihedron, if the rotation is of the type $(2p + 1) \cdot 2\pi$. On the other hand, the rotations $4p\pi$ may not have a core singularity.

All the directions of the trihedron **n**, **t**, $\mathbf{P}^0$ are not equally probable axes of rotation. In the absence of definitive observations, it may be thought that situations where space charges due to spontaneous polarization $\rho = -\operatorname{div}\mathbf{P}^0$ are, on average, zero are favoured. The field of $\mathbf{P}^0$ is analogous to that of an ordinary cholesteric (but of period equal to the pitch $2\pi/q_0$ and not the the half-pitch). The rotation dislocations about **n**, noted here as $m(\pm 2\pi)$ as in the case of Sm C, are of wedge types (equivalent to Friedel's nuclei), twist types of mixed types. It is clear that the wedge geometry is more favourable to div $\mathbf{P}^0$ tending to a zero average value (over a period of $2\pi/q_0$), while the twist geometry is not. The rotation dislocations about **t** are analogous to the $\tau(\pm 2\pi)$ (if they are compared to cholesterics) or to the $l(\pm 2\pi)$ (if they are compared with Sm C). It should be noted that in geometries where the space charge cannot be averaged at zero, typical instabilities should appear.

Note that Sm C chiralised by the addition of an active material (see Brunet, 1975) can at first display an induced polarization, but this is probably weak. Although the symmetries are the same as in pure Sm C*, it should be noted that defects can be characterized as different because of their different physical instability properties.†

### 7.3.3.2 *P disclinations*

The direction of the spontaneous polarization $\mathbf{P}^0$ is an axis of symmetry of order 2 for the configurations of the director *and* for the configurations of the layers, even if the period $\pi/q_0$ of the cholesteric coiling of the vector **t** is not commensurable with the thickness of the layers $d_0$. The ensemble of vectors $\mathbf{P}^0$ and layers has in fact the same symmetry as a helix carrying periodically arranged marks; it can certainly be stated that there is an axis of symmetry of order 2 passing through each mark and resting on the axis of the helix, also an axis of symmetry of order 2 half-way between each mark, whatever the commensurability of the periods of the helix and the marks (Frank). The directions $\mathbf{P}^0$ and the directions bisecting two directions $\mathbf{P}^0$ and $\mathbf{P}'^0$ in two adjacent layers, passing midway between the points of attachment of $\mathbf{P}^0$ and $\mathbf{P}'^0$ are therefore directions of symmetry of order 2 around which disclinations may be constructed by the Volterra process.

### 7.3.3.3 *d dispirations*

The layers are invariant through a small translation symmetry $d_0$, but this is not a symmetry operation for the configuration of directors. Over this same distance $d_0$, the director turns about **n** through an angle $\theta_0$, which is small, but is not a sub-multiple of $2\pi$. The object attained by the Volterra process (rotation $\theta_0$ of the trihedron on the cut surface, displacement of $d_0$) is a perfect dispiration. It can be called a wedge–screw dislocation or a twist–edge dislocation, whichever is

---

† M. Brunet (private communication) observes many twist lines $m(+2\pi)$ in a Sm C chiralized by the addition of active materials. It would be interesting to know if such objects are as frequent in Sm C* in the proper sense.

appropriate. The simplest object is obviously the wedge–screw dislocation, and it may be thought that it is frequently present in Sm C*.

### 7.3.3.4 *Observations*

If the specimen is introduced between two glass plates with strong planar anchoring, a non-twisted region forms near the boundaries, which is separated from the twisted bulk by a lattice of pairs of line defects (*m*–type) which affects the molecular distribution but not the distribution of layers, which remain perpendicular to the glass plate (Brunet and Williams, 1978). See also, on the same topic, Brunet and Parodi (1982) and Bourdon *et al.* (1982).

Other observations relate to the transformations of focal domains in a Sm A–Sm C* transition (Bourdon, 1980; Pĕrez, 1980).

## 7.4 SMECTICS B. LAYER DEFECTS AND RELATIONS BETWEEN ORDERED PHASES

Ordered lamellar phases appear in a great variety of forms, differing in the nature of the order in the layers and in the nature of interactions between layers, whether the systems are lyotropic or thermotropic. A detailed description of the systems known today will not be given here, as they have not yet been systematically studied from the point of view of their defects. However, purely geometric concepts will be introduced which are linked to the nature of order in these phases, and must lead to the description of some essential defects to be found in them.

### 7.4.1 Decorrelations between layers: epitaxial dislocations

De Gennes and Sarma (1972) have shown that, depending on the extension of correlations between layers, our attention can be turned to two quite different models of Sm B phases. In the first model, the layers are ordered but independent, gliding over each other with negligible friction. Apart from the interaction phenomena between the curvature of the layer and molecular organization in the layers which will be discussed later, the system behaves like a true smectic A and possesses focal conic textures (see Ribotta, 1976). In the second model, the molecular orientations in the layers are correlated over a distance of several layers: de Gennes and Sarma have suggested that this is then a plastic crystal, where the molecules would occupy a quasi-three dimensional lattice, but possessing degrees of freedom of rotation about their axes. As we have already stated, recent X-ray investigations seem to favour the second model.

Helfrich (1976) used somewhat analogous models to describe the transition between $L_\alpha$ phases (fluid lamellar) and $L_\beta$ phases (ordered lamellar) of dipalmitoyl-lecithin (D.P.L.). This lyotropic compound is formed by ordered bilayers separated by layers of water. The $L_\beta$ phase undergoes a *pretransition* at 34°C towards the higher temperature phase. This pretransition is characterized by the *loss of coherence* between the two monolayers without the order of the fatty chains disappearing. At 41°C there is a free transition towards the smectic phase with fluid layers ($L_\alpha$ phase), the chains then losing their order in each layer.

The pretransition of the $L_\beta$ phase is described by Helfrich as due to the appearance of lines *of bad adjustment* between the lattices of ordered chains of the monolayers, i.e. obviously *epitaxial dislocations*. Much earlier, Frank suggested (in a private communication) that the finite correlations (decorrelations) between layers of Sm B must be described in the same way, by distributions of dislocations situated in the plane of the layers, whose Burgers vector is itself a translation vector of the layer. It should be pointed out here that these are dislocations which establish a vernier effect between the adjacent layers, but do not perturb the long range topology. They are taken from the Frenkel–Kontorova model (see Nabarro, 1967) rather than from the model of dislocations of three dimensional media, and they must be characterized by the number of lattice parameters $n$ of a layer which occupy the same distance as $n + 1$ parameters in the adjacent layer. These lines are always present to the extent where their entropy has lessened the energy of the crystal more than their own energy increases it. $n$ is therefore a quantity which decreases with temperature. For $n = 1$ the layers are entirely decorrelated.

### 7.4.2 Defects of a single strongly curved layer

In the planar layer state, the molecules describe a two-dimensional lattice, determined, for example, by their projections on the medium surface of the layer.

It should first be stated that the energy of the deformed layer is reduced to a curvature energy if the distances between molecules and the angles between molecular arrays measured over the deformed layer $\Sigma'$ are the same as on the planar layer $\Sigma$. This requires the production of an *isometry* between $\Sigma$ and $\Sigma'$ and is only possible if $\Sigma$ and $\Sigma'$ have the same gaussian curvature $\sigma_1\sigma_2$ at all the corresponding points. $\Sigma'$ must therefore be a developable surface.

This case is very restricted. A more general situation is the conformal transformation which, of all the transformations from one surface to another, is closest to an isometry. It conserves the angles globally and the distances locally; it therefore produces on $\Sigma'$ a locally perfect crystal. Deviations from metric perfection can be relaxed by the presence of an excess or lack of sites, which, by the very definition of conformal transformations, does not create long range distortions. Therefore this is obviously a transformation analogous to that described by Nye for a three-dimensional crystal (see subsection 6.2.3).

Let us define firstly a dislocation density on a curved two-dimensional lattice: let there be a surface $\Sigma'$ defined by the quadratic formula

$$ds^2 = A^2\,du^2 + B^2\,dv^2, \tag{7.36}$$

and let us suppose that the lines $u = $ const (see Figure 7.4) are the images of the lines of the rectangular lattice of $\Sigma$. Let there be an elementary rectangle $PP'MM'$ (see Figure 7.4).

218

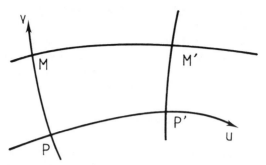

Figure 7.4. Reticular directions traced on a surface (deformed two-dimensional crystal) and Nye's definition of dislocation densities

Let us also suppose that the molecules are distributed on the lines $u = \text{const.}$, $v = \text{const.}$ with the same periodicity as on $\Sigma$. By definition, the number of edge dislocations which pierce the surface $MM'PP'$ and whose Burgers vectors is along $PP'$, is defined by

$$\frac{1}{\rho_u} = \lim \left( -\frac{PP' - MM'}{PM \cdot MM'} \right), \tag{7.37}$$

and the same for $\rho_v$. These quantities are the geodesic curvatures of the lines $u = \text{const.}$ and $v = \text{const.}$,

$$\frac{1}{\rho_v} = \frac{1}{AB} \frac{\partial B}{\partial u}; \qquad \frac{1}{\rho_u} = -\frac{1}{AB} \frac{\partial A}{\partial v}. \tag{7.38}$$

The geodesic curvatures are therefore the densities $\alpha_{31}$, $\alpha_{32}$ (taking the axis 3 along the normal to the layer) and we have:

$$\alpha_{11} = \alpha_{22} = \alpha_{12} = \alpha_{21} = 0,$$
$$\alpha_{31} = \rho_u^{-1} \qquad \alpha_{32} = \rho_v^{-1}.$$

A distribution of dislocations of this type may create distortions parallel to $\Sigma'$ for it is not possible to maintain a distance on $\Sigma'$ between adjacent molecules which, whatever the direction joining these molecules, is equal to the distance between the same molecules on $\Sigma$. But this property can be brought about if the transformation which makes the lines of the lattice on $\Sigma$ pass to lines of the lattice on $\Sigma'$ is conformal. According to Darboux, it is sufficient to choose coordinate lines on $\Sigma'$ such that

$$A = B = \lambda. \tag{7.39}$$

This same author demonstrates that this choice is always possible for *whatever* $\Sigma'$, and in an infinity of ways. These coordinate systems are called isothermic systems. The curved two-dimensional crystal thus defined, whose lattice is arranged along isothermic lines, possesses geometric properties analogous to those of the plane crystal obtained by the double orthogonal glide described by

Nye (1953), by replacing simply the curvature of the plane lattice lines by the goedesic curvature of the curved lattice lines on $\Sigma'$. We have, according to equations (7.38) and (7.39):

$$\alpha_{31} = +\frac{\partial(1/\lambda)}{\partial v}; \qquad \alpha_{32} = -\frac{\partial(1/\lambda)}{\partial u}. \tag{7.40}$$

In the case described by Nye, where $\Sigma'$ is therefore plane, we have indeed

$$\alpha_{31} = \frac{\partial\phi}{\partial s_u} \qquad \alpha_{32} = \frac{\partial\phi}{\partial s_v} \tag{7.41}$$

formulae which defined $\alpha_{31}$ and $\alpha_{32}$ as the respective curvatures of the lines $u = \text{const.}$, $u = \text{const.}$, $\phi$ being an angle of rotation. We therefore have

$$\frac{\partial\alpha_{31}}{\partial s_v} = \frac{\partial\alpha_{32}}{\partial s_u} \quad \text{(Nye).} \tag{7.42}$$

Here this formula is modified because of the curvature of $\Sigma'$. In fact its analogue is $\theta_{33} = 0$, which is written here

$$\frac{\partial\alpha_{31}}{\partial s_v} - \frac{\partial\alpha_{32}}{\partial s_u} = \sigma_1\sigma_2.$$

In conclusion, it is always possible to transform an ordered planar layer into any surface, without introducing defects other than Nye dislocations: this operation requires a double glide over the dense orthogonal rows of the planar layer, accompanied by layer curvatures.

### 7.4.3 Relations between layers

Finally, suppose that we have a stacking of parallel ordered layers, each arranged according to an isothermic distribution of dense directions. If the layers are strongly correlated, this condition is not sufficient to ensure a minimal presence of defects. One condition which can be fulfilled is the absence of screw dislocations parallel to the layers. We have discussed such a possibility in section 6.2.2.2 and shown that this requires the layers to be curved in *isothermic* surfaces, i.e. the principal lines forming an isothermic system. The only surfaces of this type are Dupin cyclides. This is probably not enough to insure that the Nye dislocations discussed in the foregoing section be continuous from one layer to the next, a condition which must be fulfilled if the correlations between layers are strong. However, such a geometry might favour the presence of focal domains, contrary to what intuition tells us. But it is clear that the growth of such textures would require energetically exorbitant cooperative deformations. These textures cannot therefore appear except in small sizes, certainly fragmented, and as elements of decomposition of grain boundaries. It would therefore be interesting to research the existence of walls in the mosaic structure. Another possibility is to obtain the focal domains by cooling in the B phase of a phase with focal conics but with liquid layers.

# Chapter 8
# Rotation dislocations in solids

Rotation dislocations exist in solids. Two types with very different natures and occurrences can be distinguished.

(1) The angle of rotation $\Omega$ is a *rotation symmetry* of the crystal; the line energies are in this case very great (see section 2.3 and following): this is therefore a rare, if not non-existent case, except perhaps if the lines are coupled. These pairs are known to be equivalent to translation dislocations if they are formed by pairs of opposite signs; Friedel (1964) suggests their existence in graphite. They could lead to focal domains if they couple lines of the same sign.

(2) The disclination limits a grain boundary, or a twin: the angle $\Omega$ can be any angle and it is known as the *angle of rotation of the limited wall*. This is a particularly frequent occurrence: it is known that because of Weingarten's theorem (section 2.2) there is no discontinuity of the $e_{ij}$ strains, and therefore of the $\sigma_{ij}$ stresses, on the wall. Apart from grain boundaries and twins, each time a physical phenomenon is associated with the presence of walls (ferromagnetic, ferroelectric materials, etc.) the line which limits them or along which they are joined can be analysed as a rotation dislocation. These are all good reasons for the detailed study of the topology and linear elasticity of rotation dislocations in crystalline solids.

Although rotation dislocations are most naturally introduced from the concept of the *limited wall* in three-dimensional crystalline solids, with rotation dislocations limiting such walls, it must not be forgotten that *perfect disclinations* (as opposed to the limited wall) have been observed in other types of ordered media. Thus the protomers of certain bacterial virus capsids (Harris and Scriven, 1971) or the membrane of certain bacteria (Thiéry *et al.*, 1968) form two-dimensional hexagonal crystals where perfect wedge disclinations are made necessary by the actual topology of the envelope, if it is homeomorphic to a sphere, or perhaps play a role in their growth. Following the same line of thought, Nabarro (1969) attributes to rotation dislocations a role in muscular contraction (a muscle is composed of a two-dimensional lattice of two types of filaments, one formed of actin and the other of myosin). Note that outside the biological field perfect rotation dislocations have been observed in two-dimensional lattices of vortex lines (Träuble and Essman, 1968).

Topological properties of disclinations, especially perfect disclinations, will be studied first. Elasticity will be developed essentially for the case of small angles of rotation, i.e. imperfect disclinations. A very complete presentation of this problem will be found in the recent monograph of Likhatchev and Khairoff (1975).

## 8.1  TOPOLOGICAL PROPERTIES

### 8.1.1  Perfect disclinations

These can only be wedge disclinations, with the axis identical to the axis of rotation $\Omega$: twist disclinations would necessarily be linked to a two-dimensional lattice of translation dislocations. de Wit (1971) and Harris and Scriven (1970, 1971) have studied the relationships between an isolated wedge disclination and translation dislocations. Their results are easily analysed using the concepts put forward in section 2.3.

(a) Wedge disclinations are sources or wells of translation dislocations: this results from the operational definition of segments of twist disclinations on a wedge disclination (Figure 2.16).

(b) The Burgers vector of a dislocation situated in the vicinity of a wedge line and not attached to this line is not defined in a unique manner (Figure 8.1). Sleeswyk (1966) made the same observation for a dislocation situated in a Moebius crystal, i.e. deformed by the presence of a twist line.

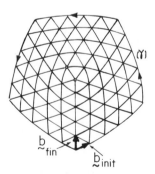

Figure 8.1. An edge dislocation changes its Burgers vector when it effects a complete turn about a disclination. The Burgers vector will be followed around the circuit $\gamma$ to show that it has turned through an angle $\Omega$ (here $\pi/3$)

(c) Suppose that a dislocation $L$ of Burgers vector $\mathbf{b}$ starts from a jog AA' on the wedge line. $\mathbf{b}$ is perpendicular to $\Omega$ and to AA'. The glide plane of $L$ is perpendicular to $\Omega$. Let us make $L$ turn about $\Omega$; the jog turns with $L$, at the same time as the local Burgers vector, which undergoes a rotation $\Omega$ for a complete turn. After completion of this turn of $L$ the crystal has undergone a relative

222

rotation $\Omega$ on both sides of the shear plane. As $\Omega$ is a rotation symmetry of the lattice, there is no need for a twist boundary to accommodate this shear.

(d) Consider a dislocation loop surrounding a wedge line (Figure 8.2). The Burgers vector **b** must undergo a discontinuity along the length of the loop. This is equal to

$$\Delta \mathbf{b} = (\Omega - 1) \cdot \mathbf{b} \tag{8.1}$$

where $\Omega \cdot \mathbf{b}$ represents the effect of a rotation $\Omega$ on **b**. $\Delta \mathbf{b}$ is quite obviously a translation symmetry of the medium. This is the Burgers vector of a segment of dislocation joining $L$ to the rotation dislocation, which is joined by a jog. The same result is true for a twist line.

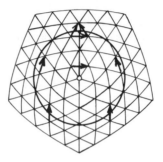

Figure 8.2. Wedge disclination surrounded by a translation dislocation: there must be a segment of dislocation of Burgers vector $\Delta \mathbf{b}$ joining the loop to the wedge line

(e) If a translation dislocation is displaced in a plane perpendicular to the disclination line and crosses it, it brings with it a segment of translation dislocation of a Burgers vector $\Delta \mathbf{b}$, linked by a jog to the rotation dislocation (Figure 8.3).

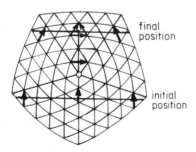

Figure 8.3. Edge dislocation crossing a disclination

### 8.1.2 Rotation dislocations, grain boundaries and twins

Several works refer to the observation or analysis of partial wedge disclinations. (Figure 2.17).

De Wit (1972b) (Figure 8.4) has shown that in a face centred cubic crystal, twins (111) can be joined together along a wedge line. Figure 8.4a illustrates the joining together of five perfect crystals, each making a dihedral angle of 70°32'. The remaining angle is 7°20'. By joining together the two free faces, a wedge line of quite low energy, of force $\Omega = 7°20'$ is obtained. This configuration has been observed on many occasions (Melmed and Hayword, 1959; Wentorf, 1963; Ogburn et al., 1964; Nohara and Imura, 1969; Galligan, 1972a; Hall, 1973, etc.). It is certainly much more likely than that shown in Figure 8.4b where the remaining angle $\Omega$ is large.

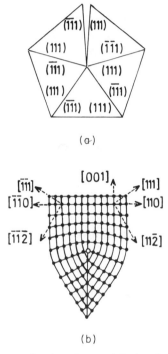

(a)

(b)

Figure 8.4. (a) Pseudopentagonal twinned crystal (before joining): $\Omega = 7°20'$, (b) Twist wall terminating on a wedge line: $\Omega = +70°32'$. This situation is infinitely less probable than the former

Marcinkowski, Das and Sadananda (1973) have analysed the behaviour of partial wedge disclinations in ordered and disordered crystals of NaCl structure, and noted their relationships to antiphase walls and grain boundaries. They have studied the stability of dipoles and quadrupoles of wedge lines, on the basis of a calculation of interaction between lines used by Kroupa and Lejcek (1972). Their considerations are limited to the choice $\Omega = 53°6'$ (along $\langle 100 \rangle$), which corresponds to a coincidence boundary (Bollman, 1970; see also Saada, 1976).

Deeper analyses, made for any angle $\Omega$ and linked to experimental observations, would be welcome. It would also be interesting to extend the results of subsection 8.1.1 to partial rotation dislocations; if $\Omega$ is any angle, the vector $\Delta \mathbf{b}$ cannot be a perfect Burgers vector. This must either restrict the possible values of $\Omega$, or necessarily lead the rotation lines to couple by pairing of opposite $\Omega$'s (modulo a rotation of symmetry of the lattice).

Das and Marcinkowski (1971) have made interesting assumptions on the passage of edge dislocations through a twist boundary, pointing out the creation of a *twist* dislocation loop in the plane of the boundary, with an angle of rotation equal to the angle of the grain boundary, but of opposite sign. This leads the two grains on both sides of the boundary to coincide inside the loop, and allows the edge dislocation to glide from one grain to the other inside the loop.

It is known that in body centred cubic metals in a certain temperature range plastic deformation is accompanied by the formation of walls of dislocations. It has been suggested by Vesely (1975), following electron microscope observations of molybdenum, that the idea that they are attended by internal stress fields due to rotation dislocations plays a role in the formation of these walls. In the first stage of plastic deformation, the primary glide planes are organized in limited parallel disorientation walls, with a well defined axis of rotation, the rotation increasing with the deformation. With strong deformations, stresses due to rotation dislocations are relaxed through the formation of walls transversal to the primary walls, and this leads to the formation of entirely closed low angle boundaries. Vesely has also suggested that the limits of the walls are not clearly defined, but are spread to reduce the long range stresses, and also that they should be analysed as densities of disclination loops.

### 8.1.3 Conservative and non-conservative motion

Das *et al.* (1973a, b) have analysed the motion of a dislocation line $L$ set up by the Volterra process

$$\mathbf{d} = \Omega \wedge (\mathbf{r} - \mathbf{r}_0), \tag{8.2}$$

and recognized the possibility of three types of motion:

(a) displacement of $L$ at $\Omega$ and $\mathbf{r}_0$ fixed;
(b) $L$ fixed, displacement of $\mathbf{r}_0$ by bringing $\Omega$ parallel to itself; and
(c) a motion combining the first two.

All these possibilities do not carry the same importance, for energy considerations lead us always to take $\Omega$ along the line $L$ (if $\Omega$ is large) or to use models of rotation dislocations which are sometimes more complex than that of equation (8.2), in particular for twist disclinations (see discussion in section 8.2). The discussion here will therefore be restricted to some obvious cases.

#### 8.1.3.1 *Motion of a wedge line in a direction parallel to itself*

This motion is necessarily non-conservative, whether the line is displaced by moving $\Omega$ (case (c)) or with a fixed $\Omega$ (case (a)), i.e. there is in the two cases addition or removal of matter in the vicinity of the line. It is useful to state, however, that in case (a), the displacement takes place by stacking of edge dislocations, whose motion out of their glide plane is non-conservative, while in case (c), there is annihilation or creation of dislocations. This can only take place by an emission of *jogs* along the line (see Figure 8.5) to which segments of edge dislocations are attached whose nucleation and growth require a diffusion of vacancies or interstitials.

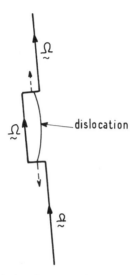

Figure 8.5.   Jog on a wedge disclination, to which is attached an edge dislocation (non-conservative displacement)

### 8.1.3.2   *Motion of a twist line*

Two cases can be distinguished, depending on whether the line is displaced in a plane parallel to $\Omega$ by the formation of jogs (Figure 8.6a) or in a plane perpendicular to $\Omega$ by the formation of kinks (Figure 8.6b). It is supposed that $\Omega$ is displaced with $L$. Figure 8.6a shows nucleation and growth of an edge dislocation and non-conservative motion, as in Figure 8.5; Figure 8.6b shows nucleation and growth of a screw dislocation which does not need a diffusion of vacancies or interstitials. The motion is therefore conservative.

$L$ can also be curved in the plane perpendicular to $\Omega$: the screw dislocations necessary for this phenomenon form a twist boundary. Das *et al.* have supposed that the line $L$ can be blocked around two wedge segments (Figure 8.6a) whose (non-conservative) motion is difficult and around which it turns like a windmill (analogous to the Frank–Read windmill for dislocations), which permits the multiplication of twist rotation dislocations.

226

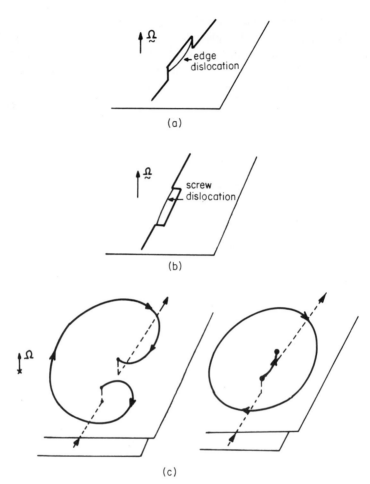

Figure 8.6. (a) Jog on a twist disclination: nucleation and growth of an edge dislocation and non-conservative motion of the line. (b) Kink on a twist disclination: nucleation and growth of a screw dislocation and conservative motion of the line. (c) Frank–Read windmill for twist disclinations

## 8.2  MATHEMATICAL FORMULATION OF THE GEOMETRY OF DISCLINATIONS AND LINEAR ELASTICITY

This section contains an extension of the theory of chapter 6 to include $e_{ij}$ strains in a form adapted to linear phenomena in solids. The $e_{ij}$ strains, like the rotation gradients, are supposed small. The non-commutativities of rotations will be disregarded. The material in this section is inspired by several works, without following them very closely (Kröner, 1958; de Wit, 1970a, b, c, 1973; Mura, 1972a, b; Kléman, 1974a). Anthony (1970a, b) has developed a non-Riemannian formulation of the same problem, and has extended elasticity by introducing the effect of torques into the expression of stresses (torque–stress theory). This attempt must be tied up to the previous discussion of Cosserat media (Gunther, 1958; Kröner, 1962, 1963; Mindlin and Tiersten, 1962; Schaeffer, 1967); also, see the collection of articles in Kröner (1968). But as observed by McClintock *et al.* (1958) the role of torques in a crystalline solid is certainly negligible compared to that of symmetrical stresses. The simplest point of view will therefore be adhered to.

### 8.2.1 Plastic distortions and elastic deformations (Kröner, 1958, 1963)

Kröner's definition of densities of dislocations will be used here, as it conforms perfectly with certain types of problems, for example the analysis of magnetostriction stresses (See Chapter 9). Plastic distortion $\beta^P_{ij}(\mathbf{r})$ of a perfect crystal is defined as the result of the following operation: the crystal is cut into infinitely small elements of volume; if the element centred on $\mathbf{r}$ undergoes the distortion $\beta^P_{12}(\mathbf{r})$, this means shearing it in a plane perpendicular to the direction 1, along the direction 2, by a quantity equal to $\beta^P_{12}(\mathbf{r})$ per unit length along the direction 1. The distortion $\beta^P_{11}(\mathbf{r})$ consists of a dilation $\beta^P_{11}(\mathbf{r})$ along the direction 1 (Figure 8.7). These operations are carried out while maintaining the orientation of the lattice. Figures 8.7c and 8.7d represent the result of operations $e^P_{ij}$ and $\omega^P_{ij}$.

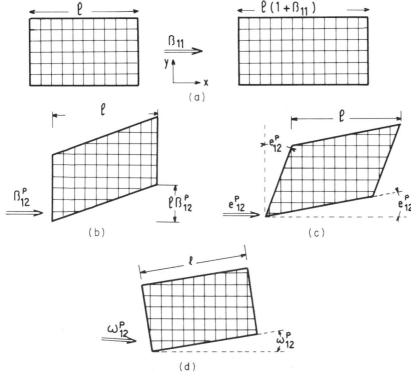

Figure 8.7. Plastic distortions: (a) $\beta^P_{11}$; (b) $\beta^P_{12}$; (c) $e^P_{12} = \frac{1}{2}(\beta^P_{12} + \beta^P_{21})$; (d) $\omega^P_{12} = \frac{1}{2}(\beta^P_{12} - \beta^P_{21})$.

The elements of volume distorted in this way cannot be put back without recovery or vacancies. To describe this phenomenon a density of translation dislocations is introduced which can be represented quantitatively in the following manner. Let there be a closed circuit $\gamma$ in the perfect crystal. To an element $\mathrm{d}x_i$ of this circuit corresponds in the distorted medium an element $\mathrm{d}x_i + \beta_{ji}^{\mathrm{P}}\,\mathrm{d}x_j$ (summation over $j$). The closures defect is therefore

$$\varDelta b_i = \oint_\gamma \beta_{ji}^{\mathrm{P}}\,\mathrm{d}x_i = \int_S \varepsilon_{jkl}\beta_{li,k}^{\mathrm{P}}\,\mathrm{d}S_j. \tag{8.3}$$

and we can define a dislocation density as:

$$\alpha_{ij}^{\mathrm{P}} = \varepsilon_{ikl}\beta_{lj,k}^{\mathrm{P}}. \tag{8.4}$$

By definition, plastic distortion does not result in elastic deformation. It must be understood that it can be obtained by making edge dislocations pass through volume elements, either by glide ($\beta_{12}^{\mathrm{P}}$) or by climb ($\beta_{11}^{\mathrm{P}}$). It is not necessary to make screw dislocations pass through them to obtain the most general distribution of densities of translation dislocations. To give back a 'compatible' form to a plastically distorted medium each element is taken back *elastically* to its primitive form by applying adequate forces to the surface. Elastic deformations and rotations are therefore introduced:

$$\begin{aligned} e_{ij}' &= -e_{ij}^{\mathrm{P}}, \\ \omega_{ij}' &= -\omega_{ij}^{\mathrm{P}}. \end{aligned} \tag{8.5}$$

The elements are then put back together and the applied force relaxed. Supplementary strains and rotations $e_{ij}''$, $\omega_{ij}''$ appear. The final state is completely compatible, does not suffer any applied force, but contains *internal stresses* $\sigma_{ij}^+ = C_{ijkl}e_{kl}^+$ and lattice rotations $\omega_{ij}^+$:

$$\begin{aligned} e_{ij}^+ &= e_{ij}' + e_{ij}'' = -e_{ij}^{\mathrm{P}} + e_{ij}'', \\ \omega_{ij}^+ &= +\omega_{ij}' + \omega_{ij}'' = -\omega_{ij}^{\mathrm{P}} + \omega_{ij}''. \end{aligned} \tag{8.6}$$

The total distortion

$$\beta_{ij}^{\mathrm{T}} = \beta_{ij}^{\mathrm{P}} + \beta_{ij}^+ \tag{8.7}$$

is compatible,† and we can put

$$\alpha_{ij}^+ = \alpha_{ij} = +\varepsilon_{ikl}\beta_{lj,k}^+ = -\varepsilon_{ikl}\beta_{lj,k}^{\mathrm{P}} = -\alpha_{ij}^{\mathrm{P}}. \tag{8.8}$$

From this point on the superscript plus sign will be suppressed in the expressions of elastic quantities.

In the total distortion $\beta_{ij}^{\mathrm{T}}$, the strains $e_{ij}$ and rotations $\omega_{ij}^{\mathrm{P}}$ are of very different natures:

(a) The strains $e_{ij}^{\mathrm{P}}$ and $e_{ij}$ do not involve the crystallographic nature of the material. They can be defined for an amorphous material without any

---

† Compatibility is the standard term meaning that $\beta_{ij}^{\mathrm{T}}$ is the derivative of a displacement $u_j^{\mathrm{T}}$; $\beta_{ij}^{\mathrm{T}} = u_{j,i}^{\mathrm{T}}$.

modification. In this sense, the $e_{ij}^P$ and $e_{ij}$ are geometric quantities having the same nature.

(b) On the other hand, the quantities $\omega_{ij}^P$ and $\omega_{ij}$ are very different in nature. Only $\omega_{ij}$ represents a real rotation of the lattice, while $\omega_{ij}^P$ is the rotation of an imaginary lattice traced in the perfect material before the operation $\beta_{ij}^P$ takes place (see Figure 8.7).

Let us put $\omega_j = \frac{1}{2}\varepsilon_{jpq}\omega_{pq} \cdot \boldsymbol{\omega}$ is the angle of the rotation vector. Its gradient, using equation (8.8), needs:

$$K_{ij} = \omega_{j,i} = \varepsilon_{jpq}e_{qi,p} + \tfrac{1}{2}\delta_{ij}\alpha - \alpha_{ji}, \tag{8.9}$$

where $\alpha = \alpha_{11} + \alpha_{22} + \alpha_{33}$.

This formula shows clearly the different origins of the rotations of the lattice. The quantity $\frac{1}{2}\delta_{ij}\alpha - \alpha_{ji}$ corresponds to the contributions of the dislocations themselves, independently of the elastic deformation which they cause. This is the contortion tensor of Chapter 6 (denoted here by $K_c$), corresponding to the generalized process of glide (equation (6.35)). $\varepsilon_{jpq}e_{qi,p}$ is the elastic contribution. We can write also:

$$\alpha_{ij} = K\delta_{ij} - K_{ji} + \varepsilon_{ipq}e_{qj,p} \tag{8.10}$$

### 8.2.2  Expression for the elastic distortion field

The stresses $\sigma_{ij} = C_{ijkl}e_{kl}$ obey the equilibrium relations

$$\sigma_{ij,i} + f_j = 0, \tag{8.11}$$

where $f_j$ is the applied force per unit volume. We have by definition

$$e_{kl}^T = e_{kl} + e_{kl}^P,$$

where, $e_{kl}^T$ being compatible, we can put

$$e_{kl}^T = \tfrac{1}{2}(u_{l,k}^T + u_{k,l}^T).$$

These expressions are then substituted into the equilibrium equations, using the symmetry of the coefficients $C_{ijkl}$. This yields:

$$C_{ijkl}u_{l,ki}^T + f_j = C_{ijkl}e_{kl,i}^P. \tag{8.12}$$

It can be seen that the plastic deformations play the role of a fictitious force. To solve the above equation, we introduce Green's function $G_{jn}$

$$C_{ijkl}G_{jn,ik}(\mathbf{r}/\mathbf{r}') + \delta_{ln}\delta(\mathbf{r} - \mathbf{r}') = 0 \tag{8.13}$$

$G_{jn}$ is the displacement in the direction $j$, at point $\mathbf{r}$, due to a point force situated at $\mathbf{r}'$ in the direction $n$, obeying the same boundary conditions as equation (8.11) (for internal stresses we must have on the surface $\sigma_{ij}v_i = 0$). This yields:

$$u_n^T(\mathbf{r}) = + \int G_{jn}(\mathbf{r}/\mathbf{r}')\{f_j(\mathbf{r}') - C_{ijkl}e_{kl,i}^P(\mathbf{r}')\}\, dV' \tag{8.14}$$

230

(for this derivation compare with the discussion of an analogous problem in smectics, sub section 5.6.6).

This expression is quite general; in its derivation no hypothesis has been made on the $e_{ij}^P$, and the $\omega_{ij}^P$ rotations are not involved (indifference to torques). It can then be used for translation dislocations as well as for defects of a more general nature.

From equation (8.14) we obtain $e_{ij}^T$ (therefore $e_{ij}$) and $\omega_{ij}^T$ (therefore $K_{ij}$). It seems from this equation that the existence of a generalized glide (i.e. a distortion without $e_{ij} = 0$ strain) is determined in the absence of the bulk forces $f_i$, not from geometric but from purely energetic considerations (presence of the $C_{ijkl}$). In fact this is not true; apart from the very particular case where the plastic deformations obey $C_{ijkl}e_{kl,i}^P = 0$ it is necessary (and evidently sufficient) to have

$$e_{ij}^P = 0, \qquad \omega_{ij}^P \neq 0 \tag{8.15}$$

to obtain generalized glide. This condition can be expressed on $\alpha_{ij}^P$ in the following manner (Eshelby, 1956)

$$\eta_{ij} \equiv -\tfrac{1}{2}(\varepsilon_{jnl}\alpha_{il,n}^P + \varepsilon_{inl}\alpha_{jl,n}^P) = 0 \tag{8.16}$$

$\eta_{ij}$ is the *incompatibility tensor*, a purely geometric quantity, defined by the expression:

$$\eta_{ij} = -\varepsilon_{ikl}\varepsilon_{jmn}e_{km,ln}^P = \varepsilon_{ikl}\varepsilon_{jmn}e_{km,ln} \tag{8.17}$$

and of which it will now be shown that it can be expressed as a function of the $\alpha_{ij}^P$ as indicated by equation (8.16). The condition of generalized glide created by dislocations of density $\alpha_{ij}^P$ requires $\eta_{ij} = 0$.

Let us write that $K_{ij}$ is a gradient, i.e. that an integral of the type $\int_A^B K_{ij}\,dx_i$ does not depend on the path followed between $A$ and $B$. This yields:

$$\theta_{ij} = \varepsilon_{ikl}K_{lj,k} = 0$$

This expression reads also, because of equation (8.9):

$$\theta_{ij} = \varepsilon_{ikl}\varepsilon_{jpq}e_{ql,pk} + \tfrac{1}{2}\varepsilon_{ikl}\alpha_{,k} - \varepsilon_{ikl}\alpha_{jl,k} = 0. \tag{8.18}$$

The first term of this equation is symmetrical and equal to $\eta_{ij}$. By taking the symmetrical part of the second term this finally yields equation (8.16).

As Eshelby (1956) and Kröner (1955, 1958) have shown, the $\eta_{ij}$ tensor defined by equation (8.17) which is of a purely geometrical nature, is the source of internal stresses. The solution of the equilibrium equation $\sigma_{ij,i} = 0$ can be expressed indeed with the help of a tensorial potential (also called Airy's function) $\chi_{ij}$ in the form:

$$\sigma_{ij} = \varepsilon_{ipq}\varepsilon_{jmn}\chi_{pm,qn} \tag{8.19}$$

(this expression satisfies $\sigma_{ij,i} = 0$). If Hooke's law is then written in the inverse form:

$$e_{ij} = \frac{1}{2\mu}\left\{\sigma_{ij} - \delta_{ij}\frac{v}{1+v}(\sigma_{11} + \sigma_{22} + \sigma_{33})\right\}, \tag{8.20}$$

and if the incompatibility of each term is calculated, this yields:

$$\nabla^4 \chi'_{ij} = \eta_{ij}, \qquad (8.21)$$

with

$$\chi'_{ij} = \frac{1}{2\mu}\left(\chi_{ij} - \delta_{ij}\frac{\nu}{2+\nu}\chi\right), \qquad (8.22)$$

where the solution of (8.21) must obey the gauge condition $\chi'_{ij,i} = 0$ (as in theory of the vector potential, supplementary conditions can be imposed on the tensorial potential $\chi'$). Equation (8.21) proves our assertion: in the case of an infinite medium, its solution is

$$\chi_{ij}(\mathbf{r}) = -\frac{1}{8\pi}\int \eta_{ij}(\mathbf{r}')\,|\mathbf{r} - \mathbf{r}'|\,d^3\mathbf{r}'. \qquad (8.23)$$

and the internal stresses depend indeed only on $\eta_{ij}$.

The concept of fictitious forces introduced above deserves a historical comment: these forces have been used for a long time to describe the internal stresses due to a non-homogeneous temperature distribution in an isotropic material (thermo-elasticity theory of the Cosserat brothers): in this case they are very simply linked to the temperature gradient. This is an energy concept; these forces involve indeed the stiffness coefficients $C_{ijkl}$. But the above statement indicates that the important concept is of a geometrical nature: actually it is these $\alpha_{ij}$ *dislocations* and $\eta_{ij}$ *incompatibilities*, associated with plastic distortion $\beta^P_{ij}$

$$\beta^S_{ij} = \delta_{ij}\alpha(T - T_0)$$

which are the origin of internal stresses created by the inhomogeneous distribution of $T$, independently of elastic quantities. The problem of magnetostriction is of the same type. If the word 'fictitious' had to be used, it would be reasonable to apply it to these dislocations, distortions and incompatibilities introduced by the Kröner process and which do not figure topologically in the geometric balance of the medium (they are non-existent, in a way, since the medium is left compatible at the end of the process). However, the terms of quasi-plastic distortion, quasi-dislocations and quasi-incompatibilities will be used in the discussion which follows, since they are used in practice.

### 8.2.3 Discrete dislocations; walls; Somigliana dislocations

First, the symbolism necessary for the study of discrete objects will be introduced. The notations used will be those used by Kunin (1965) and de Wit (1973a).

The Dirac distributions on lines $L$, surfaces $S$, volumes $V$, are defined as follows:

$$\delta_i(L) = \int_L \delta(\mathbf{r} - \mathbf{r}')\,dL'_i,$$

$$\delta_i(S) = \int_S \delta(\mathbf{r} - \mathbf{r}') \, dS'_i, \tag{8.24}$$

$$\delta(V) = \int_V \delta(\mathbf{r} - \mathbf{r}') \, dV'.$$

It is obvious that $\delta_i(L)$ behaves like a field of unit vectors tangential to $L$, and $\delta_i(S)$ as a field of unit vectors normal to $S$. From these definitions we obtain:

$$\int_V \delta_i(L') \phi(\mathbf{r} - \mathbf{r}') \, dV' = \int_L \phi(\mathbf{r} - \mathbf{r}') \, dL'_i, \tag{8.25}$$

and the analogous relations for $\delta_i(S)$ and $\delta(V)$. If an oriented line $L$ passes through an oriented surface $S$ at a point $\mathbf{r}$, one obtains:

$$\int_S \int_L \delta(\mathbf{r} - \mathbf{r}') \, dL'_i \, dS_j = \frac{t_i v_j}{\mathbf{t} \cdot \mathbf{v}}, \tag{8.26}$$

where $\mathbf{t}$ is the unit tangent to the line and $\mathbf{v}$ the unit normal to the surface. From this we obtain:

$$\int_S \delta_i(L) \, dS_i = \int_L \delta_i(S) \, dL_i. \tag{8.27}$$

These quantities are equal to zero if there is no intersection between $L$ and $S$. If there is an intersection they are equal to $\pm 1$, the sign depending on the relative orientation.

The notion of derivative is easily extended to these Dirac distributions:

$$\delta_{i,j}(L) = \int_L \delta_{,j}(\mathbf{r} - \mathbf{r}') \, dL'_i, \quad \text{etc.}$$

$$\tag{8.28}$$

$$\int_V \delta_{i,j}(L') \phi(\mathbf{r} - \mathbf{r}') \, dV' = -\int_L \frac{\partial \phi(\mathbf{r} - \mathbf{r}')}{\partial x'_j} \, dL'_i.$$

Also the set of equalities

$$\int_V \delta_{,i}(V) \phi(\mathbf{r}) \, dV = -\int_V \phi_{,i}(\mathbf{r}) \, dV = \int_{S_0} \phi(\mathbf{r}) \, dS_i$$

$$\tag{8.29}$$

$$= -\int_V \delta_i(S) \phi(\mathbf{r}) \, dV,$$

leads us to write symbolically

$$\delta_{,i}(V) = -\delta_i(S_0), \tag{8.30}$$

where $S_0$ is the *closed* surface which forms the boundary of $V$. We have in the same way:

$$\varepsilon_{ijk} \delta_{i,j}(S) = -\delta_k(L), \tag{8.31}$$

and for a closed surface:

$$\varepsilon_{ijk}\delta_{i,j}(S_0) = 0. \tag{8.32}$$

Finally, for a closed curve, we have:

$$\delta_{k,k}(L_0) = 0. \tag{8.33}$$

For a discrete dislocation of Burgers vector $b_i$, situated on a line $L$, there is a corresponding density of dislocations which can be written:

$$\alpha_{ij} = \delta_i(L)b_i, \tag{8.34}$$

since it is obvious that the integral $\int_{Sl}\alpha_{ij}\,dS_i$ is indeed equal to $b_i$ for every surface $S$ pierced by $L$ once, and oriented in the same way as $L$. In the same way, the condition of continuity $\alpha_{ij,i} = 0$ is proved if $L$ is a closed line. One will check that the density in equation (8.34) derives from the plastic distortion

$$\beta^{P}_{ki} = -\delta_k(S)b_i. \tag{8.35}$$

Furthermore, in this plastic distortion the actual expression of the Volterra process can be recognized: displacement of the two lips of the surface $S$ by the quantity $-\mathbf{b}$ (in the plastic state) therefore $+\mathbf{b}$ (in the state in which the lips of the cut surface are reglued and the whole material elastically relaxed).

Now consider surface defects, restricting the discussion to two types: those characterized by a distribution $\beta^{P}$ in the form of Dirac's function on $S$, and those characterized by a discontinuity of $\beta^{P}$ on $S$. More general cases can be envisaged, but those we are going to deal with correspond to the most frequent physical cases.

### 8.2.3.1  $\beta^{P}$ localized on S (Somigliana dislocations)

This is written:

$$\beta^{P}_{ij} = -d_{ijk}\delta_k(S), \tag{8.36}$$

and is the simplest generalization of the linear defect. It is obviously a defect obtained by displacing the two lips of the cut surface by a quantity $\mathbf{d}$ which varies with the usual point on $S$. Let us take a circuit $\gamma$ oriented in the same way as $S$ and cutting $S$ orthogonally at $M$. We have:

$$d_j(M) = -\int_\gamma \beta^{P}_{ij}\,dx_i = d_{ijk}(M)v_k v_i. \tag{8.37}$$

This defect was first introduced by Somigliana (1914).† It is not equivalent, in the most general case, to a wall of dislocations spread over $S$. In fact we have

$$\alpha_{ij} = \varepsilon_{ipq}\{d_{qjk,p}\delta_k(S) + d_{qjk}\delta_{k,p}(S)\}, \tag{8.38}$$

---

† The name Somigliana dislocation is frequently given to the line bounding $S$, and by linguistic inconsistency, deformation fields and stresses due to $S$ are attributed to this line.

234

and this quantity does not satisfy $\alpha_{ij}v_i = 0$, i.e. the density of lines piercing $S$ is not zero. However, the continuity condition $\alpha_{ij,i} = 0$ is assured.

According to Weingarten's theorem, the only case where there is no discontinuity of internal stresses on $S$ is when $\mathbf{d}$ is a rigid displacement, the sum of a translation and a rotation. It is easy to show that $\beta_{ij}^P$ then necessarily has the form:

$$\beta_{ij}^P = -\{b_j + \varepsilon_{jpq}\Omega_p(x_q - x_q^0)\}\delta_i(S), \tag{8.39}$$

which results in

$$\alpha_{ij} = \delta_i(L)\{b_j + \varepsilon_{jpq}\Omega_p(x_q - x_q^0)\} + \delta_j(S)\Omega_i - \delta_{ij}\Omega_k\delta_k(S). \tag{8.40}$$

It can be seen here that the condition $\alpha_{ij}v_i = 0$ is satisfied. The only physical case of a Somigliana defect of interest here is where the displacement obeys Weingarten's theorem, i.e. is the sum of a translation dislocation $\mathbf{b}$ and rotation dislocation $\mathbf{\Omega}$.

If the cut surface is analysed in terms of dislocations (equation (8.40)) it falls into two distinct parts. Firstly there is a surface part, which corresponds to the relative rotation of the crystal from one side of $S$ to the other, formed by dislocations in the sense of Nye (generalized glide), describing a disorientation wall (diremption).† Since this is bounded by $L$, an array of variable translation dislocations on $L$ compensates the twist parts of the diremption on $L$. De Wit called this model of a rotation dislocation the 'dislocation model of disclination'. This is surely not the most general model of a limited diremption compensated by translation dislocations. For these dislocations are here along $L$, while the topological conditions of compensation of the twist parts simply require that they end on $L$.

What we are describing here is in fact a rotation dislocation, in the Volterra sense (equation (8.39) reproduces equation (8.2)), but note that we have no discontinuity of rotations on $S$ ($K_{ij}$ is integrable). In fact these discontinuities are not introduced automatically by a displacement in the form of equation (8.2), which is only a sum of translations not affecting the lattice (see Friedel's remark reported in Chapter 2, equation (2.15)).

On the other hand, the $\alpha_{ij}$ distributions of the type

$$\alpha'_{ij} = +\delta_i(L)\varepsilon_{jpq}\Omega_p(x_q - x_q^0),$$

or

$$\alpha''_{ij} = +\delta_j(S)\Omega_i - \delta_{ij}\Omega_k\delta_k(S),$$

correspond to a discontinuity of $\omega_i$ on $S$. Another example of $\alpha_{ij}$ distribution having discontinuities of this type will be shown before formulating the general theory. $\alpha'_{ij}$ and $\alpha''_{ij}$ do not satisfy the continuity condition $\alpha_{ij,i} = 0$.

---

† We will refer to a Nye's distribution of dislocations on a wall when the strain field $e_{ij}$ of this distribution is zero. We also call such an object a diremption.

### 8.2.3.2  $\beta^P$ discontinuous on $S$

Suppose that a discontinuity of $\beta^P_{ij}$ is introduced on a surface $S$:

$$\Delta\beta^P_{ij} = \Delta e^P_{ij} + \Delta \omega^P_{ij} = E_{ij} + \varepsilon_{ijk}\Omega_k. \tag{8.41}$$

If it supposed that $S$ goes to infinity and separates two regions, in each of them the $\beta^P_{ij}$ would be constant, densities of surface dislocations can be defined by integrals of the type

$$\bar{\alpha}_{ij}(\mathbf{r}) = \int_{-\infty}^{+\infty} \alpha_{ij}\,\mathrm{d}l \tag{8.42}$$

where the integration is made over a line piercing $S$ orthogonally at $\mathbf{r}$. With

$$\Delta\beta^P_{ij} = \beta^P_{ij}(\mathrm{II}) - \beta^P_{ij}(\mathrm{I}) \tag{8.43}$$

and the normal $\mathbf{v}$ to $S$ oriented from I towards II, we have

i.e.
$$\begin{aligned}\bar{\alpha}_{ij} &= -\varepsilon_{ikl}\Delta\beta^P_{lj}v_k, \\ \bar{\alpha}_{ij} &= \Omega_i v_j - \Omega_k v_k \delta_{ij} - \varepsilon_{ikl}E_{lj}v_k. \end{aligned} \tag{8.44}$$

This quantity is a good representation of a density of dislocations situated on $S$, as it can be proved that the density of lines piercing $S$, i.e. $\bar{\alpha}_{ij}v_i$, is zero.

In using Kunin's notation, it is preferable to introduce the distribution (in the Dirac sense) $\alpha_{ij}$:

$$\alpha_{ij} = \Omega_i\delta_j(S) - \Omega_k\delta_k(S)\delta_{ij} - \varepsilon_{ikl}E_{lj}\delta_k(S), \tag{8.45}$$

rather than $\bar{\alpha}_{ij}$. If the $\beta^P_{ij}$ are variable, this expression is still valid, but only represents the surface part of the total density $\alpha_{ij}$, which is continuous outside $S$.

In expression (8.45) two parts must be distinguished:

(a) The part corresponding to $\Omega$ describes a *diremption* (disorientation wall). The corresponding dislocations cannot create long range stress. This object was met above (equation (8.40)).

(b) The part corresponding to $E_{ij}$ describes the junction of two crystals of different parameters, i.e. a distribution of epitaxial dislocations. It is tempting to call this a *distranslation*. Because of Weingarten's theorem, there is a corresponding discontinuity of stresses on $S$.

A surface object such as that in expression (8.45) cannot be easily defined except where $S$ effectively divides the space into two disjointed parts. In fact, if we suppose that $S$ is bounded by a line $L_0$, the continuity condition is not satisfied throughout:

$$\alpha_{ij,i} = \delta_i(L_0)(E_{ij} + \varepsilon_{ijk}\Omega_k) = \delta_i(L_0)\Delta\beta^P_{ij}. \tag{8.46}$$

There is no problem where $L_0$ is rectilinear and along an eigen direction (on the left) of $\Delta\beta^P_{ij}$. The condition satisfied in this case $\alpha_{ij,i} = 0$ simply means that on $L_0$ the epitaxial dislocations and the diremption dislocations satisfy a conservation law on the Burgers vectors, analogous to the one satisfied by translation dislocations and diremption dislocations in equation (8.40). But here it is convenient to give a general interpretation of equation (8.46), which interpretation covers the case of diremptions and rotation dislocations corresponding to a localized plastic distortion $\beta_{ij}^P$.

### 8.2.4 Rotation dislocations (continuous approach and discrete approach)

The preceding discussion introduced several cases of rotation dislocations bounding a wall (de Wit's model of a rotation dislocation, limited diremption, distranslation etc.). In Chapter 6, which dealt with situations where $e_{ij} = 0$, it was shown that a rotation dislocation can be defined as the site of singularities in a contortion field $K_{ij}$. The application of this definition in the case of crystalline solids leads to the concept of a limited diremption, but ignores the other types of rotation dislocations which we have just met. A wider definition is therefore necessary here, including in particular the fact that certain singularities of the strain field $e_{ij}$ can also be considered as rotation defects (the same applies to de Wit's model), without the contortion tensor $K_{ij}$ itself displaying singularities.

We will note

$$\theta_{ij} = \varepsilon_{ikl}K_{lj,k}, \qquad (8.47)$$

the *tensor curl of the contortion field*. This is a density of rotation dislocations in the same sense as in Chapter 6.

When $\theta_{ij}$ is different from zero, the incompatibility tensor $\eta_{ij}$ (see subsection 8.2.2) will be defined by equation (8.17) reproduced here

$$\eta_{ij} = -\varepsilon_{ikl}\varepsilon_{jmn}e_{km,ln}^P = \eta_{ji}. \qquad (8.17)$$

The relations between $\eta_{ij}$ and $\theta_{ij}$ can be obtained by taking the curl of equation (8.9). We obtain:

$$\eta_{ij} = \theta_{ij} - \tfrac{1}{2}\varepsilon_{ikj}\alpha_{,k} + \varepsilon_{ikl}\alpha_{jl,k}, \qquad (8.48)$$

which reads also, by putting into evidence the symmetrical and antisymmetrical parts of $\theta_{ij}$:

$$\varepsilon_{ijp}\theta_{ij} + \alpha_{kp,k} = 0, \qquad (8.49a)$$

$$\theta_{ij}^S = \eta_{ij} - \tfrac{1}{2}\{\varepsilon_{ikl}\alpha_{jl,k} + \varepsilon_{jkl}\alpha_{il,k}\}. \qquad (8.49b)$$

These nine equations link six components of the $\eta_{ij}$, nine components of the $\alpha_{ij}$, and nine components of the densities of rotation defects $\theta_{ij}$. Fifteen components can therefore be fixed to start with.

Kröner's plastic process begins with the data of the $\beta_{ij}^P$; these permit the definition of both the $\alpha_{ij}$ ($= -\alpha_{ij}^P$) and the $\eta_{ij}$ ($= -\eta_{ij}^P$). All the remaining quantities are then perfectly defined by equations (8.49). It is clear that the $\theta_{ij}$ cannot be anything else than Dirac distributions or their derivatives. It is reasonable to expect that, for a $\beta_{ij}^P$ field which is regular almost throughout, the $\theta_{ij}$ are Dirac derivatives, i.e. weak singularities obtained by short range coupling of opposing disclinations. There only remains the possibility of singular $\eta_{ij}$: these are the sources of stresses both of isolated translation dislocations and of walls of translation dislocations. These can lead to fields of singular rotations, as Eshelby first stresses. However, it is well known that this is not the only case of a singular rotation field.

Instead of beginning with the $\beta_{ij}^P$ field as Kröner does, let us take the $\alpha_{ij}^P$ field as a fundamental field, i.e. the field of translation dislocations. The arrangement of $\alpha_{ij}^P$ fields thus defined is certainly more extensive than that of Kröner, defined by equation (8.8). It is evident from equation (8.49), that it is then always possible to choose the $\theta_{ij}$ so that the sources of stresses $\eta_{ij}$ equate to zero. This surprising result is good evidence of the weakness, long recognized by Kröner, of an approach to the theory of rotation dislocations by a formalism of continuous densities (see, for example, the discussions of the Gaithersburg conference, 1969†). For if it is supposed that, the $\alpha_{ij}^P$ being given, the $\theta_{ij}$ which result from them are those which correspond to the annulment of the sources of stresses, then the $\theta_{ij}$ have no reason to correspond to the rotation symmetries of the lattice, and there are necessarily surface discontinuities of rotations, which the continuous approach ignores. It remains no less true that the present analysis reveals various sources for rotation dislocations, which can be analysed in two ways.

(a) In an amorphous medium: the continuous approach is then correct (equation (8.49)) and it is possible to start with $\alpha_{ij}^P$ and $\theta_{ij}^S$ being given. The $\theta_{ij}^S$ quantities represent a local rotation without translation ($P_G$, equation (2.15)) which can be calculated on a surface element

$$\int \theta_{ij}\,dS_i = \Delta\Omega_j$$

while the $\alpha_{ij}$ represent local translations without rotation ($T$, equation (2.15))

$$\int \alpha_{ij}\,dS_i = \Delta b_j.$$

Because of equation (8.49a), the Burgers vector $\Delta b_j$ is not defined in a unique manner: it depends on the choice of the surface element. However, the total process $P_V = P_G + T$ obeys elastic incompatibilities $\eta_{ij}$, and therefore is submitted to a strain field $e_{ij}$, independent of the choice of $S$. This independence reflects the arbitrary choice of the reference lattice in an amorphous material. The continuous elastic theory of amorphous media has been developed by Mura, de Wit and Anthony. In spite of the differences in presentation from the present discussion, there is little difference in the results and the interpretations are the same.

(b) In a crystalline medium; if $\theta_{ij}$ is to correspond to a lattice symmetry $\Omega$, we must put

$$\theta_{ij} = \delta_i(L)\Omega_j.$$

† Proceedings of the Conference on *Fundamental Aspects of Dislocation Theory*, Natl. Bur. Std. (U.S.) Spec. Pub. 317 (1970), held in Gaithersburg (1969).

We are then given nine $\alpha_{ij}^P$ (subject to the three conditions in equation (8.49a)) since the $\theta_{ij}$ are already given. The $\eta_{ij}$, and therefore the $e_{ij}^P$, are deduced from equation (8.49b). Kröner's plastic distortion process must then be replaced by the following: the plastic distortion $\beta_{ij}^P$ are reduced to the plastic strain $e_{ij}^P$; a plastic contortion $K_{cij}^P = \alpha_{ji}^P - \delta_{ij}\alpha^P$ must be added to it. The rejoining and relaxation processes are not modified. In the final result no particular cut surface appears, except perhaps if the $\eta_{ij}$ which results from the given quantities possesses surface singularities of the type already mentioned. If $\theta_{ij}$ does not correspond to a lattice symmetry, it is necessary to specify the surface $S$, resting on $L$, on which the discontinuity takes place. This can be introduced as above, by fixing the surface $\alpha_{ij}^P$. Some examples are given.

### 8.2.4.1 De Wit's model of a rotation dislocation

With the distribution in equation (8.40), it is possible to take $\theta_{ij} = 0$, since $\alpha_{ij}$ obeys the relations (imposed by equation (8.49) if we want $\theta_{ij} = 0$):

$$\varepsilon_{ikj}\alpha_{,k} + \varepsilon_{jkl}\alpha_{il,k} = \varepsilon_{ikl}\alpha_{jl,k}. \tag{8.51}$$

The Kröner process can therefore be used. It is equivalent here to the process including the distortion $e_{ij}^P = \frac{1}{2}(\beta_{ij}^P + \beta_{ji}^P)$ and the lattice rotation $K_c^P$ (plastic contortion). We also have:

$$\eta_{ij} = \delta_i(L)\Omega_j + \delta_j(L)\Omega_i.$$

### 8.2.4.2 Limited diremption

$$\alpha_{ij} = -\Omega_i\delta_j(S) + \Omega_k\delta_k(S)\delta_{ij}. \tag{8.52}$$

We have here, after equation (8.49a),

$$\theta_{ij}^A = \frac{1}{2}\{\Omega_j\delta_i(L) - \Omega_i\delta_j(L)\}.$$

We can then impose the constraint:

$$\theta_{ij} = \Omega_j\delta_i(L), \tag{8.53}$$

which leads to, following equation (8.49b):

$$\eta_{ij} = 0. \tag{8.54}$$

Mura and de Wit were the first to introduce this defect, under the names (respectively) of *impotent disclination line* and *compensated disclination line*.† The limited diremption is also the sum of the preceding rotation dislocation and of the following (translation) defect:

---

† Nye's wall, which is found in ferromagnets, is close to this diremption, but has a rather simpler physical meaning.

### 8.2.4.3  *Rotation dislocation in the Volterra sense*

$$\alpha_{ij} = \delta_i(L)\varepsilon_{jpq}\Omega_p(x_q - x_q^0), \tag{8.55}$$

which corresponds to

$$\begin{aligned}
\theta_{pq} &= \delta_p(L)\Omega_q, \\
e^P_{pq} &= -\tfrac{1}{2}(\delta_p(S)\varepsilon_{qrl} + \delta_q(S)\varepsilon_{prl})\Omega_r(x_l - x_l^0) \\
\eta_{pq} &= \delta_p(L)\Omega_q + \delta_q(L)\Omega_p.
\end{aligned} \tag{8.56}$$

### 8.2.4.4  *Distranslation*

$$\alpha_{ij} = \varepsilon_{ikl}E_{lj}\delta_k(S). \tag{8.57}$$

This defect, which occurs in magnetostriction, corresponds to

$$\theta_{pq} = \tfrac{1}{2}\varepsilon_{jpq}E_{ij}\delta_l(L) \tag{8.58}$$

i.e. to an antisymmetrical rotation dislocation, associated with sources of internal stresses:

$$\eta_{ij} = \tfrac{1}{2}\varepsilon_{ikp}\varepsilon_{jql}E_{pl}(\delta_{k,q}(S) + \delta_{q,k}(S)). \tag{8.59}$$

### 8.2.5  Distortions and line energies

De Wit (1973c) has calculated the distortions and gradients of rotation $K_{ij}$, associated in isotropic elasticity with different types of lines (except distranslations) discussed in the preceding section.

Reference may also be made to the calculations of Huang and Mura (1970, 1972), Chou *et al.* and Li and Gilman (1970), concerning rotation dislocations obeying equation (8.55). Zhukovskii and Vladimirov have recently calculated the stress field created by a segment of rotation dislocation bounded at its two extremes by singular points.

Examples relating to phenomena due to magnetostriction will be discussed in the next chapter.

# Chapter 9

# Rotation dislocations and magnetism

There are two very different types of singularity that occur in magnetism.

(a) *Singularities of the spin system:* in accordance with the previous chapters, the term *spin lattice* will be used to indicate that the magnetization distribution has properities of symmetry (see Chapter 2). The nature of the singularities can be defined from an analysis of these symmetries.

(b) *Quasi-dislocations (translation and rotation)* representing the sources of internal stresses of *magnetostriction*, due to an inhomogeneous distribution of spins. These fictitious defects, introduced in the preceding chapter, are located essentially on spin singularities (walls, wall junctions and dislocation lines). They are therefore involved in the free energy of the latter. However, the stress fields which they create interact also directly with lattice dislocations of the *supporting* perfect magnetic crystal; therefore they play an important role in the plasticity of magnetic materials. Although thoroughly investigated (e.g. Trauble, 1962, 1969) the plasticity of materials has not been studied from this angle. This is unfortunate, but it will not be discussed here either. The reader should refer to the articles by Brown Jr., (1940, 1941), Vicena (1954, 1955), Malek (1957a–e), Krönmuller and Seeger (1961), Pfeiffer (1967a–d) and to the present author's review article (published in Nabarro, 1980).

It will be supposed that the crystal being considered here is perfect apart from spin singularities.

This study cannot be undertaken here without some knowledge of the fundamental concepts of the theory of Weiss domains. A short summary of this will therefore be given first. A detailed account will be found in the article by Kittel and Galt (1956) and, for example, the works of Chikazumi (1966), Craik and Tebble (1956), Herpin (1968) or Kneller (1956). This will be followed by an account of the theory of magnetostriction, applied to walls and wall junctions with the restriction that they are considered to be infinitely thin. The end of the chapter deals with curvilinear spin singularities (topological and energy properties).

Most of the results concern ferromagnets.

## 9.1 FUNDAMENTAL CONCEPTS OF THE THEORY OF DOMAINS

### 9.1.1 Weiss domains and Bloch walls

A domain is a region of uniform magnetization $\mathbf{M}$, of constant intensity $|\mathbf{M}| = M_S$. In a crystal, $\mathbf{M}$ tends to follow *directions of easy magnetization*, for which the magnetocrystalline energy is minimal. In a cubic crystal, the density of magnetocrystalline energy is written:

$$\rho f_K = K_1(\alpha^2\beta^2 + \beta^2\gamma^2 + \gamma^2\alpha^2) + K_2\alpha^2\beta^2\gamma^2 + \cdots \qquad (9.1)$$

$\alpha$, $\beta$ and $\gamma$ are the direction cosines of $\mathbf{M}$ in the quaternary axes of the cube. $\rho f_K$ is invariant under the symmetries of the cube and under the transformation of $\mathbf{M}$ to $-\mathbf{M}$.

|  | $K_1$ | Easy axis |
|---|---|---|
| Fe | $4.8 \times 10^5 \, \text{erg/cm}^3$ | $\langle 100 \rangle$ |
| Ni | $-4.5 \times 10^4 \, \text{erg/cm}^3$ | $\langle 111 \rangle$ |

In iron and nickel, the sign of $K_1$ determines the easy axis; the term in $K_2$ and the higher order terms are smaller.

*Bloch wall.* This separates two domains magnetized along the easy directions; the magnetization turns across the wall avoiding the appearance of magnetic 'space charges' (div $\mathbf{M} = 0$), which makes the component of $\mathbf{M}$ normal to the wall remain constant. Walls at 180° and 90° in iron, and walls at 180°, 71° and 109° in nickel can be distinguished (Figure 9.1).

The energy is obtained by minimizing the sum of the anisotropic energy $\rho f_K$ (equation (9.1)) and the exchange energy:

$$\rho f_{Ex} = A(\nabla \mathbf{n})^2, \qquad (9.2)$$

where $\mathbf{M} = M_S\mathbf{n}$ and $A$ is a constant of the order of $10^{-6} \, \text{erg/cm}$ (note the similarity of equation (9.2) with the energy density of a nematic in the case of isotropic elasticity). In the sum $\rho f_{Ex} + \rho f_K$ appears the length:

$$\delta_w = \left(\frac{A}{K_1}\right)^{1/2}, \qquad (9.3)$$

which is the typical width of a wall. The typical energy is

$$\gamma_w = 2(AK_1)^{1/2} \qquad (9.4)$$

For a quick derivation of expressions (9.3) and (9.4), we can use the following single argument: the exchange energy is of density $A/\delta^2$, according to equation (9.2), for a typical variation of the magnetization extending over a width $\delta$. The contribution of $\rho f_{Ex}$ to the total energy is therefore $A/\delta^2 \times \delta = A/\delta$. In the same

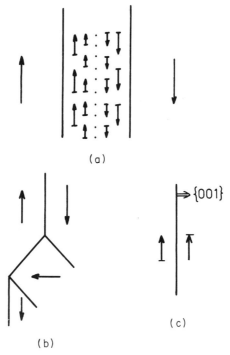

Figure 9.1. 180° and 90° walls. (a) Scheme of the distribution of **M** in the 180° wall. We have adopted the symbol of the 'nail' already used for nematics, upon which an arrow has been superimposed to indicate the direction of magnetization. (b) Current arrangement of 180° and 90° walls in iron. It will be noted that the 90° walls are parallel to planes {011}; the magnetization makes an angle of 45° with the plane of the wall. (c) 90° wall {001}: this situation is not frequently found in iron

way the contribution of anisotropy is of the order of $K_1\delta$. The sum $(A/\delta + K_1\delta)$ is minimal for $\delta = \delta_w$, and its value is then $\gamma_w$.

For iron, we obtain $\delta_w \sim 1000$ Å, $\gamma_w \sim 1.5\,\text{erg/cm}^2$. The value of $\delta_w$ indicates that the wall is extended over a hundred sites: this justifies the *continuous hypothesis* used in the calculations. This hypothesis is still valid for Co and Ni, but it is not the same for rare earth metals. For these, where magnetocrystalline anisotropy is twenty to fifty times greater than that of the preceding compounds, the thickness of the wall is of a few atomic dimensions only. Calculations involving the discrete nature of crystalline solids are therefore necessary. Barbara (1973) and Egami (1973) should be consulted on this subject.

### 9.1.2 Action of a magnetic field

In the presence of a constant applied magnetic field **H**, the free energy density becomes:

$$\rho f = A(\nabla \mathbf{n})^2 + K(\alpha^2\beta^2 + \beta^2\gamma^2 + \gamma^2\alpha^2) - \mathbf{H}\cdot\mathbf{M}. \tag{9.5}$$

Let us begin with a previously demagnetized sample, i.e. such that the distribution of different domains ensures that the total flux coming out is zero, and apply to it a weak magnetic field. The domains oriented along the field grow, while the domains magnetized in the opposite direction shrink. This is brought about by a movement of walls conserving their orientation, without modification of the direction of magnetization of the domains. For $H \gtrsim H_K = K_1/M_S$ (anisotropy field), the magnetization begins to turn towards the direction $\mathbf{H}$, until the sample is saturated. During this stage, the characteristic length is no longer $\delta_w$ but $\delta_H = (A/HM_S)^{1/2}$. This simplified scheme (see Néel, 1944) only describes *reversible* phenomena along the magnetization curve. The existence of structural defects (dislocations, grains and precipitates) and of the actual network of Bloch walls, which are of finite dimensions, imposes irreversible phenomena (Barkhausen jumps, coercive field and hysteresis).

The existence of demagnetizing fields $\mathbf{H}_d$ due either to surface magnetic moments $\sigma = \mathbf{M} \cdot \mathbf{v}$ (where $\mathbf{v}$ is the normal to the layer oriented towards its exterior) or to bulk moments ($\rho = -\operatorname{div}\mathbf{M}$), constitutes one of the great difficulties of the theory of domains, because of the non-local character of these fields. $\mathbf{H}_d$ is obtained from Maxwell's equations ($\operatorname{rot}\mathbf{H}_d = 0$; $\operatorname{div}\mathbf{B} = \operatorname{div}(\mathbf{H}_d + 4\pi\mathbf{M}) = 0$) which reads here:

$$\mathbf{H}_d = -\nabla\phi,$$

$$\nabla^2\phi = 4\pi\operatorname{div}\mathbf{M}, \qquad \text{bulk equation,} \qquad (9.6)$$

$$\left(\frac{\partial\phi}{\partial v}\right)_e - \left(\frac{\partial\phi}{\partial v}\right)_i = 4\pi\mathbf{M}\cdot\mathbf{v}, \qquad \text{surface equation,} \qquad (9.7)$$

$$\phi_e = \phi_i.$$

It can be shown (Brown, 1966) that the energy density associated with the demagnetizing field is in the form

$$\rho f_d = -\tfrac{1}{2}\mathbf{H}_d \cdot \mathbf{M} \qquad (9.8)$$

## 9.1.2.1 Néel walls and Bloch walls

The emergence of a surface Bloch wall necessarily creates surface charges. Suppose that the easy direction is parallel to the surface, and that the sample is a slab whose thickness is of the order of magnitude of the width of the wall $\delta_B$ (see Figure 9.2b). The surface charges form a dipole, the energy of which can be calculated as that of an ellipse of minor axis $\delta_B$ and of major axis $D$, uniformly magnetized. We obtain (Néel, 1945):

$$E_B = \frac{\pi\delta_B^2 M_S^2 D}{\delta_B + D}. \qquad (9.9)$$

This quantity is of the order of $\pi M_S^2\delta_B^2$ if $D$ is large enough compared to $\delta_B$. The surface demagnetizing field, of the order of $-4\pi(\mathbf{M}\cdot\mathbf{v})\cdot\mathbf{v}$, tends to bring

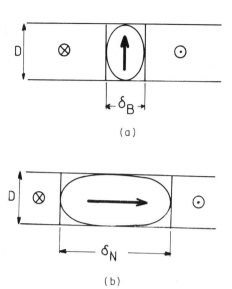

Figure 9.2. Energy diagram for walls in a thin slab: (a) Bloch; (b) Néel

the magnetization back into the plane of the slab. This is an important effect, as soon as the surface demagnetizing field overcomes the anisotropic field $\mathbf{H}_K$, i.e. for $q = K_1/2\pi M_S^2 < 1$. It is therefore interesting, in the media known as *weakly anisotropic* ($q < 1$), to consider the extreme situation where the magnetization is parallel to the slab (Figure 9.3) even in the interior of the wall. In this situation, every variation of $\mathbf{M}$ is accompanied by bulk charges. The energy of the corresponding *Néel wall* is calculated by assimilating the wall to an elliptical cylinder of minor axis $D$ and major axis $\delta_N$ (Figure 9.2b). It is clear that the width of the Néel wall $\delta_N$ is large compared to $\delta_B$, because of the tendency of the magnetic charges to spread out to diminish their density (see Néel, 1945; Prutton, 1964; Aharoni, 1966; La Bonte, 1969). We therefore have:

$$E_N = \frac{\pi \delta_N D^2 M_S^2}{\delta_N + D} \sim \pi M_S^2 D^2. \tag{9.10}$$

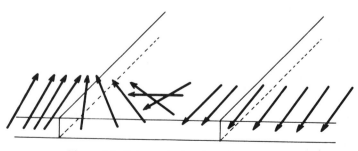

Figure 9.3. Schematic drawing of a Néel wall

The contributions of anisotropy and exchange must be added to the quantities $E_N$ and $E_B$ which are of magnetostatic origin (see Figure 9.4). But roughly, equations (9.9) and (9.10) already indicate that below a thickness of $D \sim \delta_w$, Néel walls are favoured. The energy of the wall is still diminished if it is formed by *segments of alternating polarity* (cross ties) separated by linear singularities (crossed or circular Bloch walls which are also wedge rotation dislocations $S = \pm 1$) (Middelhoek, 1963; Huber, Smith and Goodenough, 1958) (Figure 9.5).

In thick layers, at the emergence of a wall in the surface, the demagnetizing field modifies the distribution of **M** in the wall in a way intermediary between the Bloch and Néel distributions. This may be called a *surface disclination* by analogy with surface disclinations in nematics, which as we have seen can be considered as resulting from the intersection of a wall with the surface.

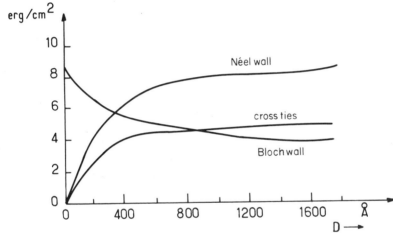

Figure 9.4. Compared energies of Bloch and Néel walls and of cross ties in a thin slab (after Middelhoek, 1963)

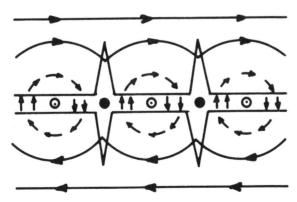

Figure 9.5. Scheme of a wall with alternating polarities (Néel segments separated by Bloch type disclination lines)

246

In weakly anisotropic media ($q < 1$), the tendency towards the disappearance of div $\mathbf{M} \neq 0$ zones is likely. ($4\pi M_S > H_K$). Hubert (1969) has made a systematic study of div $\mathbf{M} = 0$ geometries. The concept of *closure domain* applies to these situations (Figure 9.6.).

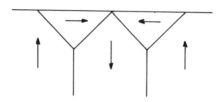

Figure 9.6. Closure domains in a weakly anisotropic material, when the easy directions are parallel and perpendicular to the slab (see Lifshitz, 1944)

On the other hand, in strongly anisotropic media ($q > 1$), the charges remain spread out on the surface, and it is expected that wall junctions are either strongly charged zones, or are sequences of segments of alternating polarity separated by *singular points*, since it is difficult to distribute the magnetization in zones of zero divergence.

### 9.1.3  Magnetostrictive effects (uniform magnetization)

The coupling between elastic distortions $e_{ij}$ and magnetization leads to the following energy density (Kittel and Galt, 1956):

$$\rho f_{\text{ms}} = B_1(e_{xx}\alpha^2 + e_{yy}\beta^2 + e_{zz}\gamma^2) \qquad + 2B_2(e_{xy}\alpha\beta + e_{yz}\beta\gamma + e_{zx}\gamma\alpha) \quad (9.11)$$

for a cubic ferromagnet (the $e_{ij}$ and $\alpha$, $\beta$, $\gamma$ are written in the quaternary axes of the cube). For other crystalline systems, Callen and Callen (1965) should be referred to. $B_1$ and $B_2$ are the two constants of magnetostriction, which are usually expressed using the dimensionless coefficients $\lambda_{100}$ and $\lambda_{111}$:

$$
\begin{aligned}
B_1 &= -\tfrac{3}{2}(C_{11} - C_{12})\lambda_{100}, \\
B_2 &= -3C_{44}\lambda_{111},
\end{aligned}
\qquad (9.12)
$$

where the $C_{ij}$ are elastic coefficients of the cubic system

$$
\begin{aligned}
\sigma_{xx} &= C_{12}(e_{xx} + e_{yy} + e_{zz}) + (C_{11} - C_{12})e_{xx}, \\
\sigma_{xy} &= \sigma_{yx} = 2C_{44}e_{xy}.
\end{aligned}
\qquad (9.13)
$$

*Table of numerical values*

|    | $\lambda_{100}$ | $\lambda_{111}$ |
|----|---|---|
| Fe | $20.7 \times 10^{-6}$ | $-21.2 \times 10^{-6}$ |
| Ni | $-45.9 \times 10^{-6}$ | $-24.3 \times 10^{-6}$ |

Consideration of material magnetized along a fixed direction $\alpha$, $\beta$, $\gamma$, extending to infinity, not submitted to any external stress field, clarifies the physical significance of these different quantities. The (uniform) deformation of such a medium is calculated by minimizing the density

$$\rho f_{\text{magel}} = \tfrac{1}{2}\sigma_{ij}e_{ij} + \rho f_{\text{ms}},$$

which leads to the expressions:

$$\begin{aligned}
e_{xx}^0 &= \tfrac{3}{2}\lambda_{100}\alpha^2 + K, \\
e_{xy}^0 &= \tfrac{3}{2}\lambda_{111}\alpha\beta,
\end{aligned} \tag{9.14}$$

where $K$ is a constant independent of $\alpha$, $\beta$, $\gamma$, which will be made equal to zero, since it corresponds to a pure dilation of the material. Equations (9.14) inform us that $\lambda_{100}$ and $\lambda_{111}$ are, but to a numerical factor, the deformations along the $\langle 100 \rangle$ directions (if the material is magnetized along $\langle 100 \rangle$) or $\langle 111 \rangle$ (if the material is magnetized along $\langle 111 \rangle$), with respect to a reference state which need not be precisely defined (it depends on $K$) but may be either the statistical state of magnetostrictive deformation when $\mathbf{M}$ takes all possible directions (see Chikazumi, 1966) or the paramagnetic state. The latter point of view is more promising; moving away from the Curie point, the cubic material takes a local symmetry which depends on the direction of local magnetization. As long as the magnetostriction coefficients are small, this change of symmetry can be treated as a perturbation of the cubic crystal, described by the $e_{ij}(\mathbf{r})$ of equation (9.14).

Let us substitute the values of equation (9.14) into $f_{\text{magel}}$. This yields:

$$\rho f_{\text{magel}} = \left(\frac{B_1^2}{C_{11} - C_{12}} - \frac{B_2^2}{C_{44}}\right)(\alpha^2\beta^2 + \beta^2\gamma^2 + \gamma^2\alpha^2). \tag{9.15}$$

The magnetostriction of a uniformly magnetized material is analogous, energetically, to an anisotropy term, which is included in the result of any measurement. $B_1$ and $B_2$ must therefore be measured separately. It appears that in ferromagnets of the type Fe, Co or Ni, the coefficient which appears in equation (9.15) is only a weak part of $K_1$. On the other hand, in rare earth metals, this may be the predominant term.

## 9.2 THEORY OF MAGNETOSTRICTION

The state of distortion described by equation (9.14) is actually an illustration of the Kröner process, the $e_{ij}^0$ being the *plastic* strains $e_{ij}^P$ (here they will be called quasi-plastic, to distinguish them from true plasticity), and the $\omega_{ij}^P$ being zero. The $e_{ij}^0$ are the strains undergone by the paramagnetic phase when, cut into infinitesimal volume elements, it is cooled below the Curie point without variation in the direction of $\mathbf{M}$, which is fixed in advance. Infinitesimal densities of dislocations

$$\alpha_{ij} = -\varepsilon_{ikl}e_{lj,k}^0, \tag{9.16}$$

supplementary strains $e_{ij}^+$, lattice rotations $\omega_{ij}^+$, and internal stresses $\sigma_{ij}^+$ can therefore be associated with the $e_{ij}^0$.

It is clear that in expressions (9.11) and (9.14) of the free energy density the strains of $e_{ij}$ are the sum $e_{ij}^0 + e_{ij}^+$. Let us suppose moreover that the sample is submitted to applied stress $\sigma_{ij}^A$ and internal stress $\sigma_{ij}^D$ (due to lattice dislocations for example). It is necessary to add these terms to equation (9.14). This then yields:

$$
\rho f_{\text{magel}} = \left| \frac{B_1^2}{C_{11} - C_{12}} - \frac{B_2^2}{C_{44}} \right| (\alpha^2 \beta^2 + \beta^2 \gamma^2 + \gamma^2 \alpha^2)
$$
$$
+ \tfrac{1}{2} e_{ij}^+ \sigma_{ij}^+ + \tfrac{1}{2} e_{ij}^A \sigma_{ij}^A + \tfrac{1}{2} e_{ij}^D \sigma_{ij}^D \qquad (9.17)
$$
$$
+ e_{ij}^+ \sigma_{ij}^D.
$$

First of all a term of anisotropy can be recognized in this expression, then three terms describing the energies appropriate to the three systems of stresses and finally a term relating to the interaction of *two* systems of internal stresses. Interaction terms between applied stress and internal stress have not been brought in, as such a quantity is known to have a zero integral over the volume of the sample. In fact we have for example:

$$
\int \sigma_{ij}^+ e_{ij}^A \, dV = \int \sigma_{ij}^+ u_{j,i}^A \, dV = \int \sigma_{ij}^+ u_j v_i \, dS - \int \sigma_{ij,i}^+ u_j \, dV = 0.
$$

This last quantity is zero since the equilibrium of stresses necessitates $\sigma_{ij,i}^+ = 0$ and $\sigma_{ij}^+ v_i = 0$ for internal stresses. The $\sigma_{ij}^+$ are true internal stresses, and by virtue of this formula analogous to those used in Chapter 3 (one constant elasticity for nematics) can be applied to them. In particular it can be shown, (see J. Friedel, 1964) that the interaction energy of a dislocation line $L$, of Burgers vector $\mathbf{b}$, with the system of magnetostriction stresses reads:

$$
W^1 = \int_S b_i \sigma_{ij}^+ \, dS_j. \qquad (9.18)
$$

Peach and Koehler's formula can also be applied to obtain the configurational force exerted on the line by $\sigma_{ij}^+$:

$$
f_i = \varepsilon_{ijk} b_l \sigma_{jl}^+ t_K, \qquad (9.19)
$$

where $\mathbf{t}$ is the unit tangent to the oriented line.

We will deal here with the infinite plane wall and the rectilinear junction of several walls; Bloch and Néel lines will be discussed at the same time as their topology.

### 9.2.1 Infinite plane wall (Rieder, Kléman and Schlenker)

Here we are only interested in the long range effects of the wall. It is therefore supposed infinitely thin. This approximation, which identifies Bloch and Néel walls and does not take into account possible accidents of structure of the wall (Bloch and Néel lines), is true for distances from the wall larger than $\delta_w$.

The magnetostrictive effects of the wall are therefore analogous to those of a surface density of dislocations. Applying formula (8.44) we have:

$$\bar{\alpha}_{1'1'} = -\Delta e^0_{3'1'}; \quad \bar{\alpha}_{3'3'} = -\bar{\alpha}_{1'1'},$$
$$\bar{\alpha}_{1'3'} = -\Delta e^0_{3'3'}; \quad \bar{\alpha}_{3'1'} = +\Delta e^0_{1'1'}, \qquad (9.20)$$
$$\bar{\alpha}_{1'2'} = -\Delta e^0_{3'2'}; \quad \bar{\alpha}_{3'2'} = -\Delta e^0_{1'2'},$$

where $x'_2$ designates the axis perpendicular to the wall and $e^0_{i'j'}$ the variations of $e^0_{i'j'}$ between domain I ($x'_3 < 0$) and domain II ($x'_3 > 0$);

$$\Delta e^0_{i'j'} = e^0_{i'j'}(\text{II}) - e^0_{i'j'}(\text{I}), \qquad (9.21)$$

the $e^0_{i'j'}$ being by definition constant in one domain.

The corresponding $e^+_{ij}$ and $\omega^+_{ij}$ can easily be calculated. If we suppose the sample bound by two planes parallel to the wall, on which the boundary conditions read:

$$\sigma^+_{2'2'} = 0 \qquad (9.22)$$

we get (isotropic elasticity):

$$\Delta e^+_{1'1'} = -\Delta e^0_{1'1'}; \quad \Delta e^+_{2'2'} = \frac{v}{1-v}(\Delta e^0_{1'1'} + \Delta e^0_{3'3'}),$$
$$\Delta e^+_{3'3'} = -\Delta e^0_{3'3'}; \quad \Delta e^+_{1'3'} = -\Delta e^0_{1'3'}, \qquad (9.23)$$
$$\Delta \omega^+_{1'} = \Delta e^0_{3'2'}; \quad \Delta \omega^+_{3'} = -\Delta e^0_{1'2'}.$$

It is recognized in these equations that the stresses are in general discontinuous on the wall. The wall, in the sense of Chapter 8, constitutes a distribution. Using Rieder's terminology, we distinguish walls of type I, which create stress fields displaying discontinuities on the walls and extending to long range, and walls of type II, for which the $\Delta e^+_{ij}$ are zero, but the $\Delta \omega^+_i$ are not, and which are therefore analogous to disorientation boundaries. It is proposed (Kléman and Schlenker, 1972) to call these walls:

(a) Nye walls, for walls of type II; and
(b) Somigliana walls, for walls of type I.

These names and this classification will continue to be used (subsection 9.2.2 will clarify this point). But it will be noted that the walls analysed here enter into the framework of those analysed in subsection 8.2.4 as resulting from discontinuities of plastic distortions (distranslations) and not from the Volterra process.

180° *Wall.* This separates domains oriented in opposite directions; as magnetostriction is quadratic in magnetization, the $e^0_{ij}$ are zero. This wall has no magnetostrictive effects. It is not visible in X-ray topography (Lang method).†

† The principle of the Lang method is as follows: a beam of monochromatic X-rays is shone on the crystal at the Bragg incidence. The intensity of the transmitted diffracted beam depends on the elastic deformations of the crystal; the diffracted beam is registered on a photographic plate on which an 'image' of the crystal appears when the plate and crystal under the fixed incident beam are displaced. Thus, dislocations and domain walls can be seen (see Bowen and Hall, 1975; Tanner, 1976).

250

[110]90° *Wall in iron.* In the axes indicated in Figure 9.7a we have:

$$\Delta e^0_{3'2'} = \tfrac{3}{2}\lambda_{100},$$ (9.24)

all the other components being zero. This yields:

$$\bar{\alpha}_{1'2'} = -\tfrac{3}{2}\lambda_{100}; \qquad \Delta\omega^+_{1'} = \tfrac{3}{2}\lambda_{100}.$$ (9.25)

This can be analysed as a density of edge dislocations along the direction 1', and is therefore again a case of a tilt wall (Nye wall).

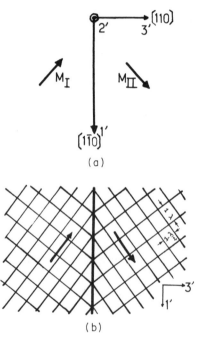

(a)

(b)

Figure 9.7. (a) [110] 90° wall. Choice of coordinates. (b) The 90° wall as a coherent twin boundary

Let us note that a crystal uniformly magnetized along a direction [100] can be analysed as a crystal of tetragonal symmetry. The wall can then be described as a coherent twin boundary in a tetragonal crystal (Figure 9.7b). This is the description adopted by Schlenker *et al.* (1968) and Polcarova and Gemperlova (1969).

### 9.2.1.1 Zigzag decomposition of the 90° wall in iron

[110] 90° walls in iron constitute the walls of Landau closure domains (Figure 9.6). These walls are not in fact plane, but zigzagging, the edges of the zigzags being along straight lines bisecting the directions of magnetization of each side of the wall (Chikazumi and Suzuki, 1955; Kaczer and Zeleny, 1960). These zigzags,

whose periodicity and amplitude are always of the order of 10 to 20 $\mu$m, have also been shown on walls located inside the sample, by Lang's method (Polcarova, 1969). It is clear experimentally that the 90° walls are practically always split into zigzags (Figure 9.8). Various authors (Chikazumi and Suzuki, 1955; Kléman and Schlenker, 1972; Labrune, 1976) have shown that this is an effect of magnetostriction.

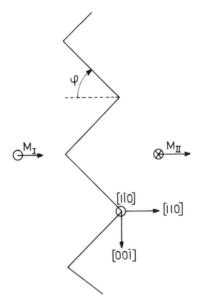

Figure 9.8. Decomposition of a [110] 90° wall into zigzags: scheme

Suppose that the two directions of magnetization [100] and [010] of the domains I and II are given. The projections of the oriented directions [100] and [010] on the direction [11h] are equal: a [11h] wall separating domains I and II, magnetized along directions $M_I = M_S$ [100] and $M_{II} = M_S$ [010], carries magnetic charges $\pm M$. [11h] is equal and opposite, therefore zero in total. The walls of normals [11h] are therefore of zero magnetostatic energy. This condition above all must be satisfied in iron, where this term of energy, of the order of $-\frac{1}{2}H_d \cdot M \sim 2\pi M_S^2$ very much prevails over any other ($2\pi M_S^2/K_1 \sim 100$). The energy of the wall [110] corresponding to $h = 0$ does not include the elastic term (Nye wall); but by comparing the terms of anisotropy and exchange for different values of $h$, we find that these terms attain their maximum values for $h = 0$. For $h \neq 0$, an elastic term appears: the wall is in fact of the Somigliana type. The elastic term has a constant spatial density, and leads to an infinite energy.† If the wall is zigzagging around the mean direction [110], the different elements of walls carry quasi-dislocations whose total stresses equate to zero at long range. The

† It can be relaxed by dislocations of the lattice of modulus equal to that of $\bar{\alpha}_{ij}$ dislocations, but of opposite signs.

total energy can then be lower than that of the plane wall [110]. The complete calculation requires an elastic analysis of each wall element bounded by the edges of the zigzag. In accordance with the discussion in the preceding chapter, the edges play the role of quasi-rotation dislocations. On the other hand, magnetostatic terms necessarily appear on the edges, which carry space charges div **M**. A complete analysis of the problem can be found in Labrune (1976), from whom Figure 9.9 is borrowed. Labrune finds a total energy of 0.96 erg per $cm^2$ of plane [110], an angle of $\phi_0 \sim 65°$ and a periodicity of $d_0 \sim 16\,\mu m$, in the most favourable energy situation. These values are very close to experimental values.

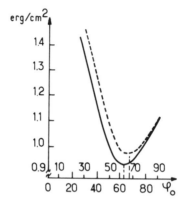

Figure 9.9. Variation of energy of a [110] 90° wall in iron. Solid line: $\gamma/\sin\phi$; $\phi$: orientation of the zigzag segments with respect to the plane $\{110\}$; $\cos^2\phi = h^2\cdot/2 + h^2$). $\gamma$ takes into account anisotropy and exchange only. Dotted line: total energy

For $h \to \infty$, the wall tends towards the [110] plane. This Somigliana wall is of minimal $\gamma(h)$. Its energy can still be reduced by giving it a zigzag structure which reduces the magnetoelastic energy at the expense of $\gamma(h)$. However, the zigzagging [001] wall retains a mean *Somigliana* characteristic (while the zigzag [110] wall keeps a mean *Nye* characteristic), and such walls can therefore be imagined only in finite media. It is probable that they have been observed in samples of iron undergoing bend. (Labrune and Kléman, 1974). Their detailed analysis can be found in Labrune (1976).

### 9.2.1.2  *Periodic domains and one-dimensional configurations*

Periodic arrays of domains bounded by plane parallel walls are frequently found. They have been studied particularly in iron (Bourret and Dautreppe, 1966; Bourret and Kléman, 1967), in iron and nickel (de Blois, 1965, 1968), in cobalt and strongly anisotropic uniaxial materials (Kooy and Enz, 1960; Jakubovicz 1966; the Czech school of Kaczer *et al.*). These walls are very frequently Somigliana walls, whose stresses are relaxed by image forces, but above all because the successive walls are of opposite signs, therefore the total stresses equate to zero at long range. These one-dimensional situations can be calculated exactly (if the sample is supposed infinite) on the basis of Brown's (1965) equations of micromagnetism, which are obtained as follows. Start from the free energy:

$$\rho f = \rho f_K + \rho f_{Ex} + \rho f_{magel} \tag{9.26}$$

written here in the absence of applied fields and space charges. This energy depends on two series of variables, those of magnetization $\alpha, \beta, \gamma$, and those of distortion $\beta_{ij}$. The minimization of $\rho f$ with respect to the elastic variables leads to the equilibrium equations:

$$\sigma_{ij,i}^+ = 0, \tag{9.27}$$

which have already been written and discussed. The minimization with respect to the magnetization variables leads to Brown's equations:

$$2A\nabla^2\alpha - 2K_1\alpha(\beta^2 + \gamma^2) + g\alpha = 2\{B_1\alpha e_{xx} + B_2(\beta e_{xy} + \gamma e_{xz})\}, \tag{9.28}$$

the other equations being obtained by circular permutation. $g$ is a Lagrange multiplier relating to the condition $\alpha^2 + \beta^2 + \gamma^2 = 1$.

Systems (9.27) and (9.28) are coupled; but in the one-dimensional case equation (9.27) can be solved independently of equation (9.28). We have, by supposing that the variation occurs along $Oz$,

$$e^+ = \begin{pmatrix} -e_{11}^0(Z) & -e_{12}^0(Z) & 0 \\ -e_{12}^0(Z) & -e_{22}^0(Z) & 0 \\ 0 & 0 & \dfrac{v}{1-v}(e_{11}^0 + e_{22}^0) \end{pmatrix} + \varepsilon_c, \tag{9.29}$$

where $\varepsilon_c$ is a constant tensor depending on the boundary conditions. From these equations (9.29) one can find again the results for the plane wall (equation (9.23)). In this case the choice of $\varepsilon_c$ is immediate ($\varepsilon_c = 0$). But the case of a periodic medium is more delicate. We can then use the property of internal stresses of having a zero mean value (see Friedel, 1964, p. 452).

$$\int \sigma_{ij}^+ \, dV = 0. \tag{9.30}$$

In the periodic case this is reduced to an integration over a period. Kléman (1969) has made a calculation in this way for a system of periodic walls in iron and nickel. The results indicate a strong sensitivity to the constants of magnetostriction of the domain widths and of the angle swept by $\mathbf{M}$ on either side of the walls ($\mathbf{M}$ does not necessarily reach the easy direction). It would be useful to take up these calculations again in precise experimental cases.

### 9.2.2 Wall junction; finite plane wall (Kléman, 1974a)

Magnetic walls are distranslations. Putting

$$E_{ij}^0 = +\Delta e_{ij}^0 \tag{9.31}$$

they correspond, when they are limited, to the distribution of defects:

$$\alpha_{ij} = \varepsilon_{ikl} E^0_{ij} \delta_k(S),$$
$$\theta_{pq} = -\tfrac{1}{2} \varepsilon_{jpq} E^0_{lj} \delta_l(L), \tag{9.32}$$

formulae which reproduce (except for the sign) equations (8.57) and (8.58). Stress fields and distortion fields can then be calculated from the incompatibilities

$$\eta_{ij} = -\tfrac{1}{2} \varepsilon_{ikp} \varepsilon_{jql} E^0_{pl} \{\delta_{k.q}(S) + \delta_{q.k}(S)\}. \tag{9.33}$$

These are distranslations of a reference state defined 'before magnetization'. We understand by this that the $E^0_{ij}$ are applied to a paramagnetic state; there is a constant dilation term which is not taken into account.

The idea that these walls are *distranslations* neglects an essential part of the physical reality. It does not allow us to distinguish between Nye walls (which are coherent twin boundaries of the state 'after uniform magnetization') and Somigliana walls (which are incoherent twin boundaries in the same state.) Nye walls can be considered as diremptions, rather than particular distranslations, and it is important to distinguish in Somigliana walls a differential rotation effect from one domain to the other, and an effect of internal stresses.

Furthermore, the analysis of *wall junctions* shows that this is a case of particular distranslations. Remember that in the crystal under consideration, lattice dislocations (translation or rotation) do not exist, but only walls and wall junctions. There are therefore two compatibility conditions to fulfill on the junctions: (a) concerning rotations and (b) concerning translations. These conditions are well known in grain boundaries (Amelinckx, 1957). In the language of distranslations, these conditions read:

$$\sum_a \theta^{(a)}_{pq} \delta_p(L) = 0, \tag{9.34a}$$

where $\theta^{(a)}_{pq}$ represents the density of rotation defects in the wall $(a)$ ending on $L$. Equation (9.34a) says that the junction is not a rotation dislocation line of the lattice.

$$\varepsilon_{ipq} \sum_a \theta^{(a)}_{pq} = 0, \tag{9.34b}$$

i.e.

$$\sum_a E^{0(a)}_{li} \delta_i(L) = 0.$$

Equation (9.34b) is equivalent to the condition $\sum \alpha^{(a)}_{ij,i} = 0$.

It is clear that if condition (9.34b) is obeyed, condition (9.34a) is automatically obeyed, since it brings about $\sum_a \theta^{(a)}_{pq} = 0$. The junction line $L$ must be along an eigen direction of the symmetrical tensor $\sum_{(a)} E^{0(a)}_{ij}$. It is therefore a straight line, or an array of segments of a straight line. Only the very particular case when $\sum_{(a)} \mathbf{E}^0$ is degenerate produces curved lines. Therefore if the junctions are curved, it must be admitted that translation dislocations of the lattice end there. Condition (9.34a) is then not automatically verified, and it is possible that rotation dislocations of the lattice are coupled to $L$. But although these new possibilities are interesting, they will be left aside.

Conditions (9.34a) and (9.34b) will be expressed in the language of coherent or incoherent diremptions. To do this, as suggested by formulae (9.23), two parts can be distinguished in $E_{ij}^{0(a)}$:

and
$$E_{ij}^{(a)} = -\Delta e_{ij}^{+(a)}$$
$$\Omega_i^{(a)} = -\Delta \omega_i^{+(a)} \tag{9.35}$$

(see equations (9.31) and (9.32)). Let us note that formulae (9.32) constitute a similar (but not identical) decomposition.

When different walls join, the total distortion is not the sum of distortions (9.35) and quasi-plastic distortions $e_{ij}^{0(a)}$. Let us forget the latter for the moment. It can be considered that the supplementary distortions $e_{ij}^+, \omega_i^+$ which we are seeking relax the distortions (9.35). We may indeed write:

$$\alpha_{ij} = \sum_{(a)} \alpha_{ij}^{(a)} = -\varepsilon_{ikl} \sum_{(a)} E_{lj}^{0(a)} \delta_k(S_a)$$

$$= -\varepsilon_{ikl} \sum_{(a)} E_{lj}^{(a)} \delta_k(S_a) + \sum_{(a)} \Omega_i^{(a)} \delta_j(S_a) - \delta_{ij} \sum_{(a)} \Omega_k^{(a)} \delta_k(S_a), \tag{9.36}$$

where the expected decomposition appears on the right-hand side. As on each wall $(a)$ we have $\Omega_k^{(a)} \delta_k(S_a) = 0$ ($\boldsymbol{\Omega}^{(a)}$ is in the plane of the wall in a cubic ferromagnet), we may also write:

$$\alpha_{ij} = \sum_{(a)} \Omega_i^{(a)} \delta_j(S_a) - \varepsilon_{ikl} \sum_{(a)} E_{lj}^{(a)} \delta_k(S_a). \tag{9.37}$$

Condition (9.34b) reads here:

$$\boldsymbol{\Omega} \wedge \mathbf{L} + \sum \mathbf{E}^{(a)} \cdot \mathbf{L} = 0, \tag{9.38}$$

where $\boldsymbol{\Omega}$ designates the total rotation vector and $\mathbf{L}$ a unit vector tangential to $L$:

$$\Omega_i = \sum_{(a)} \Omega_i^{(a)} \tag{9.39}$$

Finally, condition (9.34a) ($\theta_{ij} = 0$) can always be achieved because $\alpha_{ij}$ satisfies the relations (8.51) as soon as relation (9.38) is obeyed.

The stress sources which appear on the merging of the $(a)$ domains can therefore be separated into two series of quite distinct physical meanings. Let us put

$$\alpha_{ij} = \alpha'_{ij} + \alpha''_{ij}; \qquad \theta_{ij} = 0 = \theta'_{ij} + \theta''_{ij}.$$

### 9.2.2.1  Nye walls

$$\alpha'_{ij} = \sum_{(a)} \Omega_i^{(a)} \delta_j(S_a),$$

$$\theta'_{ij} = \tfrac{1}{2} \sum_{(a)} \{ \Omega_i^{(a)} \delta_j(L) - \Omega_j^{(a)} \delta_i(L) \}. \tag{9.40}$$

This distribution is analogous to that of limited diremptions (equations (8.52) *et seq.*), but differs from it in that in $\theta'_{ij}$ only the antisymmetrical part of the 'density of rotation dislocations' tensor of the limited diremption is retained. We then have:

$$\eta'_{ij} = +\tfrac{1}{2}\{\Omega_i\delta_j(L) + \Omega_j\delta_i(L)\}. \tag{9.41}$$

Contrary to the limited diremption (which is also, as we have emphasized, de Wit's compensated disclination line), the set of defects described by (9.40) gives a non-zero field of distortions. But there is an advantage in bringing about the decomposition (9.40) rather than the limited diremption:† in fact, if the components of the $\mathbf{E}^a$ are all zero, i.e. if all the walls constituting the junction are of the Nye type, then obviously $\alpha_{ij} = \alpha'_{ij}$ and $\theta'_{ij} = 0$. The distribution (9.40) therefore contains all the information on the rotations from one part of the wall to the other, and their role in final distortions and final rotation gradients. These are discontinuous on $S_a$.

### 9.2.2.2 *Somigliana wall*

$$\alpha''_{ij} = -\varepsilon_{ikl}\sum_{(a)} E^{(a)}_{lj}\delta_k(S_a),$$

$$\theta''_{ij} = -\tfrac{1}{2}\varepsilon_{pij}\sum_a E^{(a)}_{lp}\,\delta_l(L) = \tfrac{1}{2}\sum_a \{\Omega_j\delta_i(L) - \Omega_i\delta_j(L)\}. \tag{9.42}$$

The equality between the two expressions of $\theta''_{ij}$ is demonstrated using (9.38). This is the distranslation studied in the preceding chapter (equation (8.57) *et seq.*). We have:

$$\eta''_{ij} = \tfrac{1}{2}\varepsilon_{ikp}\varepsilon_{jql}\sum_a E^{(a)}_{pl}\{\delta_{k,q}(S_a) + \delta_{q,k}(S_a)\}. \tag{9.43}$$

This distribution of incompatibilities gives rise to discontinuities of the $e_{ij}$ on $S_a$.

### 9.2.3 Formulae showing the deformations and gradients of rotation for finite plane walls

These formulae are taken from Kléman (1974a)

#### 9.2.3.1 *Wedge disclination*

This is the case when $\theta'_{ij} = \theta''_{ij} = 0$. It corresponds to the physically important case of the junction of Nye walls. We have, in polar coordinates:

$$e_{\rho\rho} = \frac{\Omega_3}{4\pi(1-v)}\{(1-2v)(\ln\rho + 2) - 1\},$$

---

† In our publication (Kléman, 1974a) we used a decomposition where one of the terms is a limited diremption. The decomposition used here seems more natural.

$$e_{\theta\theta} = \frac{\Omega}{4\pi(1 - v)}(1 - 2v)(\ln \rho + 2),$$

$$K_{\theta 3} = -\sum_a \Omega_3^{(a)}\delta_\theta(S_a) + \frac{\Omega_3}{2\pi\rho}.$$

To obtain the total strains computed with respect to the reference state 'before magnetization', it is convenient to add to these quantities the quasi-plastic strain $e_{ij}^0$. The total quantities $e_{ij} + e_{ij}^0$ are compatible. A displacement function can then be chosen which has been calculated by Miltat and Kléman (1973) for the classical $Y$ junction of iron (Figure 9.10) formed by a 180° wall and two 90° walls. These quantities are discontinuous on the walls: the walls can be seen by Lang's method because of the discontinuity in the rotation gradients. The rest of the contrast is due to global displacements resulting from the $e_{ij}$ and $e_{ij}^0$.

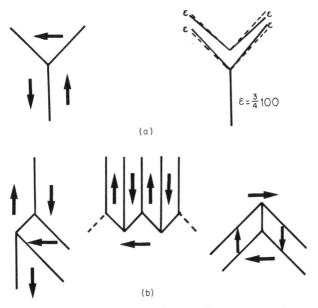

Figure 9.10. (a) $Y$ junction in iron. $\Omega_3 = -3\lambda_{100}$. (b) Typical associations of junctions of opposite signs

The energy of the zone of radius $\rho_0$ around the junction is of the order of (see Friedel, 1964):

$$W = \frac{\mu\Omega_3^2\rho_0^2}{16\pi(1 - v)} \tag{9.45}$$

(within the present case $\Omega_3 = \pm 3\mu\lambda_{100}$). This is a very considerable energy. This explains why the junctions are often observed by pairs of opposing rotation vectors (Figure 9.10).

If $\Omega_3 = \sum_a \Omega_3^{(a)}$ vanishes, the junction does not create any elastic strain. There only remain discontinuities in the rotations. It is clear that a Néel line on a *plane* Bloch wall is a junction of this type. It cannot therefore be analysed by the present formalism, which does not take into account the distribution of magnetization across the wall. But it is clear that the strains created by such an object can only extend over a short distance (of the order of $\delta_w$).

### 9.2.3.2 Twist disclination

The formulae for a half-wall $(a)$, bounded along the line $L$, taken as the $Oz$ axis, are given here. The axes are those of Figure 9.11. To work out these formulae, we have used de Wit's (1973c) calculations. We suppose $\Omega_3^{(a)}$ to be zero:

$$e_{11}^{(a)} = \frac{\phi^{(a)}}{2\pi} E_{11}^{(a)} - \frac{2}{4\pi v} \sin \phi^{(a)} \cos \phi^{(a)} E_{22}^{(a)},$$

$$e_{22}^{(a)} = \frac{\phi^{(a)}}{2\pi} E_{22}^{(a)} + \frac{1}{4\pi v} \sin \phi^{(a)} \cos \phi^{(a)} E_{22}^{(a)},$$

$$e_{33}^{(a)} = \frac{\phi^{(a)}}{2\pi} E_{33}^{(a)},$$

$$e_{12}^{(a)} = \frac{1}{4\pi} (E_{11}^{(a)} - E_{22}^{(a)})(1 + \ln \rho) - \frac{E_{22}^{(a)}}{8\pi v} \cos 2\phi^{(a)}, \quad (9.46)$$

$$e_{23}^{(a)} = \frac{E_{31}^{(a)}}{4\pi} (1 + \ln \rho),$$

$$e_{31}^{(a)} = \frac{\phi^{(a)}}{2\pi} E_{31}^{(a)} - \frac{\Omega_1^{(a)}}{4\pi} (1 + \ln \rho);$$

$$K_{11}^{(a)} = \frac{1}{2\pi\rho} (\cos \phi^{(a)} E_{31}^{(a)} - \sin \phi^{(a)} \Omega_1^{(a)}),$$

$$K_{21}^{(a)} = \frac{1}{2\pi\rho} (\sin \phi^{(a)} E_{31} + \cos \phi^{(a)} \Omega_1^{(a)}),$$

$$K_{13}^{(a)} = -\frac{1}{2\pi\rho} \cos \phi^{(a)} E_{11}^{(a)}, \quad (9.47)$$

$$K_{23}^{(a)} = -\frac{1}{2\pi\rho} \sin \phi^{(a)} E_{11}^{(a)},$$

$$K_{31}^{(a)} = -\Omega_1^{(a)} \delta_3(S_a).$$

The existence in the $e_{ij}^{(a)}$ of a discontinuity on $S_a$ and in the $K_{ij}^{(a)}$ of a discontinuity $\Omega_1$ in the rotation on either side of $S_a$ will be noticed.

This formalism has been applied by Miltat (1976a, b) to the study of *fir-tree domains* in samples of iron–silicon oriented approximately along a plane (100) (Figure 9.12). The bottom of these domains is made of a curved Somigliana wall

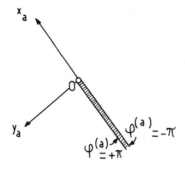

Figure 9.11. Coordinate axes describing the half-wall (*a*) ending on the junction *Oz*

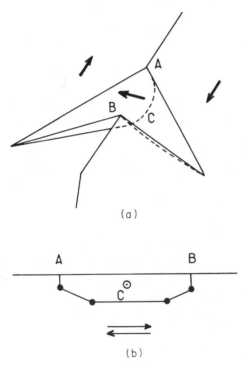

(a)

(b)

Figure 9.12. (a) Fir-tree patterns; (b) elastic outline (J. Miltat, 1976b)

which Miltat analysed as an array of wall segments joined by twist quasi-disclinations. Miltat made a complete calculation of the strain field of these walls, taking image forces into account, and he showed that the field of displacements which results from it accounts reasonably well for the contrast in Lang's method.

This study, and the preceding one on *Y* junctions, constitute the first experimental observation of the deformation fields of an elastic rotation dislocation.

### 9.2.4 Sinusoidal antiferromagnets and helimagnets

These materials are not uniformly magnetized in their ground state. Thus, the phenomenological theory of magnetostriction leads to assigning internal stresses to the perfect crystal. These stresses have the periodicity of the spin lattice and are at the origin of a periodic modulation of the parameter of the lattice (Nourtier, 1973).

## 9.3 TOPOLOGICAL AND ENERGETIC STUDY OF THE SINGULARITIES IN THE SPIN LATTICE

### 9.3.1 Topology of ferromagnets

The majority of known ferromagnets have *discrete* easy magnetization axes (direction $\{100\}$ for iron, $c$ axis for cobalt and rare earth metals). This implies that the fundamental defect of the medium is a wall (Bloch or Néel as the case may be). Gadolinium is an exception: in the temperature range 0 to 240 K (Corner and Tanner, 1976) its anisotropy is conic, and the easy axis makes an angle $\theta_0(T)$ with the $c$ axis, the azimuthal angle not being fixed. In this case, expected defects are *walls* and *disclinations* in the spin lattice, the latter corresponding to rotational symmetries multiple of $2\pi$ about the $c$ axis for the basal non-zero component of magnetization. Finally, amorphous ferromagnets with zero anisotropy should be introduced here: their essential defects are singular points.

This classification of defects as a function of the symmetries of the magneto-crystalline anisotropy will be clarified in Chapter 10: here we have only discussed defects which are *topologically stable* with regard to angular variations of magnetization which take place without modification of the anisotropy energy. In fact, the other energy contributions (stray-fields, magnetostriction, etc.) lead to the consideration in each case of other stable defects. They can generally be linked either to new symmetries introduced by core structures of topologically stable defects (Bloch or Néel lines in materials with discrete anisotropy, singular points in the same materials and certainly also in materials with conic anisotropy), or to surface effects (disclination line and walls in amorphous materials, etc.). These various phenomena will be described first.

#### 9.3.1.1 *Bloch wall or Néel wall*

Along the length of the same wall there may be Bloch segments and Néel segments. The transition is made through a disclination of strength $|\frac{1}{2}|$ (Figure 9.13); this does not contradict the fact that only disclinations of integral strength are allowed, since it must be considered here that this is a defect built in the *helimagnetic* inner region of the wall, which has a rotation symmetry $\pm n\pi$ along every direction of spin. The line is therefore a $\lambda(\pm\frac{1}{2})$ for the Bloch part, or a half $S = \pm 1$ for the Néel part, and has no core singularity. A Néel segment can therefore be transformed continually into a Bloch segment.

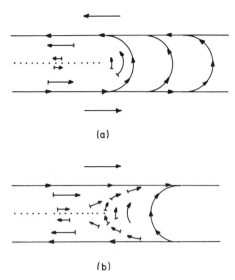

(a)

(b)

Figure 9.13. Two possibilities of transformation from a Néel segment to a Bloch segment:
(a) $\lambda(+\frac{1}{2})$; (b) $\lambda(-\frac{1}{2})$

### 9.3.1.2 *Bloch lines*

These are rotation dislocations of 180° Néel walls. They can be created from the preceding model by reducing the Bloch segments to zero. Circular Bloch lines (wedge $S = 1$) and crossed Bloch lines (wedge $S = -1$) can be recognized (Figure 9.4) (see Huber *et al.*, 1958; de Blois and Graham, 1958). Bloch lines on 90° walls have not been observed.

### 9.3.1.3 *Néel lines*

These are rotation dislocations of 180° Bloch walls (for various observations see Williams and Goertz, 1952; de Blois and Graham, 1958; Carey and Isaac, 1966; Schön and Buchenau, 1972, have studied the anisotropic coupling of the direction of the line with the crystal). Nabarro (1972) has studied their topology from a very general point of view. Figure 9.14 indicates two possible configurations for these lines, without core singularities.

### 9.3.1.4 *Other disclination lines*

Bloch (or Néel) lines are found in walls; but isolated disclination lines can also exist: for example in gadolinium with conic anisotropy and (less probably) in amorphous materials. Disclination lines can be classified by their strength (always integral), the sign of magnetization (see Figure 9.15) and the singular or non-singular character of their core.

262

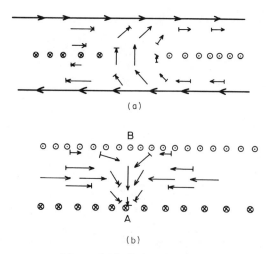

(a)

B

A

(b)

Figure 9.14. Twist Néel lines

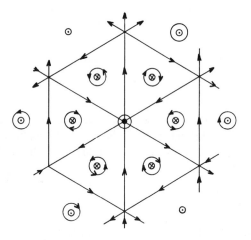

Figure 9.15. Hypothetical distribution of wedge lines in gadolinium, allowing a regular filling of the plane, and weak magnetostatic and magnetostrictive energies. We have supposed that there is no applied field: the $c$ axis is perpendicular to the diagram. The magnetization passes from a direction $C$ (at the core of the non-singular $S = 1$ lines) to a direction situated on the cone of the easy directions (projected here along the sides of the equilateral triangles in the diagram). The apexes of these triangles represent $S = -2$ lines

### 9.3.1.5 Singular points

Singular points on Bloch or Néel lines are similar to singular points on lines without a singular core in nematics (Chapter 3). These points have been described by Feldtkeller (1964, 1965) and seem to have been observed by Harrison and Leaver (1973), using geometries of very thin specimens (therefore they observe surface lines rather than wall lines). Döring (1968) described singular points within the limit where anisotropy is zero (amorphous): these are therefore topologically stable defects.

### 9.3.1.6 *Surface lines*

The intersection of a 180° wall with a surface must be considered as a disclination (imperfect because the surface section is marked by the wall) of order $\pm\frac{1}{2}$. The rearrangements of spin in the vicinity of the surfaces are essentially due to magnetostatic effects. By supposing div $\mathbf{M} = 0$, Hubert has shown that the configurations are antisymmetric. More precise calculations by La Bonte (1969) do not give significantly different results. The verifications made by Harrison and Leaver on thin specimens (observed by electron microscopy) confirm these models.

Recent studies of charged walls in implanted bubble materials have shown that these walls are in fact highly charged wedge surface lines, located in the implanted zone of strength $|S| = \frac{1}{2}$. The sign of the line is directly imaged by the interaction of the charges with the Bitter powder deposited on the specimen: they repel or attract the magnetic colloid according to their sign. It seems that these lines play a role in the mobility of the bubbles by a process in which the $S = \frac{1}{2}$ lines surround the bubbles (like a lassoo) and drive them away (Kléman and Puchalska, 1980).

### 9.3.1.7 *Wall junctions*

It is always possible to consider the junction of Bloch walls as a $\lambda(-\frac{1}{2})$ split in imperfect disclinations along the walls (Figure 9.16a). Through *conicization* in the direction of $\mathbf{M}$ around the normal to the wall, the situation without anisotropy (Figure 9.16a) changes into the situation characteristic of a $Y$ junction (Figure 9.16b). The core then has the topology of a $\lambda(+\frac{1}{2})$ which is transformed continually at long range into that of a $\lambda(-\frac{1}{2})$. In the same way the core in Figure 9.16c is of the type $\lambda(+1)$ while at long range we have $\lambda(-1)$.

Pairings of lines of opposite signs have been observed in striped domains. This gives rise to edge dislocation configurations (Figure 9.17) (see Hirth and Wells, 1970; Bourret and Kléman, 1967).

### 9.3.2 Lines in magnetically hard materials

Materials such as magnetoplumbite, orthoferrite and uniaxial garnets have strong uniaxial anisotropies. They are characterized by a quality factor $q = K/2\pi M_s^2$ (ratio of the effects of anisotropy and stray-fields) which is large compared to unity. In samples cut along two faces perpendicular to the easy axis, the magnitude of anisotropy prevents the formation of domains of closure. *Bubble materials* are obtained in this way (Kooy and Enz, 1960; Bobeck and Della Torre, 1975). These materials have domains in the form of cylinders parallel to the easy axis. The dynamics of these bubbles under an applied field have been closely studied, and Néel lines which restrain their movement (hard bubbles) have

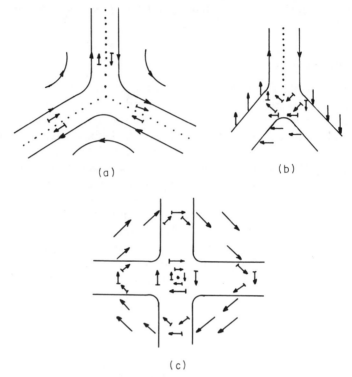

Figure 9.16. Junction of Bloch walls. (a) Without anisotropy. (b) With the introduction of anisotropy: the core is transformed into a $\lambda(+\frac{1}{2})$. (c) Junction of four walls

been shown to exist. The properties at equilibrium are determined by two lengths $\delta_w = (A/K)^{1/2}$ and $\delta_s = (A/2\pi M_s^2)^{1/2}$. The magnetostatic effects are weak $(\delta_s \gg \delta_w)$, and the effects of exchange determine entirely the interactions between lines (Figure 9.18); they have been studied by Slonczewski and Malozemoff (1973) and Thiele (1974) among others.

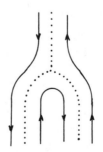

Figure 9.17. Pairing of two $\lambda$ of opposite signs

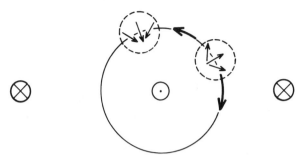

Figure 9.18. Scheme of a bubble. The arrows indicate the direction of magnetization in the middle of the wall. Néel lines are shown as dashed circles

### 9.3.3 Lines in magnetically soft materials (q ≪ 1)

#### 9.3.3.1 *Bloch lines wedge*

Exchange energy being the dominant term in magnetism, it is reasonable to look for configurations of strong exchange energy, then to deal with other terms as perturbations, taking into account their relative importance.

Exchange introduces into the free energy a term resembling that studied in nematics. It can therefore be concluded that, except in the vicinity of the Curie point, non-singular core structures are favoured. There is therefore a line energy of the order of (see equation (3.49): note the equivalence $K_1 \leftrightarrow 2A$):

$$W_{Ex} = 4\pi|S|A. \tag{9.48}$$

Magnetostatic charges must be totally avoided in soft materials. This virtually forbids the presence of radial $S = 1$ lines. On the other hand, *circular Bloch lines* ($S = 1$), which are solenoidal (div $\mathbf{M} = 0$), are favoured. Their size is therefore determined by the competition between the anisotropy energy and exchange energy, i.e. is of the order of $\delta_w$ if $K \neq 0$.

Cross Bloch lines ($S = -1$) are not solenoidal, but the total magnetization is zero. The radius is therefore essentially determined by an interaction between the effect of stray-fields (quadrupolar) and the exchange effect, and is therefore of the order of $\delta_s$.

These order of magnitude arguments lead to the same results as those obtained by more complete calculations (Feldtkeller and Thomas, 1965; Baürich, 1966).

Terms of magnetostriction have still not been taken into account. Consider the case of a planar circular Bloch line (with singular core). A calculation using equation (9.2) leads to a density of quasi-dislocations:

$$\alpha_{3\theta} = -\frac{3}{2r}\lambda_S \tag{9.49}$$

in cylindrical coordinates. ($\lambda_S$ is a coefficient of isotropic magnetostriction.) $\alpha_{3\theta}$ can be interpreted as a *density* of tilt walls ending on the line, and giving rise to a total rotation vector

$$\Omega_3 = \int \alpha_{3\theta} r \, d\theta = -3\pi\lambda_S. \tag{9.50}$$

The stresses due to the dislocation density $\alpha_{3\theta}$ become singular at the origin; it is clear that their contribution to the total energy is reduced by making the core non-singular. But fundamentally the line behaves like a wedge disclination. If $r_0$ is the radius of the extension of the line, the appropriate energy of this central region is of the order of

$$W_{\text{me}} = \mu r_0^2 \Omega_3^2. \tag{9.51}$$

This quantity is low compared to $W_{\text{Ex}}$ as long as $r_0$ is less than the characteristic length

$$\delta_{\text{me}} = \frac{1}{\lambda_s} \left(\frac{A}{\mu}\right)^{1/2} \gg r_0. \tag{9.52}$$

The study of a cross Bloch line leads to analogous results. The density $\alpha_{ij}$ is certainly more complex, but as long as $\delta_{\text{me}} \gg \delta_s$, a characteristic dimension of a cross line, the approximation remains valid.

This discussion assumes the geometries around the line to be perfectly cylindrical and of well-determined dimensions. In fact, it is convenient to distinguish two situations, which can be characterized using a parameter

$$p = \frac{\delta_w^2}{\delta_{\text{me}}^2} = \mu \frac{\lambda_s^2}{K}.$$

(a) $p > 1$. The walls are very wide and have no physical meaning in the limit $K = 0$. It is therefore magnetostriction which controls the size of circular lines. We have $r_0 \sim \delta_{\text{me}}$. Cross Bloch lines are of size $\delta_s$ (we suppose $\delta_s < \delta_{\text{me}}$). It can be imagined that the lines are assembled regularly paving the plane with square arrays of size $\delta_{\text{me}}$, the apexes being circular lines, the centres cross lines.

(b) $p < 1$. Walls have a physical meaning. The location of the lines on the walls has the effect of screening at short range the stress sources located on these lines, by stress sources of opposite signs, carried by the edges of the walls. Expression (9.49) is then no longer valid, and it is necessary to repeat the calculation using a model of the type shown in Figure 9.19, where the rapid variation of the curvature of the lines of force of magnetization on the axis $x = 0$ leads to sources of stress of a new type. They can be evaluated by supposing that this is an abrupt variation of $\mathbf{M}$ along an element of the wall along $Oy$, the lines of force of $\mathbf{M}$ being circles on either side of $Oy$. This yields:

$$\alpha_{31} = \frac{3\pi}{4d} \lambda_s \sin \frac{\pi y}{d} - \tfrac{3}{4}\lambda_s \sin \frac{\pi y}{d} \delta(x),$$

$$\alpha_{32} = \frac{3\pi}{4d} \lambda_s \cos \frac{\pi y}{d} \{U(x) - U - x)\}, \tag{9.53}$$

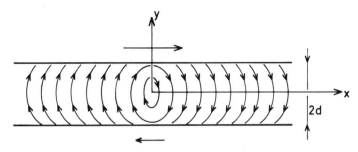

Figure 9.19. Model of circular Bloch line used to calculate the magnetostriction stresses

where $U(x)$ is the unit step function ($U(x > 0) = 1$; $U(x < 0) = 0$). $\alpha_{31}$ and $\alpha_{32}$ are zero outside the wall.

Stresses at long distance can be completely evaluated. The method (Kléman, 1976b) consists in splitting each half-wall on either side of $Oy$ into elements of walls of infinitesimal thickness $dy_e$, centred on $y_e$, constituting stress sources $\alpha_{31}(y_e)dy_e$ and $\alpha_{32}(y_e)dy_e$. These are densities of Somigliana walls. But it is clear that here the stress source is contained in the Dirac distribution of $\alpha_{31}$; it is a dislocation dipole situated at $x = 0$ which can be evaluated by integrating $\alpha_{31}$ over a half-thickness. This yields:

$$b_x = -\tfrac{3}{4}\frac{d}{\pi}\lambda_s. \tag{9.54}$$

With $d \sim 10^{-5}$ cm and $\lambda_s \sim 10^{-5}$, we obtain $b_x \sim 10^{-9}$ cm; i.e. a value only ten times less than that of a Burgers vector of a lattice dislocation. The linear tension, by using the Nabarro (1952) formula for a dipole, is

$$W_{\text{me}} = \frac{9\mu}{32\pi^2}\frac{d^2\lambda_s^2}{1-v}\left(\ln\tfrac{4}{3}\frac{\pi}{\lambda_s} - 1\right). \tag{9.55}$$

The stresses decrease as a function of $r^{-2}$. The successive Bloch lines, whether cross or circular, act as dipoles of opposite signs. We must take $d = \delta_w$ for circular lines, $d = \delta_s$ for cross lines.

### 9.3.3.2 Néel lines (twist)

The first calculations (Shtrikman and Treves, 1960a, b; Janak, 1966) did not take magnetostriction into account, refer to magnetically soft materials or suppose that the wall is essentially plane. Recent observations (Puchalska and Sadoc, 1976) in amorphous materials seem to indicate that zigzag walls can be controlled by magnetostatic and magnetostrictive interactions between the Néel lines of the edges (Figure 9.20). These zigzags have also been observed in thick samples (Craik and Tebble, 1956; see Duda et al., 1974; with regard to observations on monocrystalline iron and yttrium YiG garnets). By splitting the half-walls into a

268

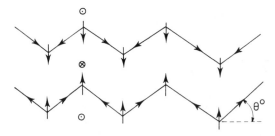

Figure 9.20. Zigzag 180° wall: the magnetization in the middle of the wall is shown. The rapid variations of magnetization at the edges of the wall create stresses and long-range lattice curvature if $\theta_0 \neq 0$

density of Somigliana walls (Kléman, 1976b), we arrive at the following strain field (cylindrical coordinates):

$$e_{11} = \frac{3}{2}\frac{\lambda_s d}{\pi}\sin\theta_0\,\frac{\sin\phi}{r} = -e_{22},$$

$$e_{33} = 0,$$

$$e_{12} = -\frac{3}{4\pi}\lambda_s d\sin\theta_0\,\frac{\cos\phi}{r},$$

$$e_{31} = -\frac{3\lambda_s d^2}{2\pi^2}\frac{1}{\cos\theta_0}\frac{\sin 2\phi}{r},$$

$$e_{32} = +\frac{3\lambda_s d^2}{4\pi^2}\frac{1}{\cos\theta_0}\frac{\cos 2\phi}{r^2},$$

(9.56)

where $d$ is the thickness of the wall.

It can be seen that for $\theta_0 = 0$ (plane wall) the only existing strains are in $r^{-2}$ (the dipolar term). But the zigzag wall has terms in $r^{-1}$ analogous to those created by a dislocation. They cannot therefore be neglected, even for $\lambda_s$ as small as those found in YiG, amorphous materials or iron. Probably their importance is essential in the case of strongly magnetostrictive materials such as rare earth metals.

Note that for $\theta_0 = \pi/2$, the dipolar term becomes catastrophic: a double wall with a Néel line at the end is unstable under magnetostriction. A relaxation process consisting of the formation of a *bubble* at the end of the double wall can be imagined (Figure 9.21). It seems that the process has been observed and can in certain cases be at the origin of the formation of *hard* bubbles, i.e. containing Néel lines (see Grundy et al., 1972).

### 9.3.4 Singular points

Feldtkeller and Döring noted that, in spite of the singularity of magnetization at a singular point, the exchange energy does not diverge. The results obtained here

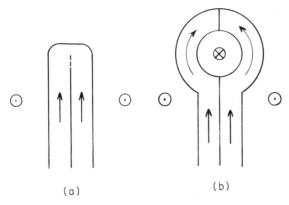

Figure 9.21. Swelling of the magnetostrictive origin at the end of a double wall facilitates the formation of hard bubbles

are the same as those obtained for nematics:

$$W = 8\pi A |S|. \tag{9.57}$$

### 9.3.5 Topology of helimagnets (Kléman, 1970)

Helimagnets, apart from their symmetries, must have the same types of singularities as cholesterics. But they differ from them in three ways:

(a) here the periodicity is equal to the pitch (and not the half-pitch);[†]
(b) the two chiralities are equally probable; and
(c) the anisotropic interaction of spins with the crystal lattice in general forbids spins from leaving the basal plane (except under very high fields).

The third difference constitutes a limitation with regard to cholesterics. This probably makes $\chi$ lines more likely than others. The second difference leads to the consideration of a type of defect new to cholesterics: *spin twin-boundaries* perpendicular to the $\chi$ axis.

#### 9.3.5.1 *Wedge $\chi$ lines*

These are necessarily of integral order and have configurations resembling those shown in Figure 2.18. In fact, configurations of the same order but with spins of opposite signs must be distinguished from each other. Configurations in a right-handed helimagnet must also be distinguished from configurations in a left-handed helimagnet.

---

[†] The direction of **M** remains a direction of symmetry of rotation of order 2 (but not the direction perpendicular to **M**, which is of order 1). Therefore in a helimagnet $\lambda(\pm\frac{1}{2})$ lines can be found. See also subsection 9.3.1.1.

The definition of the $\chi$ in terms of helicoidal generation given in subsection 4.4.3 (equation (4.19)) can here be extended without difficulties:

$$S = 1 - n, \tag{9.58}$$

where $n$ is a positive integer if the direction of rotation of the pattern has the same chirality as the helimagnet in the pattern, negative in the opposite case (see Figure 9.22). This definition of strength does not allow for a distinction between the chiralities and the sense of spins.

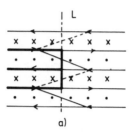

a)

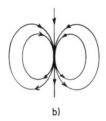

b)

Figure 9.22   Generation of a wedge $\chi$ by a helicoidal rotation (here $n = -1, S = 2$); right-handed helimagnet

### 9.3.5.2   *Twin boundary and twin disclination*

Let us suppose that a line $L$ along $\chi$ is divided into two parts, $L_1$ and $L_2$, along which exist two singularities of strength

$$S_1 = 1 - n_1 \quad \text{and} \quad S_2 = 1 - n_2,$$

situated in helimagnets $H_1$ and $H_2$ of opposite chiralities. Is there continuity of the singularity line $L$ if $n = n_1 = n_2$? According to the first definition of strength it is sufficient to construct $L$ beginning with a configuration $S = 1 - n$ in a basal plane, and to turn it helicoidally along $L$ first in one direction (in $H_1$) then in another (in $H_2$). There is therefore no discontinuity except on $L$. The plane of contact between $H_1$ and $H_2$ can be defined as a twin boundary.†

---

† The term 'twin boundary' rather than 'wall' is preferred here since there is an obvious analogy with the twin boundary by reticular merohedrism (G. Friedel, 1929) which takes place when the lattice is more symmetrical than the pattern. Here the lattice is for example an array of parallel planes of period $p$. The symmetry lacking in the pattern is a centre of symmetry.

Definition (9.58) is therefore compatible with the existence of two chiralities. Remember that a $\chi$ wedge disclination can be identified with an edge dislocation. We have, after equation (4.20):

$$b_1 = -p_0(1 - n_1),$$
$$b_2 = -p_0(n_2 - 1),$$

(9.59)

where $p_0 = +p_1 = -p_2$ ($p_0 > 0$). Formulae (9.59) imply that $L_1$ and $L_2$ are oriented in opposite directions (Figure 9.23). We therefore have for $n_1 = n_2$, $b_3 = 0$.

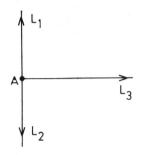

Figure 9.23. Twin disclination along $L$

If $n_1$ and $n_2$ are of opposite signs, or different, a line $L_3$ of Burgers vector

$$b_3 = \pm p_0(n_1 - n_2)$$

(9.60)

must start from $A$ in the twin boundary. Therefore this twin displays a step which is a twin disclination (Figure 9.24).

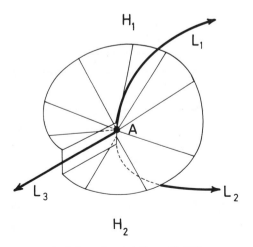

Figure 9.24. Spin twin and Cottrell–Bilby process

272

One recognizes here the possibility of a process of displacement of the twin boundary analogous to that proposed by Cottrell and Bilby (1951) for the displacement of coherent twin boundaries in crystals. By turning $L_3$ about $L_1$ and $L_2$, the twin is displaced along $L_1$ or $L_2$. The motion of $L_3$ is a *climb* in the twin plane, since $b_3$ is along the $\chi$ axis (Figure 9.25).

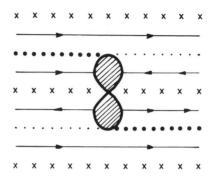

Figure 9.25. Junction of two $\chi$ lines on a twin disclination

### 9.3.5.3 *Twist $\chi$ lines*

Because of the very strong anisotropy, the distribution of spins is probably of the type shown in Figure 9.26. It can be seen that this singularity can be defined as a line on a Bloch wall parallel to the $\chi$ axis, situated entirely in $H_1$ (or $H_2$). Figure 9.26 supposes that this wall is of finite thickness: this can only be due to the presence of a certain anisotropy *in* the base plane; when this anisotropy is zero, the faces of the wall are plane and the wall is of infinite thickness at infinity.

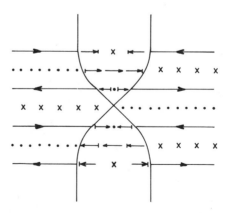

Figure 9.26. Twist $\chi$ in a strongly anisotropic helimagnet

# Chapter 10

## Classification of singularities in ordered media using homotopy groups. Generalization of the Burgers circuit. Comparison with the Volterra process

### 10.1 LIMITATIONS OF THE CLASSIFICATION OF DEFECTS BY THE VOLTERRA PROCESS. POSSIBILITY OF A MORE GENERAL CLASSIFICATION

The classification of defects presented in Chapter 2 makes a correspondence between each element of H, the symmetry group of the medium, and a type of curvilinear defect constructed by the Volterra process on the cut surface.

This classification is inadequate in a number of ways.

(1) It is limited to defects of dimensionality equal to one in three-dimensional space. Thus, singular points and walls, twin boundaries, etc. escape classification. It is clear that these objects are also linked to the symmetry properties of the ordered medium (e.g. singular points (nematics) or spin walls in anisotropic ferromagnets). Note also that G. Friedel gave a classification of twins by reticular merohedrism as a function of crystalline symmetries.

More generally, in conjunction with the spirit of modern theories of phase transitions (see Ma, 1976, and Toulouse and Pfeuty, 1975) a complete classification of defects must define, in a space of any dimensionality $d$, all singularities of dimensionality $d'$. The integer $r$ will then be used, defined by

$$d' = d - r - 1, \qquad (10.1)$$

where $d' = 0$ for singular points, $d' = 1$ for lines, etc.

To designate the presence of a defect in an ordered medium, we propose indeed to use the term *broken symmetry*. This usually designates a change of phase in the Landau sense. However it is already evident from the Volterra process that a dislocation ($d' = 1$) breaks some symmetry in some sense, and this will appear true for any d', in the course of this chapter.

273

274

(2) The Volterra process itself does not enable all curvilinear defects to be obtained in a simple manner. Thus, the defect in Figure 10.1, typical of a nematic, must be described by a rotation vector $\Omega$ which is variable along the line (and tangential to the line) and the same applies to helicoidal or circular defects found in cholesterics (Chapter 4). This description can only be made at the cost of introducing infinitesimal densities of dislocations (rotation or translation) which can sometimes appear artificial.

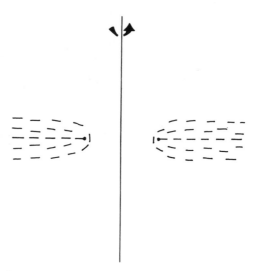

Figure 10.1. The diagram is of a revolution; the meridian section is formed by two configurations $S = +\frac{1}{2}$

(3) The problem of the topological stability of lines of integral strength was dealt with in the discussion of nematics. Remember that these are unstable under a transformation where their singular core disappears by *escape into the third dimension*. Topological stability must be distinguished from physical stability, the discussion of which touches on the calculation of distortions associated with a line and its energy. The Volteraa process provides one part of the boundary conditions (on the cut surface) for this calculation (the other part being provided by the nature of the surfaces bounding the sample), and this therefore means that the physical stability of a line can be dealt with without ascertaining the topological stability. Note also that this does not necessarily mean that the topological stability should in all cases be a finer concept than that of physical stability, which can eventually be produced under particular boundary conditions (on external surfaces, or by the action of an applied field).

The classification under discussion (Toulouse and Kléman, 1976; Kléman, Michel and Toulouse, 1977; Rogula, 1976; Volovik and Mineev, 1977) depends on the generalization in $r$ dimensions of the concept of the Burgers circuit and the introduction of the manifold of internal states $V$, which will at the same time define the *dimensionality* and *connectivity* properties of the order parameter (see

Chapter 2). The natural language of such a theory is mathematical, since the essential concepts are those of algebraic topology (see, for an elementary introduction, Zisman, 1972). These concepts will be set out with the use of many examples without rigorous mathematical accuracy.

A number of excellent review papers concerning the classification of singularities in ordered media homotopy groups have recently appeared (Mermin, 1979; Michel, 1980; Mineyev, 1980). The present chapter is not conceived as a complete review of the subject, but as an introduction with some emphasis on particular points. Mermin's paper gives a very pictorial discussion of homotopy theory, well suited for physicists who are totally unaware of the subject. Michel's paper starts with a general discussion of broken symmetry in physics, before going on to the applications of homotopy theory. Mineyev's paper contains many examples chosen from the anisotropic phases of liquid $^3$He.

## 10.2 BURGERS CIRCUIT IN THE DEFORMED MEDIUM. HOMOTOPY GROUPS OF THE MANIFOLD OF INTERNAL STATES

### 10.2.1 A first simple example: amorphous ferromagnetism

In this case the order parameter is the magnetization vector **M**, the modulus of which is a known function of temperature. The basic state is defined by the given direction of **M** (at a given temperature) in relation to an absolute reference frame: the spins are then all parallel to one single direction. The array of possible directions can be represented by the *manifold of internal states* $V = S_2$, consisting in the two dimensional sphere

$$x^2 + y^2 + z^2 = 1.$$

Each point of the sphere corresponds to the direction cosines $x, y$ and $z$ of **M**. At a given temperature the modulus of **M** is fixed, and it can be normalized to unity. We are therefore only interested in the angular variables of the order parameter.

Consider, in the usual three-dimensional space, $d = 3$, a singularity of the ordered medium of dimension $d' = 2 - r$, and surround it with a sphere $S_r$, i.e.:

(a) two points ($S_0$) in the case of a surface defect ($r = 0$; $d' = 2$);
(b) a circle $S_1$ (or every circuit homotopic to a circle in the sample) in the case of a curvilinear defect ($r = 1$; $d' = 1$); and
(c) a sphere $S_2$ (or every circuit homotopic to $S_2$) in the case of a singular point ($r = 2$; $d' = 0$).

Make a correspondence between every point $m$ of the sample included in $S_r$ and the point $M \in V$ which indicates the direction of magnetization **M** at $m$. In this way one produces a mapping

$$f : S_r \rightarrow V, \tag{10.2}$$

which makes a sub-manifold $\sigma$ of $V$, of dimension $r$ equal in general to that of $S_r$, correspond to $S_r$. A sub-manifold of dimension $r$ will be denoted $\sigma_r$. We are interested in the topological properties of these sub-manifolds and will identify in the same class all those which are homotopic to each other on $V$. For example, if $V$ is a sphere $S_2$, all the closed lines $\sigma_1$ traced on this sphere, which are mappings of $S_1$ onto $S_2$, are homotopic (they can be identified by continuous displacement and deformation on $V$). In the example chosen these lines are also homotopic to a point, but more general cases exist. Thus, the set of mappings of $S_2$ onto $V$ will be distributed in a set of *equivalence classes* (in the homotopic sense): this set possesses a natural group structure, which will be appreciated in the examples.

If $\sigma_r$ can be contracted to a point on $V$ in a continuous manner, this indicates that it is possible to continuously submit the magnetization on $S_r$ to rotations which will make it take a uniform direction; $S_r$ does not then in general surround any singularity in the sample, or surround a number of singularities compensating each other. It will be said that $\sigma_r$ is homotopic to zero (or to a point): this equivalence class forms the identity element of the natural group structure of mappings (10.2).

On the other hand, if $\sigma_r$ is not contractible, it will be said that $S_r$ surrounds a *topologically stable* singularity.

It is clear that every $\sigma_0$, like every $\sigma_1$ on $S_2$, is contractible to a point: lines and walls are not topologically stable in an amorphous ferromagnet. On the other hand, singular points are topologically stable in an infinite number of ways, for the $\sigma_2$ which cover $V = S_2$ an integral number $N$ of times are not homotopic to a point. At the end of Chapter 5 an integer $N$ was defined which corresponds to the operation described here. $N$ is known as the degree of the mapping $f : S_2 \to S_2$. It can be shown that all the possible mappings between the Burgers circuits $S_2$ traced in the deformed medium and $V (= S_2)$ are exhausted by taking $N$ as all the integral values between $+\infty$ and $-\infty$. $N = 0$ corresponds to the case where $\sigma_2$ is homotopic to a point; opposite values of $N$ correspond to two opposite *orientations* of $\sigma_2$ in the mapping, i.e. to opposing distributions of $\mathbf{M}$ (for more details on the topological significance of this concept of orientation, which generalizes the concept of direction of rotation on a circle in the multidimensional case, see Arnold, 1974). Finally, all these possibilities appear as elements of the additive group of relative integers, $Z$, since a circuit surrounding two singular points of indices $N_1$ and $N_2$ has the index $N_1 + N_2$; this reads:

$$\pi_2(S_2) \simeq Z. \tag{10.3}$$

$\pi_2(V)$ is the second homotopy group of the manifold of internal states $V$. Each of its elements corresponds to a homotopy class of the mappings of $S_2$ onto $V = S_2$. We have in the same way:

$$\pi_0(S_2) \simeq \pi_1(S_2) \simeq 0. \tag{10.4}$$

These groups designate the homotopy groups of the mappings of $S_0$ and $S_1$ onto $V = S_2$ and only contain the neutral element (or identity, denoted 0 or I as appropriate).

Therefore there are no topologically stable surface singularities in an amorphous ferromagnet. It should not be deduced that Bloch walls are not topologically stable. We shall see below (subsection 10.2.4) that they are indeed classified (as defects) in a much less trivial way.

### 10.2.2 A second simple example: the nematic phase

In this case the order parameter is the director $\mathbf{n}$. The variety of internal states can be built up from $S_2$ by *identifying* two opposing points on $S_2$, since $\mathbf{n} \simeq -\mathbf{n}$. We can still only take half a sphere, and identify the diametrically opposing points of its boundary. The manifold thus constructed is the *projective plane $P_2$*. It can again be imagined, since only the topological properties of $V$ interest us, as a full circle whose diametrically opposed points are identified: in fact such an object is homeomorphic to a half-sphere whose diametrically opposed points are identified (see Figure 10.2).

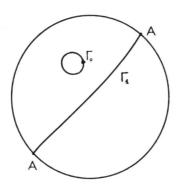

Figure 10.2. Representation of a *projective plane $P_2$* by a full circle whose diametrically opposed points are identified. The loop $\Gamma_0$ is homotopic to zero; the loop $\Gamma_1$ is not. The non-orientable nature of $P_2$ can easily be visualized. Consider an oriented loop $\Gamma_0$ and displace it towards the boundary of the representation (notice that this boundary is not that of $P_2$, which is a closed surface), up to crossing. A part of $\Gamma_0$ now appears in a diametrically opposite part of the representation, with a defined orientation. Eventually $\Gamma_0$ can be transported in its entirety by a continuous movement to this opposite part. It is easy to see that its orientation is now opposite to the one in its former position

The properties of the projective plane should be discussed at some length; reference should be made to the work of Hilbert and Cohn-Vossen which has already been quoted. It will be noted here that this is a non-orientable surface, like the Moebius ribbon, and we can also obtain $P_2$ by joining along the edge of a hole in a sphere the single border of a Moebius ribbon.

The homotopy groups of $P_2$ are the following:

$$\pi_0(P_2) \simeq 0; \qquad \pi_1(P_2) \simeq Z_2; \qquad \pi_2(P_2) \simeq Z. \qquad (10.5)$$

(a) $\pi_0(P_2) \simeq 0$. There are no topologically stable walls in $P_2$. This means, as for $S_2$, that the manifold $V$ is here connected (i.e. that it is possible to join any two points on $V$ by a line segment which stays entirely in $V$).

(2) $\pi_1(P_2) \simeq Z_2$. The first homotopy group of $P_2$ only contains two elements, the neutral element $I$, corresponding to the absence of singularities, and the element (here denoted $a$), which must necessarily be its own inverse, and the square of which must be the neutral element:

$$a = a^1; \qquad a^2 = I. \qquad (10.6)$$

$\pi_1(P_2)$ can be calculated simply from Figure 10.2. The neutral element corresponds to contractible loops of the type $\Gamma_0$ (see Figure 10.2). The loop $\Gamma_1$, which joins two diametrically opposed points $AA$ on the boundary, is a non-contractible closed loop. Let us orient $\Gamma_1$ and displace it continuously by turning $AA$ on the boundary making it complete a half-turn around this point. The final result is a loop oriented in the inverse direction. We therefore have $a^{-1} = a$. It could be demonstrated in the same way that the loop $\Gamma_2$ obtained by joining $\Gamma_1$ to a loop of the same class $\Gamma'_1$, is homotopic to zero, therefore $a^2 = I$.

To determine the homotopy class of a line $L$ in the ordered medium, it is convenient to take a circuit $\gamma$ which surrounds it, homotopic to a circle $S^1$, and to consider its mapping into $V$. The class will obviously be $a$ or $a^{-1}$, according to the orientation of $\gamma$. The first equality (10.6) tells us that the lines in nematics are not orientable. $S = \frac{1}{2}$ and $S = -\frac{1}{2}$ lines must therefore be identified in three-dimensional space.† The closed loops, whose opposing sections are of these different strengths (see Figure 2.19), can therefore be constructed so that the passage between these different configurations is continuous. This result cannot be obtained trivially from the Volterra process. Note also that the scheme of Figure 10.1 enters into the classification.

The second of the equalities (10.6) tells us that the lines of integral order are *topologically unstable*: this confirms the result in Chapter 2 and justifies our identification of the element $a$ with the class of lines of half-integral order.

(3) $\pi_2(P_2) = Z$. The same result is found here as for the amorphous ferromagnet: singular points in nematics are topologically stable, and are indexed like those in amorphous ferromagnets. However, the sign of the degree of mapping $N$ depends on the sign of the director, which is to a certain extent arbitrary: if the singular point of degree $N$ describes a circuit in the nematic medium belonging to a non-trivial class of $\pi_1(P_2)$, i.e. surrounding a line $S = \pm\frac{1}{2}$, then the director changes sign like $N$; it can be said that the homotopy group $\pi_1(P_2)$ has a non-trivial action on $\pi_2(P_2)$. For the same reason, the product of the coalescence of two singular points $|N_1|$ and $|N_2|$ depends on the relative trajectories of these two points when they approach each other and is equal to $\||N_1| \pm |N_2|\|$ (Volovik and Mineyev, 1977).

---

† It will be noted that if we are restricted to the consideration of wedge lines (see Chapter 2, Figure 2.17) it would be necessary to take a two-dimensional nematic, for which $V = S_1$, and $\pi_1(S_1) = Z$. A correspondence can then be made between each element $n$ of $Z$ and the wedge lines of order $S = n/2$. The three-dimensional case is more subtle.

### 10.2.3 Generalization of preceding examples

The above examples will be generalized for media defined in an $n$-dimensional space, the order parameter being a vector or director of dimension $n$ (in the preceding cases we had $d = n = 3$).

#### 10.2.3.1 Vectorial order parameter (normalized spin field)

The manifold of internal states is then $V = S_{n-1}$. For the calculation of homotopy groups of spheres, reference should be made to Steenrod (1951) or Hu (1959). We find:

$$\pi_i(S_{n-1}) \simeq \begin{cases} 0, & i < n-1, \\ Z, & i = n-1. \end{cases} \tag{10.7}$$

The only non-trivial case corresponds to

$$r = n - 1 = d - d' - 1,$$

i.e.

$$d' = d - n. \tag{10.8}$$

This applies to amorphous ferromagnets ($n = 3$) and to superfluids ($n = 2$): the angular part of the order parameter is the phase $\phi(0 \leq \phi \leq 2\pi)$, therefore $V = S_2$; for $d = 3$ this yields $d' = 1$ (vortex lines of $^4$He). Figure 10.3 sums up the results: on diagonal lines we have singularities of a given dimension. The zone on the right of the curve $d = n$ is non-physical (defects of negative dimensionality). Toulouse (private communication) has noted that this forbidden zone actually corresponds to the systems for which the phase transitions towards the ordered state (order of the spin field type, i.e. $V = S_{n-1}$) only occur at absolute zero (see Migdal, 1975).

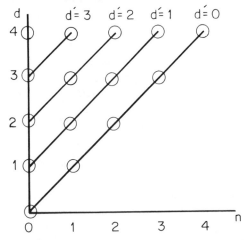

Figure 10.3. Nature of defects (points, lines, surfaces) as a function of the dimensionalities $d$ (of space) and $n$ (of the order parameter) for spin fields ($V = S_{n-1}$)

10.2.3.2 *Projective order parameter (director field); rotation group* $SO(3) = P_3$

Suppose that the manifold of internal states is $V = P_{n-1}$, i.e. the sphere $S_{n-1}$ whose diametrically opposed points are identified. We then have:

$$\pi_1(P_{n-1}) = Z_2,$$
$$\pi_r(P_{n-1}) = \pi_r(S_{n-1}), \qquad r \neq n. \tag{10.9}$$

$n = 4$ ($V = P_3$) corresponds to the A phase of $^3$He and deserves further comment. $P_3$ is in fact a *topological group*, more precisely a Lie group (see Gürsey, 1964), i.e. a manifold representative of a group; here the group of rotations in three-dimensional space of a solid invariable in form. The demonstration is simple: let us set this solid in a reference frame to which it is invariably attached, of rectangular axes $l$, $\Delta_1$ and $\Delta_2$. Let us define the motion of this frame by its rotation $(\Omega, v)$ in relation to an absolute trihedron. Let $\lambda$ and $\mu$ be the polar angles of the direction $v$. The vector $\Omega, v$ has components:

$$x_1 = \Omega \sin \lambda \cos \mu,$$
$$x_2 = \Omega \sin \lambda \sin \mu, \tag{10.10}$$
$$x_3 = \Omega \cos \lambda.$$

We therefore have

$$x_1^2 + x_2^2 + x_3^2 = \Omega^2. \tag{10.11}$$

$\Omega$ can be chosen in the determination $-\pi \leq \Omega < \pi$. Let us put $x_4^2 = \pi^2 - \Omega^2$. This yields:

$$x_1^2 + x_2^2 + x_3^2 + x_4^2 = \pi^2. \tag{10.12}$$

This equation represents a three-dimensional sphere $S_3$ embedded into a four-dimensional euclidean space. But $S_3$ is not the representative manifold of the group of rotations; two diametrically opposed points must again be identified ($\pm x_i$) which represent rotations differing by an angle $2\pi$. The definitive manifold is therefore $P_3$ (also denoted $SO(3)$: special group of rotations not containing reflections). $P_3$ can be represented in three-dimensional space as the closed ball $x_1^2 + x_2^2 + x_3^2 \leq \pi^2$ with identification of the diametrically opposed points on the boundary (Figure 10.4).

$P_3$ is obviously the manifold of internal states of $^3$He-A, whose order parameter is a trihedron $1$, $\Delta_1$ and $\Delta_2$. We have:

(1) $\pi_1(P_3) \simeq Z_2$. There are two equivalence classes of singular lines in $^3$He as in the nematic phase. One of these classes (the neutral element) corresponds to the absence of singularities, which can still be obtained as the square of any other class ($a = a^{-1}; a^2 = 0$); reference to the discussion in Chapter 6 shows that this is a case of rotation dislocations of angle $4\pi$ whose core can be abolished.

Class $a$ corresponds to rotation dislocations (vortex lines, disgyrations, here identified) of angle $2\pi$ around any direction of the trihedron.

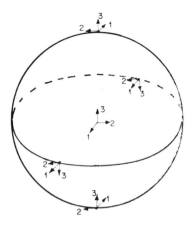

Figure 10.4. The ball $\Sigma x_i^2 \leq \pi^2$ represents the special group of rotations SO(3) (i.e. without reflections) of a solid in ordinary space, i.e. rotations of an undeformable trihedron with three unequal axes

(2) $\pi_2(P_3) \simeq 0$. There are no topologically stable singular points in $^3$He-A. If the topology of the sample makes the presence of one singular point inevitable (e.g. a spherical sample with **1** always perpendicular to the wall), this point is at one extremity of a singularity line, or is eventually split into surface lines (see

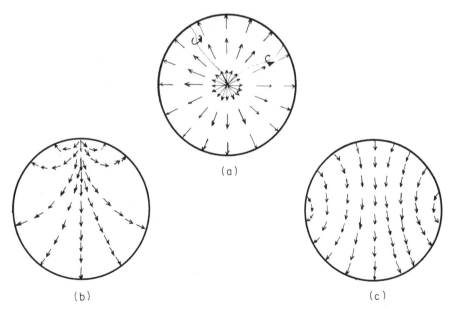

(a)

(b)                                                      (c)

Figure 10.5. $^3$He-A in a sphere, **1** normal to the surface. The obligatory central singularity is linked to two disclination lines (a) which tend to give rise to surface lines (b) and (c) (from Mermin, 1976)

Figure 10.5, borrowed from Mermin, 1976). Such singular points have been compared to Dirac monopoles (linked by chords: here singular lines; see Polyakov, 1976).

### 10.2.4 The homotopy group $\pi_d(V)$ and higher order groups

These groups correspond to singularities of negative dimensionality (see equation (10.1)). They were considered for the first time by Finkelstein and Misner in the study of the differential equation

$$\frac{\partial^2 \phi}{\partial x^2} - \frac{\partial^2 \phi}{\partial t^2} - K^2 \sin \phi = 0, \tag{10.13}$$

known as the Sine–Gordon equation. The function $\phi$ is defined on the interval $(0, 2\pi)$ and the manifold of states $V = S_1$ can therefore be introduced. The space on which $\phi = \phi(x)$ is defined is one-dimensional. $\pi_1(S_1)$ therefore corresponds to defects of dimensionality $-1$. The interpretation here is very simple, however. Suppose that the boundary conditions, at the two extremities of the $x$ axis, are identifiable (e.g. $\phi = 0$ for $x = -\infty$; $\phi = 2\pi$ for $x = +\infty$).

These boundary conditions mean that the $x$ axis can be identified with a circle $S_1$. $\pi_1(V)$ therefore represents the group of different homotopy classes of mappings of the $x$ axis (supplied with these boundary conditions) onto $V(= S_1)$. If this group is non-trivial, this means that *continuous* arrangements $\phi(x)$ exist which can be reduced to a constant $\phi_0$ in a continuous manner. It is essential here that these arrangements are continuous, which fact profoundly differentiates the topological states $d' < 0$ from the preceding states. This results from the fact that the mapping envisaged concerns *all* the space of $x$ and not only a *Burgers circuit*: therefore by definition there can be no singularities in it.

The nature of these arrangements is obviously a case of one-dimensional twist. For example, $\phi$ can vary from 0 to $2\pi$ in a continuous manner and from $x = -\infty$ to $x = +\infty$. Finkelstein (1966) called this a *kink*. This is a solution of the *soliton* type of the Sine–Gordon equation (see Scott *et al.*, 1973). There can be $1, 2, \ldots, n$ solitons on the $x$ axis affected by a minus or a plus sign according to the sign of the twist. In fact we have:

$$\pi_1(S_1) \simeq Z. \tag{10.14}$$

More generally, suppose that the medium under consideration, of dimension $d$, is submitted to uniform boundary conditions. By identifying all the points of the boundary, the medium can be considered as a sphere $S_d$. $\pi_d(V)$ is the group of *kinks* or *generalized twists*, which cannot be reduced to zero. In all the cases previously described we have $\pi_d = Z$. It is clear that this possibility describes, in the case of an amorphous ferromagnet, the presence of twist walls. In the case of a nematic this describes the chiralized nematic (Figure 4.9b without the core can be understood as representing a chiralized nematic with uniform boundary conditions). In the case of $^3$He-A this analysis points to the existence of chiralized situations with walls, which cannot be reduced to zero.

Louis Michel has recently proposed calling 'unknotted configurations' (u.c.) all the configurations pertaining to a non-trivial element of $\pi_d(V)$.

Cylindrical twist walls are easily visualized as such u.c. They are two-dimensional objects, and as such they are unknotted in the sample when we restrict ourselves to two-dimensional inhomogeneities. Hence, the relevant homotopy group is $\pi_2(V)$, which is non-trivial in an amorphous ferromagnet or a nematic phase. The physical stability of two-dimensional u.c. has been proven by Belavin and Polyakov (1975) for the case of a free energy density reduced to second-order gradient terms

$$\rho f = A(\nabla \mathbf{n})^2.$$

The various minima of energy of two-dimensional u.c. are $W = 8\pi A|m|$, where $m$ is an integer. The configuration of minimal energy is given by

$$e^{i\alpha} \operatorname{ctg} \frac{\beta}{2} = \prod_{i,j} \left(\frac{z - z_i}{\lambda}\right)^{m_i} \left(\frac{\lambda}{z - z_j}\right)^{n_j},$$

where $\alpha$ and $\beta$ are azimuthal and polar angles of $\mathbf{n}$, and $z = x + iy$. The integers $m_i$ and $n_i$ are submitted to the constraints

$$m = \sum m_i > \sum n_j,$$

and $\lambda$ is an arbitrary scale parameter. The indifference of the energy to the value of $\lambda$ makes this model unrealistic. As pointed out by Shankar (1977), any term of fourth order (like $B (\nabla \mathbf{n})^4$) in the free energy removes the neutrality with respect to scale changes. It remains to discover whether, physically, $d_0 = \sqrt{(B/A)}$ is a macroscopic length or not, compared to molecular dimensions.

An example of Belavin and Polyakov u.c. is furnished by the nematic in the capillary tube observed by Williams, Piéranski and Cladis (1972) and discussed in subsection 3.3.3. But here the u.c. is restricted (by homeotropic conditions) to a part of its spatial range of existence. It obeys the equation above with $m = 1$:

$$e^{i\alpha} \operatorname{ctg} \beta/2 = z/\lambda,$$

with $\lambda$ equal to the cylinder radius.

The stability of three-dimensional u.c. (pertaining to $\pi_3(V)$) is also dependent upon some scale $\lambda$. In the case where $V = S_2$ or $P_2$, we have $\pi_3 = Z$, and representative elements of each class $m$, called Hopf maps (see Steenrod), are given for samples of infinite volume by the formula

$$e^{i\alpha} \operatorname{ctg} \frac{\beta}{2} = \left(\frac{(\lambda^2 - r^2)^2 + 4\lambda^2 r^2 \cos^2 \theta}{4\lambda^2 r^2 \sin^2 \theta}\right)^{|m|/2}$$

$$\times \exp\left[i m \left(\tan^{-1} \frac{2\lambda r \cos \theta}{\lambda^2 - r^2} - \right)\right],$$

where $r$, $\theta$ and $\phi$ are the spherical coordinates of a point defined by vector **r**. Here too a stable size $d_0$, different from molecular dimensions, depends crucially on the existence of a $B$ term. Since the Hopf map above corresponds to $B = 0$, some change in the configuration has to be expected with $B \neq 0$, but this does not lead to any topological change in the u.c. for a given $m$.

A Bloch wall can be defined as a one-dimensional u.c. when the spin is restricted to two-dimensional variations; i.e. it pertains to a class of $\pi_1 (S_1)$. The scale is fixed by a competition between exchange terms of the type above and anisotropy terms (rather than fourth-order gradient terms), as is well known.

The physical meaning of higher order groups ($r > d$) is not clear. Let us point out that, after Finkelstein (1966), if the group $\pi_{d+1}(V)$ contains a cyclic sub-group of order 2, then the elements of the group $\pi_d(V)$ can be represented by double-valued functions which would correspond to spin properties. Groups of order higher than $d + 1$ have never been considered.

### 10.3  RELATIONSHIP BETWEEN SYMMETRY PROPERTIES OF ORDERED MEDIA AND THE MANIFOLD OF INTERNAL STATES. CRYSTALS

The foregoing section does not directly involve the symmetry of the ordered medium, but rather the notion of order parameters. This is not without interest, as the results of investigating the differential equation (10.13) can be considered in the same light. However, there are several examples of ordered media where the manifold $V$ does present great difficulties of definition by the order parameter. The method which follows, taken essentially from Michel, begins with a definition of $V$ which does not involve the concept of order parameters and leads to precise results when the symmetry group of the ordered medium is a discrete sub-group of a Lie group (this is the case of the 230 symmetry groups of crystals).

#### 10.3.1  The manifold $V$ defined as an orbit of the group $G$

In the various examples of ordered media studied in section 10.2, the medium remains globally invariant under spatial rotations: the action of any spatial rotation $g$ gives the medium of primitive orientation $\mu$ a new orientation, $g \cdot \mu$, but the physical laws of the medium are unchanged. The group of spatial rotations $(SO(3) = P_3$ if the medium is chiral; $O(3)$, containing reflections, if the medium is not) is also, in all these examples, the symmetry group $G$ of the (disordered) high temperature phase, and $H$, the symmetry group of the ordered medium, is a sub-group. G will be called the thermodynamic group.

This situation is quite general: a *thermodynamic* group $G$ can be associated with each ordered medium of symmetry group $H$ (see Landau and Lifshitz, 1958). All the cases of figures of the usual media are obtained by taking the euclidian group E(3) for $G$, a semi-direct product† of the rotation group O(3) and the group of all

---

† For this concept see Lomont (1959) or Gürsey (1964)

translations $R^3$:

$$E(3) = R^3 \square O(3).$$

The property under consideration is the following: *the manifold of internal states V is equivalent to the cosets of the quotient G/H.*

Consider a state $\mu$ of the medium, invariant under $H$, i.e. in the low-temperature state. The totality of the states $g \cdot \mu$, where $g$ is any element of $G$, defines the so-called orbit $V$ of physically equivalent states (see Michel, 1980). Each of these states $g \cdot \mu$ is invariant under the action of all the elements of the coset $Hg^{-1}$ belonging to the quotient $G/H$. We may then write:

$$G/H = V. \tag{10.15}$$

This equation identifies each coset of $G/H$ with a point of $V$. Some of the results already mentioned in section 10.2 will be obtained now from this equation. This method will also be used for the instructive case of the triclinic crystal.

### 10.3.1.1 Nematic

The symmetry group $H$ is the (semi-direct) product of the translation group $R^3$ and the group $D_{\infty h}$ of the $\pi$ rotations about every straight line perpendicular to the director (plus reflection).† The quotient $E(3)/H$ is also $O(3)/D_{\infty h} = SO(3)/D_\infty$; the reflections having disappeared in the dividend and the divisor. This quotient is equal to $P_2$. In fact, $SO(3)/D_\infty = P_3/D_\infty$ can be represented with the help of the representation of $P_3$ in Figure 10.4 (see Figure 10.6). $D_\infty$ includes the set of rotations which corresponds to a diameter of the ball

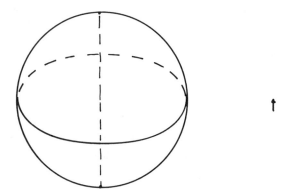

Figure 10.6   Representation of the symmetry group $D_\infty$ of a nematic, as a sub-group of $SO(3)$. The orientation of the nematic is shown on the right. The vertical diameter corresponds to the rotation about the director, the great circle corresponds to the rotations of $\pi$ about the perpendiculars to the director

† See Landau and Lifshitz (1958) for the usual notations of groups. $R$ designates the real straight line, isomorphic to the group of unidirectional translations.

and its conjugate great circle: the diameter is taken parallel to the director. All the points of $D_\infty$ must be identified in the quotient $P_3/D_\infty$. The action of the rotations g of $P_3$ which cause the director to occupy all the directions in space leads to a rotation of the representation of $D_\infty$ in Figure 10.6.

All the possibilities are clearly exhausted by making a correspondence between each point $A$ of the boundary sphere on the one hand and the diameter issuing from $A$ and its conjugate great circle on the other hand. The quotient $P_3/D_\infty$ is therefore isomorphic to this sphere, which is obviously identifiable with $P_2$, since the diametrically opposed points are equivalent.

### 10.3.1.2  $^3He\text{-}A$

Let us restrict ourselves to the case where $V$ is along $\mathbf{l}$ (see Chapter 6). The symmetry group $H$ is the (semi-direct) product of the translation group $R^3$ and the cyclic group with two elements $Z_2$ containing the neutral element and the central reflection (the trihedron $\mathbf{l}$, $\Delta_1$ and $\Delta_2$ is a pseudo-trihedron, $\mathbf{l}$ being analogous to a magnetic moment). The quotient $E(3)/H$ is equal to $O(3)/Z_2$, which is again $SO(3)/I = P_3$.

### 10.3.1.3  *Centrosymmetric triclinic crystal* (Michel)

This example is at the opposite extreme from the preceding examples. The symmetry group is the semi-direct product of the quantified translations $Z^3$ and the cyclic group with two elements $Z_2$ containing the neutral element and the central reflection (the basic trihedron of the Bravais lattice having the same genericity as that of $^3$He-A). Translations have the notation $Z^3$ because the symmetries of translation along a given direction form an Abelian group isomorphic to the group of the integers $Z$. $Z^3$ is the product of three such groups. We therefore have:

$$E(3)/H = (R^3/Z^3) \,\square\, P_3.$$

The quotient $R/Z$ is also a straight line element whose two extremities are identifiable; this is therefore a circle $S_1$. The product of three circles $S_1$ is a three-dimensional torus $T^3$. $V$ is therefore the product of two manifolds, $T^3$ and $P_3$, which play separate roles. Corresponding to the torus there are the translation dislocations:

$$\pi_1(T^3) \simeq \pi_1(S_1) \otimes \pi_1(S_1) \otimes \pi_1(S_1) \simeq Z \otimes Z \otimes Z,$$
$$\pi_i(T^3) \simeq 0, \qquad i \neq 1,$$

whose homotopy classes are isomorphic to the Burgers vectors $n_1\mathbf{b}_1 + n_2\mathbf{b}_2 + n_3\mathbf{b}_3$. Rotation dislocations analogous to those of $^3$He correspond to $P_3$.

#### 10.3.1.4 *Non-centrosymmetric triclinic crystal* (Michel)

The symmetry group is now reduced to $Z^3$, the identity element representing both the zero translations and the $2\pi$ rotations. The quotient is very different: reflections cannot be eliminated between the dividend and the divisor. $V$ is therefore the (non-trivial) product of the two manifolds $T^3$ and $O(3)$.

But since $O(3)$ is the group of rotations with reflections it is formed from *two* copies of $SO(3) = P_3$; the homotopy groups of $O(3)$ are isomorphic to those of $P_3$, except the groups of order zero: the mapping of the sphere $S_0$ (formed by two points) onto $O(3)$ is not necessarily homotopic to zero, and we have

$$\pi_0(O(3)) = Z_2. \tag{10.16}$$

*Singular planes* are therefore possible: this is obviously a case of twins by reticular merohedrism (see G. Friedel, 1926). The other singularities are of the same type as in the centrosymmetric crystal.

### 10.3.2 General theorems concerning the calculation of homotopy groups

The calculation of homotopy groups is not in general as trivial as in the above examples, and requires a deep knowledge of algebraic topology. This subsection simply summarizes, without demonstration, a number of useful results, and knowledge of algebraic topology is not necessary for the reading of the paragraphs which follow: uninterested readers can skip it. Useful introductory texts to algebraic topology are Steenrod's and Hu's books, and the lecture notes of Poenaru (1978).

#### 10.3.2.1 *H connected*

If the symmetry group is connected, it is shown (Steenrod, 1951, p. 87 *et seq.*) that the manifold $V$ is arcwise connected and that the first homotopy group $\pi_1(V)$ is Abelian (the others are also necessarily Abelian). The standard method to calculate the various homotopy groups is to use the long exact homotopy sequence for principal fibre bundles. An example will be given with the discussion of the classification of defects in Sm A phase (subsection 10.3.3.3).

#### 10.3.2.2 *H discrete*

Suppose that $G = E(3)$ (non-chiral media). $G$ contains two equal and disjointed manifolds, because of reflections. If $H$ also contains reflections, the quotient $G/H$ is then formed by a single connected manifold. Therefore $\pi_0 (G/H) = 0$; there can be no topologically stable surface singularities. On the other hand, if $H$ does not contain reflections, the $G/H$ quotient is formed by two copies of the same manifold: there are surface singularities (twins by merohedrism). There are several examples of this property in crystals. Helimagnets, which do not possess a definite chirality (but both chiralities are possible), are again an example in which, because of the absence of a centre of symmetry in $H$, there can be surface singularities (see Chapter 9).

If the medium is of well-determined chirality, $G$ is necessarily connected (we have $G = E_0$, connected component of E(3)). $H$ cannot contain reflections. The quotient $G/H$ is necessarily formed by a single connected manifold, and there can be no topologically stable surface singularities (e.g. cholesterics).

The other homotopy groups do not depend on reflections. It will therefore always be sufficient to consider $E_0/H_0$ ($H_0$ being the symmetry groups without reflections) to calculate them. The following results are obtained:

$r > 1$.

$$\pi_r(E(3)/H) \simeq \pi_r(E_0/H_0) \simeq \pi_r(E_0),$$

i.e.

$$\pi_2 = 0: \qquad \pi_3 = Z, \quad \text{etc.}$$

This property is based on the fact that $E_0$ is a covering of $V$ (for this concept see Steenrod) (in the same way $P_3$ is a covering of $S_3$, and we have $\pi_r(P_3) = \pi_r(S_3)$).

$r = 1$.  Let us introduce the concept of the *universal covering* $\bar{E}_0$ of a continuous group $E_0$. It is by definition a covering where first homotopy group is trivial. The universal covering of $E_0$ (therefore of the envisaged manifolds $V$, for which $E_0$ is a covering) is the semi-direct product:

$$\bar{E}_0 = R^3 \,\square\, SU(2), \tag{10.17}$$

where $SU(2)$ designates the Lie group of which the representative manifold is $S_3$. The fact that $S_3$ is a covering of $SO(3)$ appears in the construction which we have made of $P_3$. On the other hand, $\pi_1(S_3) \simeq 0$; $S_3$ is therefore simply connected: it is a universal covering.

The theorem is the following. The homotopy group $\pi_1(V) = \pi_1(E_0/H_0)$ is isomorphic to the image of $H_0$ in $\bar{E}_0$ in the inverse mapping $\theta^{-1}$ defined by the homomorphism of groups

$$\theta: \bar{E}_0 \to E_0,$$

which we denote by

$$\pi_1(V) \simeq \theta^{-1}(H_0) \simeq \bar{H}_0. \tag{10.18}$$

In our case, since $\bar{E}_0$ has twice as many elements as $E_0$, the first homotopy group will have twice as many elements as $H_0$.

$H_0$, like $\pi_1(V)$, may not be Abelian. This happens in most crystallographic groups. The physical significance of these non-commutativities will be studied in section 10.4. The other homotopy groups $\pi_r(V)$ for $r > 1$ are always Abelian. In what concerns $\pi_0(V)$, this is generically not a group, but a pointed set (Poenaru, 1978). However, we have no use for such subtleties.

### 10.3.3  Application to different media

A systematic exposition of the application of rules of algebraic topology to ordered media whose symmetry groups are sub-groups of the euclidean group can be found in Kléman and Michel (1978a). Here we develop our discussion for the cases of the biaxial nematic, the Sm C phase and the Sm A phase.

#### 10.3.3.1  *Biaxial nematic* $N_B$

This medium is characterized by a local trihedron with three unequal axes, each having the symmetry of a director. The symmetry group is therefore $D_2$ (Abelian dihedral group, containing four elements; the identity will be noted $I$, the rotations of $\pi$ about the three axes of the trihedron $a_1$, $a_2$, $a_3$), sub-group of $P_3$. We will not take into account translations, which are the same as those in E(3).

$D_2$ is a sub-group of SO(3), whose representative manifold is $P_3$. According to the foregoing paragraph, $\pi_1(V(N_B))$ is the transform of $D_2$ in the inverse mapping which brings $P_3$ into $S_3$, i.e. which maps each point of $P_3$ (which is half of the sphere $S_3$) into two diametrically opposite points on $S_3$, the representative manifold of the group SU(2). While $P_3$ is not simply connected (its fundamental group $\pi_1(P_3)$ is isomorphic to $Z_2$), $S_3$ is ($\pi_1(S_3) = 0$). Let us now see the physical significance of this (inverse) mapping (Toulouse, 1977).

A closed circuit surrounding a singularity line in a distorted $N_B$ sample has a natural image line in $P_3$ which, say, starts from the center $O$ and finishes in any of the points $a_i$ or in $O$ itself (see Figure 10.7). There are therefore four classes of image lines of 'Burgers' circuit in the sample according to the end-points of their natural image in $P_3$. But in fact each of this class divides in two, since there are two classes of closed loops in $P_3$. This is most easily seen by considering $S_3$, in which the two classes of closed loops in $P_3$ are represented by (i) closed loops for the

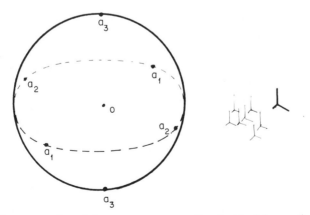

Figure 10.7. Representation of the symmetry group $D_2$ of a biaxial nematic on $P^3$. The orientation of the biaxial nematic is shown on the right

class of closed loops homotopic to zero in $P_3$ and (ii) open loops (joining two diametrically opposed points) for the class of non-trivial closed loops in $P_3$.

The homotopy classes of $\pi_1(V(N_B))$ are therefore classified by the (inverse) image of $D_2$ in $S_3$. Let us call $I$ and $-I$ the inverse images of $I$ in $D_2$, $I$ corresponding to a closed loop, $-I$ to an open loop. Call $e_i$ and $-e_i$ the two (inverse) images of $a_i$. We have the relations:

$$(-I)(-I) = I; \qquad (e_i)(-e_i) = I;$$

$$(e_i)(e_i) = -I; \qquad e_1 e_2 = e_3.$$

This group is non-Abelian (Toulouse, 1977) and isomorphic to the group of quaternions.

*Non-orientable lines* $(-I)$. A rotation dislocation of angle $\pm 2\pi$ about any direction of the trihedron corresponds to the element $-I \in \pi_1(V)$. As $-I$ is its own inverse, a closed singular line in the sample having this homotopy class is non-orientable. This means that on the same line local configurations can be found corresponding to both rotation angles $+2\pi$ and to rotation angles $-2\pi$ of the Volterra process, without it being *necessary* to introduce singular points between segments of *opposing* configurations. We have already made an analogous remark in connection with nematics (subsection 10.2.2) and indicated that in the final analysis this property corresponds to the existence of two types of lines: with diametrically opposed segments of equal configuration or not (Figure 10.1). We tend towards the same conclusion here, giving a generalized sense to the word *pinch*.†

Finally, note that the disclination $(-I)$ necessarily displays a core singularity. We have to go to disclinations with an angle of $4\pi$ $(+I)$ for this singularity to disappear, as in $^3$He.

*Orientable lines* $(\pm e_i)$. Corresponding to the elements $e_i$ and $-e_i$ there exist lines whose rotation $+\pi$ and $-\pi$ are made about the axis $Ox_i$. Thus, a closed line of $N_B$, oriented in a certain way, is in the homotopy class $e_i$ or $-e_i$, according to the orientation; the configurations on two diametrically opposed segments of this line are different and form a sort of pinch; there is no singular point anywhere on the line. It is clear that the local configurations we expect are similar to those of the $|S| = \frac{1}{2}$ lines in nematics. But the orientability of the line prevents the existence of global configurations of the type of Figure 10.1 in which there is no pinch. If such a configuration were to exist anyway, the absence of pinch would be relaxed by the appearance of a line of type $(-I)$ along the axis of revolution.

*Singular points.* These are not topologically stable, the $\pi_2(V)$ being trivial.

---

† These are general results concerning non-orientable lines: there are indeed two types of lines without singular points each time the homotopy class is its own inverse. (We will see below (section 10.4) how such lines can be distinguished by certain properties of the cut surface)

## 10.3.3.2 Smectics C (Sm C)

We will deal with this example before the smectics $A$ because here the local symmetry group is discrete (the theorems set out in subsection 10.3.2.2 can therefore be applied here). The local reference frame consists of two directors (the normal $\mathbf{n}$ to the layers and the projection $\mathbf{t}$ of the molecules on the layer) and a vector $\mathbf{N}$ ($\mathbf{N} \cdot \mathbf{t} = \mathbf{N} \cdot \mathbf{n} = 0$). The local reference frame is invariant under a rotation of $\pi$ about $\mathbf{N}$; hence the group $H_0$ is isomorphic to the product of the cyclic group with two elements $Z_2$ (rotation about $\mathbf{N}$) and of the group $Z$ (translations of the layers). The (semi-direct) product is obtained in such a way that the conjunction of a translation $t$ with a rotation $r$ (about $\mathbf{N}$) reverses $t$:

$$-t = r\,tr^{-1}.$$

This means that if an element of $H_0 = Z \square Z_2$ is represented by the symbol $(n, \alpha)$ we have the multiplication rule

$$(n, \alpha)(m, \beta) = (n + \alpha(m), \alpha\beta), \qquad (10.19)$$

where $\alpha$ and $\beta$ are elements of $Z_2$, $n$ and $m$ are relative integers ($\in Z$). $\alpha(m) = -m$ if $\alpha$ is the non-trivial element of $Z_2$. The calculation of $\pi_1$ ($V(\mathrm{Sm\ C})$) involves an inverse mapping of $H_0$ into $\bar{E}_0$, which is the double covering of $E_0$, the euclidean group without reflections

$$\bar{E}_0 = \bar{R}^3 \square S_3.$$

$\bar{E}_0$ plays here the role that $S_3$ plays for the biaxial nematics. The final result is:

$$\pi_1(V(\mathrm{Sm\ C})) \simeq Z \square Z_4. \qquad (10.20)$$

*Decomposition of $\pi_1$ into conjugation classes.* The results of section 10.4 will be anticipated in the interpretation of the elements of this homotopy group and in linking them to known (or hypothetical) defects of the Sm C phase. It will be shown in section 10.4 that a single defect (in the Volterra sense) is represented in $\pi_1$ by all the elements of a single conjugation class[†] of $\pi_1$, the classes of commutators playing a particular role.[‡] Let us denote $Z_4 = \{e, a, a^2, a^3\}$ the elements of the cyclic group with one generator: $Z_4$ ($a^4 = e$). We obtain the following classes in number $4 + \infty$:

$$
\begin{aligned}
&A : \{(2p, a)\}, &&p \in Z, \\
&A^{-1} : \{(2p, a^3)\}, &&p \in Z, \\
&B : \{(2p + 1, a)\}, &&p \in Z, \\
&B^{-1} : \{(2p + 1, a^3)\}, &&p \in Z, \qquad (10.21) \\
&q_0 : \{(q, e), (-q, e)\}, &&\text{one class for each value of } q \in Z, \\
&q_1 : \{(q, a^2), (-q, a^2)\}, &&\text{one class for each value of } q \in Z.
\end{aligned}
$$

---

† Remember that the conjugation class $C_a$ of an element $a$ of a group $G$ is the set of elements $gag^{-1}$, where $g$ is an element of $G$. If $b$ is in $C_a$, we have $C_b = C_a$.

‡ An element $a$ of $G$ is a commutator if it can be put in the form $a = \alpha\beta\alpha^{-1}\beta^{-1}$, $\alpha$ and $\beta$ being the elements of $G$. The conjugation class of a commutator is formed by commutators.

292

(*q, e*) *elements.* These are pure translation dislocations *d* (see nomenclature in Chapter 7) of smectic layers.

(0, α) *elements.* These are rotation dislocations. The disclinations $(0, e) = (0, a^4)$ are by definition topologically unstable. As with ³He, it must be interpreted as a disclination of angle $4\pi$ (or zero) constructed around any fixed direction of the base trihedron. It follows that $(0, a^2)$ represents a topologically stable rotation dislocation of angle $2\pi$ around every fixed direction of the base trihedron. But (0, α), being its own inverse, represents a *non-orientable line* (see above) (lines of the smectic layering: $\Omega$ along **N**; or lines *m* of the molecular order: $\Omega$ along **n**). This point is demonstrated in Bouligand and Kléman (1979).

The (0, *a*) element, the inverse of which is $(0, a^3)$, creates *orientable lines.* Since this is necessarily a case of rotation of angle $\pm\pi$, it represents rotation dislocations of the smectic layering or of the field **t**: ($\Omega$ along **N**) (Figure 10.8).

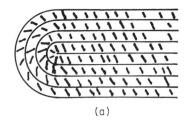

(a)

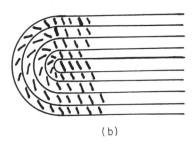

(b)

Figure 10.8. Rotation dislocation $1(+\pi)$ of the smectic layering. (a) $\Omega$ between the layers; (b) $\Omega$ inside the layers

(*n*, α) *elements.* (*n*, α) can be considered as the superimposition of a rotation dislocation (0, α) and a translation dislocation (*n*, *e*), according to the formula

$$(n, \alpha) = (n, e)(0, \alpha). \qquad (10.22)$$

This interpretation, which leads us to consider the different elements of the same conjugation class, such as class *A*, for example, as elements corresponding to different lines, is in contradiction with the fundamental property of conjugation classes, namely that all their elements represent the same object in the Volterra sense. This property is however obvious: if $a$ is the homotopy class of a line $L$ (class obtained by mapping a circuit $\gamma$ surrounding $L$ into $V$), $b$ the homotopy class of a line $L'$, $bab^{-1}$ is also the homotopy class of the line $L$. To obtain this class it is possible to consider a circuit $\delta$ surrounding $L$ once and $L'$ twice but in opposite directions. For $\delta$ finally only surrounds the singularity $L$. Another interpretation can then be given to $bab^{-1}$: let us turn $L$ about $L'$ by a complete turn, following a circuit $\gamma'$, enclosing $L'$; the homotopy class of $\gamma$ becomes $bab^{-1}$ after the return to the starting point.

(1) Thus, the $(n, \alpha)$ elements of a class $A$ or $B$ ($n$ taking all integral values) represent the same object $(0, \alpha)$ or $(1, \alpha)$ as appropriate, with $\alpha = a$ or $a^3$. This can be seen again in the following way: consider a rotation dislocation $(0, \alpha)$ with the axis of rotation (in the Volterra sense) passing through the origin, and displace the axis of rotation by a relative translation of $n/2$ layers. Thus, the cases where $\Omega$ (parallel to $N$) is between the layers (classes $A$ and $A^{-1}$) can be differentiated from cases where $\Omega$ is inside the layers (classes $B$ and $B^{-1}$) (Figure 10.8). More precisely, there is a transformation from Figure 10.8a to Figure 10.8b by the addition of a translation dislocation of odd order.

These considerations do not invalidate the interpretation of equation (10.22). They indicate that, for classes of the types considered, the decomposition (10.22) is topologically unstable and the line $(0, \alpha)$ tends to be displaced by a number $n/2$ of layers.

(2) Each of the $q$ classes contain two elements only, corresponding to the same absolute value of the *Burgers vector $q$* of the translation dislocation of the couple $(\pm q, \alpha)$. The conjugation operation does not change the Burgers vector, and it is therefore convenient here to designate the lines $(\pm q, \alpha)$ as translation dislocations of Burgers vector $\pm q$, coupled to lines of rotation dislocation $(0, e)$ or $(0, a^2)$ as the case may be. Thus, the $(q, e)$ lines are pure translation dislocations.

*Conjugation classes $q = 2n$.* In the first place there is no difference between the interpretation of the $q = 2n$ classes and the $q = 2n + 1$ classes. However, it is necessary to observe that the elements of the conjugation class $q = 2n$ are commutators of the group $\pi_1(V)$. The following example will show what this property implies. The elements of the classes $q = 2n + 1$ can also be written as the product of an element of a class $A$ or $A^{-1}$ and an element of a class $B$ or $B^{-1}$. In contradistinction, the elements of the class $q = 2n$ are products of an element of a class $A$ or $A^{-1}$ and an element of the same set $A$ or $A^{-1}$ (or the product of an element of a set $B$ or $B^{-1}$ and an element of the same set). In the latter case, this means that the two rotation dislocations of the product both have their axis of rotation (in the Volterra sense) either between the layers or inside the layers.

Let us consider the product of two rotation dislocations of opposite signs ($1(+\pi)$ and $1(-\pi)$ in this case) obeying this condition: this product is either a translation dislocation or a pinch, both representable by the same homotopy class $(2n, e)$. The pinch is known to be an evanescent singularity. On the other hand, there is no possibility of obtaining a pinch in a product of rotation dislocations leading to a class $q = 2n + 1$.

*The quotient $\pi_1/D$.* Let us call $D$ the invariant sub-group containing the classes of commutators $q = 2n$. The quotient $\pi_1/D$ represents the different classes of defects defined by the Volterra process (this will be clarified in section 10.4); the neutral element corresponds to the sub-group of commutators. We obviously have:

$$H_1 = \pi_1/D = D_2. \tag{10.23}$$

Classes $A$ and $A^{-1}$ $(=a_1)$, $B$ and $B^{-1}$ $(=a_2)$ and the set of classes $2q + 1$ $(=a_3)$ correspond respectively to the $a_1, a_2, a_3$ elements of $D_2$. It is easy to find in this homomorphism the properties of pairing of defects discussed above. It is also particularly interesting to note that the group $H_1 = D_2$ is Abelian, while the group $H_0 = Z \square Z_2$ (see equation (10.19)), which gives the Volterra classification, is not. Some general comments on such properties will be made in section 10.4.

*Singular points and walls.* The symmetry group $H = H_0 \otimes Z_2$ contains a centre of symmetry; there can therefore be no topologically stable walls in it. There are no topologically stable singular points since $H$ is discrete (see subsection 10.3.2.2).

### 10.3.3.3 *Smectic A* (Kléman and Michel, 1978a)

The local symmetry group $H_A$ is, if reflections are excluded, the semi-direct product of a group of unidirectional translations isomorphic to $Z$ and of the rotation group $D_\infty$ (symmetry group of the director). $H_A$ is non-discrete and the theorems of subsection 10.3.2 cannot be applied here. We are going to obtain the relevant quantities by beginning with the results for Sm C and by using the fact that the symmetry group $H_C$ of the Sm C phase, reflections excluded, is a sub-group of $H_A$. The method we are going to present will lead to the introduction of the concept of a fibre bundle and the use of a classic technique of algebraic topology, the exact homotopy sequence of a fibre bundle.

*First homotopy group $\pi_1(V(Sm\,A))$.* It is possible to pass continuously from a Sm C phase to a Sm A phase by making the angle of the optical axis of the Sm C with the normal to the layer tend to zero. If this geometrically acceptable operation is also physically acceptable, we have a second-order transition. The considerations which follow can therefore be used in such a case.

The geometric relationship between Sm C and Sm A can again by expressed in the following manner: suppose that the manifold of internal states $V_A = V(Sm\,A))$ is known. To every point of $V_A$ corresponds an orientation of Sm A from which an infinite number of orientations of Sm C phases can be deduced: the orientation of the layers is conserved, the optical axis is put at an angle to the normal which is different from zero. The different positions of the optical axis define a circle $S_1$. $V_A$ is then fibred by circles $S_1$ at every point of $V_A$. This geometrical construction ensures that all the orientation of the Sm C are attained once and once only: we have therefore obtained $V = V(\text{Sm C})$ as a fibre bundle over $V_A$, of fibre $S_1$ and base $V_A$. However, $V_C$ is not the direct product of $V_A$ and $S_1$. Its complete construction requires considerations of another order.

Consider, on $V_C$, a set of loops of homotopy class $(0, a^2)$: defect lines of type $m$ correspond to these loops. These lines clearly involve the geometric construction used to deduce $V_C$ from $V_A$: they are therefore homotopic to the $S_1$ fibres in the transition Sm C $\rightarrow$ Sm A. $V_A$ can therefore be obtained from $V_C$ by identifying on $V_C$ all the points of the loops of conjugation class $q_0 = 0$ and the homotopy classes which are deduced from them by inversion of the orientation and by product of the elements of the class $q_0 = 0$; i.e. the $(0, e)$ and $(0, a^2)$ elements. These elements form an invariant sub-group† in the group $\pi_1(V_C)$ which is isomorphic to $Z_2$.

The quotient $\pi_1(V_C)/Z_2$ is isomorphic to $\pi_1(V_A)$. In fact we have:

$$\pi_1(V_C)/Z_2 = (Z_2) + \alpha_1(Z_2) + \alpha_2(Z_2) + \cdots, \tag{10.24}$$

where $\alpha_1, \alpha_2 \ldots$ are the elements of $\pi_1(V_C)$ which are not in $Z_2$. All the elements of a coset $\alpha_i(Z_2)$ obviously correspond to one loop of $V_A$ by our construction, from which:

$$\pi_1(V_A) = \pi_1(V_C)/Z_2 = Z \,\square\, Z_2. \tag{10.25}$$

We also know that $(0, a^2)$ is an element which is equal to its own inverse. Geometrically this means that a loop $S_i$ of homotopy class $(0, a^2)$ can change orientation if it undergoes certain closed displacements on $V_C$ which lead it back on itself. The mappings of $S_1$ fibres onto themselves are therefore either identity mappings or mappings which change their orientation. The group of these mappings is $Z_2$ (group of the fibre bundle). This indicates clearly that the product of $V_A$ and $S_A$ cannot be a direct product.‡

*Exact homotopic sequence.* The other homotopy groups are obtained by using a classical property linking homotopy groups of the base, the fibre and the fibre bundle, namely the following sequence:

$$\rightarrow \pi_2(S_1) \rightarrow \pi_2(V_C) \rightarrow \pi_2(V_A) \rightarrow \pi_1(S_1) \rightarrow \pi_1(V_C) \rightarrow \pi_1(V_1) \tag{10.26a}$$

is a sequence of groups homomorphisms and is *exact*: this means that if three successive elements of the sequence are considered,

$$\rightarrow A \xrightarrow{i} B \xrightarrow{j} C \rightarrow \cdots, \tag{10.26b}$$

the image of the mapping $i$ is equal to the kernel of the mapping $j$

$$i(A) = \ker j, \tag{10.27}$$

† It is necessarily an invariant sub-group since it is composed of conjugation classes.
‡ A product with analogous properties is obtained by fibring a circle (the base of the fibre bundle) by a segment of a straight line which undergoes a rotation of $\pi$ along the circle. This is a Moebius ribbon.

the kernel ker $j$ designating the set of elements of $B$ which have as an image in $C$ the identity element of $C$.

The groups $\pi_n(V_A)$ are then easily calculated for $n \geq 3$. We have in fact:

$$\pi_n(S_1) \to \pi_n(V_C) \to \pi_n(V_A) \to \pi_{n-1}(S_1). \qquad (10.28)$$
$$\downarrow \qquad\qquad\qquad\qquad\qquad \downarrow$$
$$O \qquad\qquad\qquad\qquad\qquad\quad O$$

The homotopy groups of a circle $S_1$ are reduced to the zero element for $n \geq 2$. The property of exactness then reads:

$$\pi_2(S_1) \to \pi_2(V_C) \overset{P*}{\to} \pi_2(V_A) \overset{\Delta}{\to} \pi_1(S_1) \overset{i*}{\to} \pi_1(V_C) \qquad (10.30)$$
$$\downarrow \qquad\quad \downarrow \qquad\qquad\qquad\qquad \downarrow \qquad\qquad \downarrow$$
$$O \qquad\quad O \qquad\qquad\qquad\qquad Z \qquad\quad Z \,\square\, Z_4$$

of which all the groups, except $\pi_2(V_A)$, are known.

The kernel of the mapping $\pi_1(V_C) \to \pi_1(V_A)$ is $Z_2$, according to the geometric discussion of the preceding section. It is therefore also, because of the property of exactness, the image of $\pi_1(S_1)$ in $\pi_1(V_C)$. Therefore the kernel of the mapping $i*$ is formed by one element over two of the $\pi_1(S_1)$ and since the identity is always mapped onto the identity, it is formed of the even elements of $Z$. A geometrical sense can also be given to the mapping $i*$ by noting that its kernel is formed of $S_1$ fibres followed through twice; in this case these circuits are homotopic to zero: $(0, a^2)^2 = (0, I)$. However, it serves no useful purpose to stop at this interpretation. By following the exact sequence, it can be seen that the image of $\pi_2(V_A)$ in $\pi_1(S_1)$ is formed of two even elements of $Z$, and that the kernel of this mapping is the identity element of $\pi_2(V_A)$, since $\pi_2(V_C)$ includes only the identity element. $\pi_2(V_A)$ is therefore mapped isomorphically into its image by the mapping $\Delta$. Therefore $\pi_2(V_A)$ is isomorphic to the set of even elements of $Z$, which is also $Z$:

$$\pi_2(V_A) = Z. \qquad (10.31)$$

*Comments.* It is clear that the singular lines of Sm A are obtained from those of the Sm C simply by resetting the optical axis along the normal to the layers. This operation suppresses the $m$ lines of the Sm C. It will be noted that the focal conics of smectic phases are not brought into this discussion. We shall give a plausible explanation for this in subsection 10.4.4.

Sm A can display singular points (equation (10.31)). This is new with respect to Sm C. Figure 10.9 shows two examples of this. It is easy to be convinced that singular points cannot persist when the Sm $A$ phase approaches to Sm C phase; in fact $m$ lines will be necessarily appear, which may in this case terminate on the singular point. This one cannot therefore remain isolated in a Sm C. This important property has a mathematical analogy in the theory of fibre bundles, which will not be gone into here: the fibre bundle $(V_C, V_A, S_1)$ of group $Z_2$ does not possess a section which is continuous everywhere, i.e. a mapping of $V_A$ onto $V_C$ which is continuous everywhere cannot be constructed.

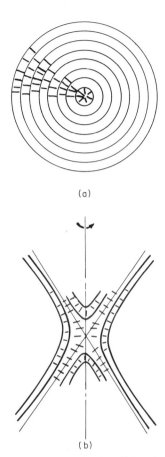

(a)

(b)

Figure 10.9    Singular points in Sm A (a) radial point; (b) hyperbolic point

## 10.4    COMPARISON BETWEEN CLASSIFICATION BY THE VOLTERRA PROCESS AND BY THE FIRST HOMOTOPY GROUP (d = 3)

### 10.4.1    Definition of the Burgers circuit

Consider a line singularity $l$ in the distorted medium, and a closed loop $\gamma$ surrounding it: this loop has been called the *Burgers circuit in the deformed medium*. The image of $\gamma$ in $G$, the thermodynamic group, will be called the *Burgers circuit*. This is generally an open loop. It is obtained in the following manner: let $M$ be a point of origin on $\gamma$; consider a neighbouring point $M + dM$ on $\gamma$, and let $dg \in g$ be the infinitesimal operation which allows the passage from the local orientation at $M$, denoted $\mu$, to the local orientation $\mu + d\mu$ at $M + dM$. $\mu$ and $\mu + d\mu$ are points in $V$, which is the space defined by the absolute orientations of the ordered medium. Two points on $V$ are always linked by at least one operation of $G$ ($V$ is a homogeneous space of $G$). If these two points are next to each other, there is a single infinitesimal operation $dg$, and we can write:

$$\mu' = \mu + d\mu = dg \cdot \mu. \qquad (10.32)$$

By following $\gamma$ from $M$ to $M$, in $G$, a circuit $Q$ is constructed from the origin (neutral element in $G$) to a point $h$ which must be an element of the symmetry group $H(M)$ of the ordered medium at $M$.†

The Burgers circuit $Q$ is obviously the generalization of the traditional Burgers circuit of the theory of translation dislocations, where $G$ is the euclidian space $R^3$ of all the translations, and $H$ the sub-group of $G$ consisting of the Bravais lattice. In this case $G/H = T^3$, the three-dimensional torus.

Note that the symmetry group at $M + dM$ is

$$H(M + dM) = dg \cdot H(M) \cdot dg^{-1}$$

which is an inner automorphism of $H(M)$; after following the circuit we have the automorphism

$$H(M) \to h \cdot H(M) \cdot h^{-1} \qquad (10.33)$$

The new symmetry group is isomorphic to the starting group: we can say that the ordered medium has undergone an operation of symmetry $h$, and the presence of the line $l$ means that the existence of this element of symmetry in $H$ can be recognized, which fact removes the degeneracy.

However the terminating point of the image $Q$ of $\gamma$ in $G$ is not uniquely fixed; it is defined up to a conjugation of $h$; for example it can be $k \cdot h \cdot k^{-1}$ where $k$ is another element of $H(M)$. Physically such a difference would correspond to the fact that a line defined by the element of symmetry $k$ would have been allowed to rotate once about the line defined by the element of symmetry $h$ (see Chapter 4, Figure 4.14). But there is another indeterminacy on $Q$, which relates to the first homotopy group of $V$. We discuss this new indeterminacy in the following subsection, as well as that one already evoked, to show that they correspond to two aspects which can be considered as belonging to the same topological phenomenon.

## 10.4.2 Relationship between the Burgers circuit, the elements of $\pi_1(V)$ and the Volterra classification

Consider all the circuits $Q(h)$ beginning at 0 and ending at $h$; they are necessarily unequivalent, if the first homotopy group of $G$ is not trivial. We have already encountered such a situation for the $N_B$ phase (subsection 10.3.3.1).

Let us represent the $n$ elements of $\pi_1(V)$ by $n$ curves closed in $G$ and passing through 0, denoted $Q_1(0), Q_2(0), \ldots, Q_n(0)$. To each of these curves correspond, by the quotient operation $G/H$, $n$ elements in $V$, belonging to $\pi_1(V)$, denoted $\alpha_i(0)$.

---

† $G$ is the space defined by the relative orientations and $V$ defined by the absolute orientations of the ordered medium.

The set $D_0 = \bigcup_i \alpha_i(0)$ is an invariant subgroup of $\pi_1(V)$. We easily get the following relationships

$$\alpha_j(h)\alpha_i(0) = \alpha_k(h)$$
$$\alpha_i(0)\alpha_j(h) = \alpha_l(h) \qquad (10.34)$$

where $\alpha_j(h)$ designates an element in $\pi_1(V)$ corresponding to one of the Burgers circuit $Q_j(h)$ by the quotient operation.† The subscripts $i$, $j$, $k$, $l$ are integers belonging to the set $[1, n]$ and are not better defined here, since there is no necessity. Now, if we consider a Burgers circuit passing successively through the points $h$, $h \cdot g$, $h \cdot g \cdot h^{-1}$ and closing at origin, we obtain

$$\alpha_k(0) = \alpha_j(h \cdot g^{-1} \cdot h^{-1})\alpha_i(h^{-1})\alpha_j(g)\alpha_i(h) \qquad (10.35)$$

Another Burgers circuit, passing through the points $h$, $h \cdot g$, $h \cdot g \cdot h^{-1}$ and terminating at $h \cdot g \cdot h^{-1} \cdot g^{-1}$ gives

$$C_k = \alpha_k(h \cdot g \cdot h^{-1} \cdot g^{-1}) = \alpha_j(g^{-1})\alpha_i(h^{-1})\alpha_j(g)\alpha_i(h) \qquad (10.36)$$

The elements of $\pi_1(V)$ which are commutators, i.e. of the form of $C_k$, play a special role. We prove that

$$\alpha_j(h)C_i = \alpha_k(g^{-1} \cdot h \cdot g)$$
$$C_i\alpha_j(h) = \alpha_l(e^{-1} \cdot h \cdot e) \qquad (10.37)$$

where $C_i$ is any commutator of $\pi_1(V)$, and $g$ and $e$ some other elements, of $\pi_1(V)$.

Using (10.35) and (10.36), we get

$$\alpha_k(g^{-1} \cdot h \cdot g) = \alpha_l(0)\alpha_i(g^{-1})\alpha_j(h)\alpha_i(g)$$
$$= \alpha_l(0)\alpha_q(h^{-1} \cdot g^{-1} \cdot h \cdot g)\alpha_r(h)$$

But $\alpha_l(0)\alpha_q(h^{-1} \cdot g^{-1} \cdot h \cdot g)$ is also a commutator, according to (10.35); therefore the theorem is proved for $C_i$ acting on the right of $\alpha(h)$. It is clear that it is true also for $C_i$ acting on the left; hence equation (10.37) is proved.

It appears clearly that equation (10.37) describes, in the first homotopy group, the indeterminacy on the terminating point of the Burgers circuit we discussed in subsection 10.4.1, while equation (10.34) describes the indeterminacy on the Burgers circuit itself, when this point is fixed. These two indeterminacies can be lumped together by recognizing that the classes of homotopy $\alpha_i(0)$ as well as the $C_i$ pertain to the same invariant subgroup $D$ of $\pi_1(V)$, the so-called commutator subgroup, since

$$\alpha_i(h \cdot g \cdot h^{-1} \cdot g^{-1})\alpha_j^{-1}(h \cdot g \cdot h^{-1} \cdot g^{-1}) = \alpha_k(0)$$

for some $k$. Therefore to any element of $D$ (even if it is not a commutator itself) corresponds a different choice in the Burgers circuit, for the same element of symmetry $g$ defining a Volterra process. This can be better expressed by

---

† We are indebted to Dr. H. R. Trebin for an enlightening correspondence on the points discussed in this subsection (see also a forthcoming review paper of this author).

300

introducing the quotient operation

$$\pi_1(V)/D = H_1(V) \tag{10.38}$$

$H_1(V)$ is called the first homology group and is in fact the set of classes of independent 1-cycles on $V$. But, more important for us, all the different Burgers circuits corresponding to the same element of symmetry $g$ (i.e. to the same element in the Volterra classification) are in the same coset of $\pi_1\ (V)/D$.

Note that $H_1(V)$ is an Abelian group (this is a property relating to the quotient of a group by its commutator subgroup).

### 10.4.3   Physical significance of commutators and conjugation classes in $\pi_1(V)$

#### 10.4.3.1   *Borromian knot* (Figure 10.10)

Two singularity lines $l_1$ and $l_2$ are linked by a circuit which encircles each of them twice in opposite directions. If we take away $l_1$, the line $l_2$ is freed from the knot $(B)$ and vice versa. The homotopy class of $(B)$ is a commutator

$$\alpha(B) = a_1^{-1}a_2^{-1}a_1a_2, \tag{10.39}$$

the exact form of which depends on the origin on $(B)$ and the orientation of $(B)$. If the origin and/or the orientation is changed, we obtain a sub-set of classes $\alpha_i(B)$ which belong to a class of commutators $C$ (which is a conjugation class), or two classes, $C_1$, $C_2$, if $C$ is not ambivalent.†

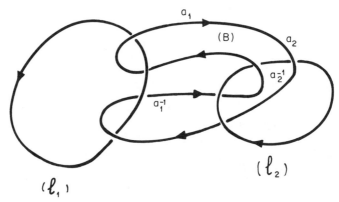

Figure 10.10. Borromian knot $(B)$ linking two singularity lines $l_1$ and $l_2$: the homotopy class of $(B)$ is a commutator of the group $\pi_1(V)$

#### 10.4.3.2   *Topological interaction between lines*

By multiplying the two sides of equation (10.39) by $a_1$, we obtain:

† A conjugation class $C$ is called ambivalent if it contains the inverses of all its elements:

$$a_1\alpha(B) = a_2^{-1}a_1a_2.\tag{10.40}$$

This quantity is the homotopy class of a circuit which surrounds $l_1$ once and $l_2$ twice in opposite directions. This is therefore a homotopy class of $l_1$. The conjugation class of $a_1$ therefore represents the set of possible topological interactions of $l_1$ with other loops present in the medium.

### 10.4.3.3 *Entangled lines* (Figure 10.11a)

It is easy to see that the homotopy class of $l_1$ and $l_2$ must commute

$$a_1^{-1}a_2a_1a_2^{-1} = 1.\tag{10.41}$$

Two lines whose homotopy classes do not commute cannot be freely entangled: they must be linked by a segment of line whose homotopy class $\alpha$ is a commutator (Figure 10.11b).

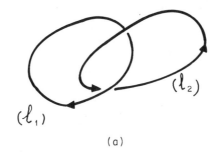

(a)

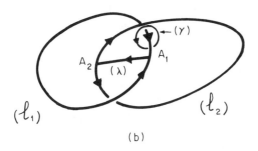

(b)

Figure 10.11. (a) The homotopy classes of two entangled lines must commute. (b) If they do not commute, the lines are linked by a singularity line whose homotopy class is a commutator belonging to an ambivalent conjugation class

$$\alpha = a_1a_2a_1^{-1}a_2^{-1}.\tag{10.42}$$

This expression can be understood as the product of class $a_1$ and $a_2a_1^{-1}a_2^{-1}$, i.e. the homotopy class of the line $l_1$ oriented in an inverse direction, after the closed circuit $\gamma_1$ has been taken along $l_1$ from one side of the knot $A_1$ to the other. The

302

product $\alpha' = a_2 a_1^{-1} a_2^{-1} a_1$ is in the same conjugation class as $\alpha$; the order of the product of the classes of $\lambda$ is not involved. But the same calculation can be made around the knot $A_2$. We then find:

$$\alpha'' = a_1 a_2^{-1} a_1^{-1} a_2. \tag{10.43}$$

$a''$ is not in the same conjugation class as $\alpha$ and $\alpha'$, but in the class containing $\alpha^{-1}$. Since this calculation was made before giving $\lambda$ a well-defined orientation, the result means that $\lambda$ must be non-orientable, i.e. belong to an ambivalent conjugation class. This is the case for example with the $-I$ element in the homotopy group $N_\mathrm{B}$: a line $\lambda(-I)$ can link two entangled lines $e_i$ and $e_j$.

### 10.4.3.4 *Relative displacement of two lines*

If a rectilinear line $l_1$ is turned about a line $l_2$ which is parallel to it, the line $l_1$ follows a circuit which is homotopic to the closed Burgers circuit surrounding $l_2$. Its homotopy class $a_1$, after a complete turn, becomes:

$$a_1' = a_2 a_1 a_2^{-1}. \tag{10.44}$$

$a_1'$ is in the same conjugation class as $a_1$. If this class is ambivalent, the line $l_1$ may, after completion of the circle, have changed sign (see, for example, Chapter 4, Figures 4.11 to 4.14).

### 10.4.4   The role of the cut surface in the boundary conditions

Let there be a line $l$ in the deformed medium. Consider a surface $\Sigma$ resting on $l$. $\Sigma$ is an image of the cut-surface of the perfect medium, before any Volterra process, and it too will be called the cut surface. It is proposed that the topology of the image of $\Sigma$ in $G$ (in a sense which will be defined below) is an essential element of the definition of the line.

$\Sigma$ is the locus of the points of orgin $M$ of the closed Burgers circuits $\gamma$ on which the homotopy classes of $l$ are calculated (Figure 10.12).

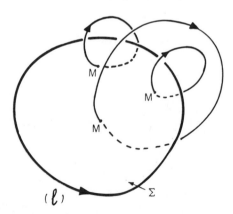

Figure 10.12. Cut surface and closed Burgers circuits originating on $\Sigma$

There corresponds to this set of circuits a set of (open) Burgers circuits $Q$ in $G$ whose extremities $h$ describe a manifold in two dimensions ($\sigma$), situated on the sub-manifold of $G$ defined as the set of conjugate points of $h_0$, the extremity of a particular direction $Q_0$. Let $N$ be this sub-manifold.

(1) If $\sigma$ can be contracted to a point, all the circuits $\gamma$ surrounding the line $l$ have images $Q$ homotopic to each other. A deformation of $\Sigma$ which conserves its topological properties (easily described in $G$) can therefore lead to a situation where the Volterra process is applied according to the definition in Chapter 1.

(2) If $\sigma$ is a non-closed manifold which can eventually be contracted to a line, it is convenient to consider the homotopy group $\pi_1(N)$ of the lines on $N$: there are then as many possible types of cut surfaces as there are classes of non-trivial lines on $N$. Take the case where $h$ corresponds to a rotation $\pi/2$ (on SO(3)). We therefore have $N = P^2$, from which $\pi_1(N) = Z_2$.

Our example corresponds, among other cases, to the nematic phase: there are two types of lines in the nematic as noted in section 10.1 (see Figure 10.1). Note the resemblance of Figure 10.1 to a focal domain of a smectic: also note that in a focal domain the hyperbola is not a true singular line: by local rearrangements the curvature of the layers can always be made to take regular forms. The ellipse should therefore be considered to be a singular line: it is obvious that it is also a line for which the Volterra process is of the type shown in Figure 10.1. Yet another observation can be made: such lines are fundamentally non-orientable (in the sense discussed above) and do not carry singular points. In conclusion, our analysis allows the differentiation of two *Volterra processes* for non-orientable lines, one leading to the classic *pinch* situation, the other to focal conics.

(3) If $\sigma$ is a closed manifold, the classification of the different $\Sigma$ is made according to $\pi_2(N)$. There is at present no example of this situation.

## 10.5   ON THE USE OF THE EXACT HOMOTOPY SEQUENCE IN VARIOUS PROBLEMS INVOLVING DEFECTS

### 10.5.1   Some remarks on defects at second-order phase transitions

In subsection 10.3.3.3 we introduced a number of properties of homomorphism, linking homotopy groups of the base, the fibre and the fibre bundle. They were used in the calculation of the homotopy groups of the Sm A phase, comparing them to the homotopy groups of the Sm C phase, whose symmetry group is a subgroup of the symmetry group of the Sm A phase. Clearly, this method, using an exact homotopy sequence (E.H.S.), can be extended to allow us to compute the homotopy groups of phases such that one of them has a symmetry group which is a sub-group of the other. In all those cases the larger manifold of internal states, $V_A$ say (corresponding to the smaller symmetry group $H_A$), is a fibre bundle over $V_B$, the base (the manifold of the larger symmetry group $H_B$). The fibre itself, along which the fibre bundle is *projected* on the base, is the quotient $F = H_B/H_A$ (see Steenrod, 1951, p. 30: the bundle structure theorem) which is a manifold; the same can be said for the quotients $V_A = G/H_A$ and $V_B = G/H_B$ (see equation (10.15)). The homotopy groups of the fibre classify new defects appearing at the transition $B \to A$. Note that $V_B$ should not be included in $V_A$: for example, the fibre bundle over a circle $V_B = S_1$, with fibre $F = Z$ the set of integers, is the real line $V_A = R$. This can be seen as follows: choose an origin of coordinates on $R$ and choose between each set of successive integral coordinates a number $0 < \phi < 1$. All points $n + \phi$, where $n$ is an integer, project on the same point (of azimut $2\pi\phi$) of the circle.

A simple exercise in exact homotopy sequence which we leave to the reader and which illustrates other possibilities of the method is the following: consider the phase transition $N \rightleftarrows \text{Sm A}$ between a nematic and a smectic A. We have here $V_A = P_2$. $V(\text{Sm A})$ is a fibre bundle over $P_2$ with fibre $F = S_1$. Using the E.H.S., it is easy to show that translation dislocations in Sm A may have a continuous nematic core because their classes of homotopy are in the kernel of $\pi_1(V_B) \to \pi_1(V_A)$, so that they can disappear without leaving any track at the Sm A $\to$ N transition, except some large distortions measured by curl $\mathbf{n}$. Conversely, the $N \to \text{Sm A}$ transition is marked by the appearance of defects classified by $\pi_1(F) = Z$, which are essentially screw dislocations (in the regions of the nematic where $\mathbf{n} \cdot \text{curl}\,\mathbf{n} \neq 0$), and edge dislocations (in the regions where $\mathbf{n} \wedge \text{curl}\,\mathbf{n} \neq 0$). Of course, the nucleation and growth of these defects is a problem far outside the realm of homotopy; the exact homotopy sequence tells us nothing about the necessity of these defects (more constraining here than in Sm A $\to$ Nem, since $\mathbf{n} \cdot \text{curl}\,\mathbf{n} \neq 0$ generally in the Nem phase), because $\pi_1(S_1)$ is the counter-image of the trivial element in $\pi_1(P_2)$, so that no defect in the Nem phase has obligation to transform to a dislocation.

## 10.5.2 Surface defects and cores of defects

### 10.5.2.1 *Surface lines*

Other situations arise in which the E.H.S. can clarify the physical situation. Examples have already been published (Mermin, 1979; Mineyev and Volovik, 1978; Volovik, 1978), either concerning the classification of defects at boundaries (where the order parameter is submitted to constraints), or when the order parameter of a given medium depends on the length on which the medium is studied. In both cases one has to introduce two manifolds, $V_1$ and $V_2$, say, such that one of them is *included* in the other ($V_2 \subset V_1$). The concept of homotopy has to be modified to one of *relative homotopy*.

Let us suppose, for example, that we want to classify point singularities on the boundary of a sample. Such singularities can be either isolated points or ends of singular lines in the bulk. We surround the point singularity by a loop, which we map on some manifold $V_2$ which is representative of the order on the boundary. Is there a relationship between $\pi_1(V_2)$, $\pi_1(V_1)$ and other homotopy groups?

Since the boundary conditions in general restrict the degrees of freedom of the order parameter, rather than enlarge them, $V_2$ is a sub-manifold of $V_1$. Hence, it is possible to compare the homotopy classes of $\pi_1(V_1)$ and $\pi_1(V_2)$. If $a$ is an element of $\pi_1(V_2)$ which is also in $\pi_1(V_1)$, then this clearly means that the corresponding surface singularity point is the end of a singularity line in the bulk. Isolated point singularities are represented by elements $b$ of $\pi_1(V_2)$ which vanish in $\pi_1(V_1)$ when the loop $\gamma$, surrounding the point of the surface is transported in a smooth manner in the bulk. This corresponds to the operation of inclusion $i$ of $V_2$ in $V_1$. This transport of $\gamma$ clearly defines a homomorphism between $\pi_1(V_2)$ and $\pi_1(V_2)$; all elements $b$ are in the kernel of this homomorphism. Finally, some surface point singularities can be generated by bulk point singularities arriving on the surface; these objects are truly point singularities and are not in ker $\{\pi_1(V_2) \to \pi_1(V_1)\}$, but are classified by the classes of homotopy of half-spheres $\sigma_2'$ embedded in the bulk and whose limiting circles are on the boundary. These are *relative homotopy groups* $\pi_2(V_1, V_2)$; the limiting circle is mapped in $V_2$. If the image of $\sigma_2'$ in the mapping is entirely in $V_2$, we contract it to a point in $V_2$; its class of homotopy is the identity element in $\pi_1(V_1, V_2)$. In fact, we have the E.H.S.:

$$\to \pi_2(V_2) \xrightarrow{i_2} \pi_2(V_1) \xrightarrow{j_2} \pi_2(V_1, V_2) \xrightarrow{\partial_2} \pi_1(V_2) \xrightarrow{i_1} \pi_1(V_1) \to . \qquad (10.45)$$

In this sequence, $\partial$ is an operation which consists of considering in $\sigma_2'$ only its boundary $\partial\sigma_2'$ on the specimen surface. It induces a group homomorphism between $\pi_2(V_1, V_2)$ in $\pi_1(V_2)$. One has, by the property of exactness,

$$\text{im } \partial_2 = \text{ker } i_1,$$

where im $\partial_2$ is the image of $\pi_2(V_1, V_2)$ in $\pi_1(V_2)$. $j$ is the operation of inclusion which consists of considering any $\sigma_2$ in $V_1$ modulo $V_2$ (i.e. reducing it to $\sigma_2'$).

Using (10.45), it can be proved that $\pi_2(V_1, V_2)$ is finally the product of two groups of elements; the first is ker $i_1$ and describes true singular surface points which are not terminations of lines; the second is the quotient group $\pi_2(V_1)/\text{im } i_2$ and describes singular surface points coming from the bulk. Surface points which are terminations of lines are in im $i_1$.

This discussion (which can be extended to any dimensionality of surface defects) shows how complex the classification of these defects can be and how powerful the homotopy method. Let us take the example of a nematic phase with parallel boundary conditions (Volovik). One has $V_1 = P_2$ and $V_2 = S_1$. Here ker $i_1 = Z$; these are the terminations of the boundary of the disclinations of integer strength which are non-singular in the bulk but cross the surface with a singular point. In what concerns $\pi_2(V_1)/\text{im } i_2$, it is equal to $Z$ also, but classifies here the bulk point singularities, which are all topologically stable on the surface. The rest of the surface singularities are the termination of half-integer lines. We have finally:

$$\pi_2(P_2, S_1) = Z \otimes Z. \qquad (10.46)$$

Let us now apply the same analysis to the dipole-locked phase of $^3$He-A for which $V_2 = S_1 \times Z_2$;† the inclusion of $V_2$ in $V_1$ is a closed loop whose homotopy class is

---

† It is interesting to notice that, although $V_2$ is a doubly connected manifold (two circles), its inclusion in $V_1$ is a unique circle. This does not affect the reasoning. Here we have assumed the same boundaries conditions as in Chapter 6, i.e. that $\Delta_1$ and $\Delta_2$ are parallel to the boundary.

the non trivial class of $\pi_1(V_1) = Z_2$. One has:

$$\pi_2(V_2) = \pi_2(V_1) = 1; \qquad \pi_1(V_2) = Z; \qquad \pi_1(V_1) = Z_2. \qquad (10.47)$$

The way $V_2$ is included in $V_1$ tells us that the induced inclusion $i: \pi_1(V_2) \to \pi_1(V_1)$ maps all multiple double loops in $V_2$ on the trivial element of $\pi_1(V_1)$, whereas any loop on $V_2$ which surrounds $V_2$ an odd number of times is mapped on the non-trivial element of $V_1$, i.e. on the non-trivial loops of $V_1$. Therefore any surface point singularity which pertains to ker $i_1$ has a configuration of *even* strength: $\Delta_1$ and $\Delta_2$ turn by angles which are multiple of $\pm 4\pi$ on the boundary. These singularities are the 'booja' of Mermin (1977). There are no other true point singularities on the surface, since $N = \pi_2(V_1)/\text{ker}\, j_2 = 1$: in other words the non-existence of topologically stable points in He-A limits the boundary singular points to the above class. Any boundary singular point having an *odd* strength must be a pinning point of a topologically stable line in the bulk of $^3$He-A. Eventually, any topologically stable singular point can split into two singular points of odd strength, linked by a singular line. Which configuration is stable is a question of energetics.

'Booja' in cholesterics have also been discussed (Stein *et al.*, 1978). We shall assume that the so-called cholesteric axis is perpendicular to the boundary; therefore $V_2 = S_1/Z_2 = S_1$. We have $\pi_2(V_1) = \pi_2(V_2) = 1$; $\pi_1(V_2) = Z$; and $\pi_1(V_1) = Q$, where $Q$ is the quaternion group. The inclusion of $V_2$ in $V_1 = \text{SU}(2)/Q$ is not easy to figure out. As in subsection 10.3.3.1, let us call $I$, $-I$, $\pm e_1$, $\pm e_2$ and $\pm e_3$ the five conjugation classes of $Q$. To each of them corresponds a certain type of loop included in $V_1$. Assume that the class $\pm e_1$ represents the $\chi$ $(\pm \frac{1}{2})$ lines in the classification of Friedel and Kléman, the class $\pm e_2$ the $\lambda(\pm \frac{1}{2})$ lines, and the class $\pm e_3$ the $\tau(\pm \frac{1}{2})$ lines. Clearly, the manifold $V_2$ is in the mapping in $V_1$ of a Burgers circuit surrounding a $\chi$ line. Its homotopy class in $\pi_1(V_2)(=Z)$ is $\pm 1$ and is $\pm e_1$ in $\pi_1(V_1)$. If $V_2$ is surrounded twice (elements of homotopy classes $\pm 2$ in $\pi_1(V_2)$), its homotopy class in $V_1$ is $-I = (\pm e_1)^2$. If it is surrounded three times ($\pm 3$ in $\pi_1(V_2)$), its homotopy class in $V_1$ is again $\pm e_1$. Only elements $\pm 4n$ in $\pi_1(V_2)$ map on $I$ in $\pi_1(V_1)$: they constitute the kernel of the inclusion $i$ and correspond to possible 'booja'. Here, as above in the $^3$He-A phase, splittings are possible which would divide the singular points on the surface in two singular points linked by a $(-I)$, or in four singular points linked by $(\pm e_1)$ lines.

## 10.5.2.2 *Cores of defects*

We have already introduced a problem of this type in subsection 10.5.1. Analogous problems happen with phases whose order parameter depends on the scale in which it is observed; an example is with $^3$He-A, $V_1$ being the dipole-free A phase, $V_2$ the dipole-locked phase (Mineyev and Volovik), for which

$$V_1 = (\text{SO}(3) \times S_2)/Z_2; \qquad V_2 = \text{SO}(3). \qquad (10.48)$$

The $V_1$ phase is coherent on distances $\xi \ll r < \xi_{\text{dip}}$, where $\xi$ is a typical atomic dimension, and $\xi_{\text{dip}} \sim 10^2$–$10^3 \xi$. The $V_2$ phase is coherent at distances larger than $\xi_{dip}$. $V_2$ is included in $V_1$.

As already stated, the order parameter of $^3$He-A can be represented by a $3 \times 3$ tensor of the form:

$$A_{\mu i} = d_\mu l_i,$$

where $\mathbf{d}$ is a real unit vector in spin space and $\Delta_1 + i\Delta_2$ is a complex nilpotent vector such that $\mathbf{l} = \Delta_1 {}_\wedge \Delta_2$ is the orbital moment of the Cooper pair. $\Delta_1 \cdot \Delta_2 = 0$. In the dipole-free phase $\mathbf{l}$ and $\mathbf{d}$ are uncorrelated: they are parallel in the dipole-locked phase.

Let us consider first lines of singularity and apply the E.H.S. (10.4.3). A number of results follow.

(a) Lines in the $V_2$ phase which are in ker $i_1$ have a core of perfect $V_1$ phase, since their homotopy classes map on the trivial class of $\pi_1(V_1)$. Hence, the core is non-singular. Notice that the core has less symmetry elements (is more ordered) than the outside region.

(b) Lines in the $V_2$ phase which are not in ker $i_1$ have a singular core. The region surrounding the core is made of $V_1$ material, and any Burgers circuit in this region belongs to im $i_1$. Reciprocally, lines in the $V_1$ phase which are in im $i_1$ are also lines in the surrounding $V_2$ phase.

(c) Lines in the $V_1$ phase which are not in im $i_1$ are not surrounded at a large distance by a 'good crystal' of the $V_2$ phase. The indication of this difficulty is in considering the mapping $j_1 : \pi_1(V_1) \rightarrow \pi_1(V_1, V_2)$; the image of the Burgers circuit surrounding the line in $V_1$ is open in $V_1$, with ends in $V_2$. Hence a *wall*, inside which the order parameter is in $V_1$, terminates on the line.

The study of points of singularity requires that the E.H.S. be extended to the left:

$$\pi_3(V_1, V_2) \xrightarrow{\partial} \pi_2(V_2) \xrightarrow{i_2} \pi_2(V_1) \xrightarrow{j_2} \pi_2(V_1, V_2).$$

We have results very similar to the former ones, replacing the word 'line' by the word 'point' and 'wall' by 'line'. One result seems worthy of emphasis: points in the $V_1$ phase which are not in im $i_2$ are ends of thread-like non-singular regions containing $V_1$ material. Let us assume that such a thread is finite and terminated by two such points of $V_1$ material, whose homotopy classes are not in im $i_2$. Draw a circuit $\Gamma$ (homotopic to a sphere) which contains these two points and is entirely in $V_1$ (this circuit is entirely situated in the thread-like region). $\Gamma$ contains a singularity of $V_1$ whose homotopy class is the product of the homotopy classes of the point singularities of its ends.

But im $i_2$ is a subgroup of $\pi_2(V_1)$. Hence, the homotopy class of $\Gamma$ is not in im $i_2$ and does not map (by $i_2^{-1}$) on any homotopy class of $\pi_2(V_2)$. This has an obvious physical consequence: it is not possible to bring the circuit $\Gamma$, by smooth movement in the ordered medium (without traversing the singular points) in a region of $V_2$ material made of good crystal, except if the homotopy class of $\Gamma$ is the trivial homotopy class. This must therefore be the case, because $\Gamma$ must be transportable smoothly to a good crystal region. Hence, the two singular points terminating the thread-like region have homotopy classes inverse one from the other, and the thread-like region is presumably unstable by collapse of the two end-points. Eventually, the thread-like region can be made of three (or more) branches with end-points whose total homotopy class is trivial. This can stabilize the whole configuration.

We have discussed a situation in which $V_1$, the larger manifold, describes the order parameter on larger scales than $V_2$, the smaller manifold. The inverse situation might happen (although we do not know of any physical example), and could be discussed in the same way as above.

# References

Abramovicz, M. and Stegun, I. (1965) *Handbook of Mathematical Functions* (Dover, New York).

Abrikosov, A. A. (1957) *Zn. Eksp. Teor. Fiz.*, **32**, 1442.

Adams, J. and Haas, W. (1971) *Mol. Cryst. Liq. Cryst.*, **15**, 27.

Aharoni, A. (1966) *J. Appl. Phys.*, **37**, 4615.

Allet, C., Kléman, M. and Vidal, P. (1978) *J. de Phys.*, **39**, 181.

Ambegaokar, V., de Gennes, P. G. and Rainer, D. (1974) *Phys. Rev.*, A, 9, 2676.

Amelinckx, S. (1957) *Physica*, **23**, 663.

Anders, W. and Biller, E. (1970) *Phys. Status Solidi*, (a) **3**, K71.

Anderson, P. W. (1973) *Phys. Rev. Lett.*, **30**, 1135.

Anderson, P. W. and Brinkman, W. F. (1975) in *The Helium Liquids*, ed. J. G. M. Armitage and I. F. Farquhar (Academic Press, London).

Anderson, P. W. and Morel, P. (1961) *Phys. Rev.*, **123**, 1911.

Anderson, P. W. and Toulouse, G. (1977) *Phys. Rev. Lett.*, **38**, 508.

Anisimov, S. I. and Dzyaloshinskii, I. E. (1972) *Sov. Phys. J.E.T.P.*, **36**, 774.

Anthony, K. H. (1970a) in *Fundamental Aspects of Dislocation Theory*, Natl. Bur. Std. (U.S.). Spec. Publ. 317, **1**, 317.

Anthony, K. H. (1970b) *Arch. Rat. Mech. Anal.*, **39**, 43.

Anthony, K. H., Essman, U., Seeger, A. and Trauble, H. (1968) in *Mechanics of Generalized Continua*, ed. E. Kröner (Springer Verlag, Berlin) 355.

D'Arcy Thompson, (1968) *On Growth and Form* (Cambridge University Press, Cambridge).

Armstrong, R. W. (1968) *Science*, **102**, 799.

Arnold, V. (1974) *Equations Différentielles Ordinaires* (Editions MIR Moscow).

Asher, S. A. and Pershan, P. S. (1979) *J. Phys.*, **40**, 161.

Baessler, H. and Labes, M. (1970) *J. Chem. Phys.*, **52**, 631.

Barbara, B. (1973) *J. Physique*, **34**, 1039.

Bartolino, R. and Durand, G. (1978) *Annal. de Phys.*, **3**, 257.

Baürich, H. (1966) *Phys. Status Solidi*, **16**, K39.

Belavin, A. A. and Polyakov, A. M. (1975) *J.E.T.P. Lett.*, **22**, 245.

Benattar, J. J., Doucet, J., Lambert, M. and Levelut, A. M. (1980) *Phys. Rev.*, A5, submitted.

Bidaux, R., Boccara, N., Sarma, G., de Sèze, L., de Gennes, P. G. and Parodi, O. (1973) *J. Physique*, **34**, 661.

Bilby, B. A. (1960) *Prog. Solid. Mech.*, **1**, 329.

Bilby, B. A. and Smith, E. (1956) *Proc. R. Soc.*, A **236**, 481.

Bilby, B. A., Bullough, R. and Smith, E. (1955) *Proc. R. Soc.*, A **231**, 263.

Billard, J. (1968) *Mol. Cryst.*, **3**, 227.

Billard, J. (1980) European Conference on Liquid Crystals, January 1980, Garmisch-Partenkirchen.

Billard, J. and Urbach, W. Z. (1972) *C. R. Hebd. Séan. Acad. Sci.*, **274B**, 1287.

Bitter, F. (1931) *Phys. Rev.*, **38**, 1903.

309

310

Blaha, S. (1976) *Phys. Rev. Lett.*, **36**, 784.
De Blois, R. W. (1965) *J. Appl. Phys.*, **36**, 1647.
De Blois, R. W. (1968) *J. Appl. Phys.*, **39**, 442.
De Blois, R. W. and Graham, C. D. (1958) *J. Appl. Phys.*, **29**, 932.
Bobeck, A. H. and Della Torre, E. (1975) *Magnetic Bubbles* (North-Holland, Amsterdam).
Bollman, W. (1970) *Crystal Defects and Crystalline Interfaces* (Springer-Verlag, New York).
Bouligand, Y., (1969) *J. Physique Colloq.*, **30**, C 4-90.
Bouligand, Y., (1972a) *J. Physique*, **33**, 525.
Bouligand, Y., (1972b) *J. Physique*, **33**, 715.
Bouligand, Y, (1973a) *J. Physique*, **34**, 603.
Bouligand, Y., (1973b) *J. Physique*, **34**, 1011.
Bouligand, Y., (1973c) *J. Microscopie*, **17**, 145.
Bouligand, Y., (1974) *J. Physique*, **35**, 959.
Bouligand, Y., (1975) *J. Physique Colloq.*, **36**, Cl-331.
Bouligand, Y. (1980) in *Dislocation in Solids*, ed. F. R. N. Nabarro, vol. 5 (North-Holland, Amsterdam).
Bouligand, Y. and Kléman, M. (1970) *J. Physique*, **31**, 1041.
Bouligand, Y. and Kléman, M. (1979) *J. Phys.*, **40**, 79.
Bouligand, Y., Cladis, P. E., Liébert, L. and Strzelecki, L. (1973) *Mol. Cryst. Liq. Cryst.*, **21**, 1.
Bouligand, Y., Derrida, B., Poenaru, V., Pomeau, Y. and Toulouse, G. (1978) *J. Phys.*, **39**, 863.
Bourdon, L. (1980) Third Cycle Thesis, Orsay.
Bourdon, L., Sommeria, J. and Kléman, M. (1982) *J. Physique*, **43**, 77.
Bourret, A. and Dautreppe, D. (1966) *Phys. Stat. Solidi*, **13**, 559.
Bourret, A. and Kléman, M. (1967) *Phys. Status Solidi*, **23**, 207.
Bowen, D. K. and Hall, C. R. (1975) *Microscopy of Materials* (Macmillan, London).
Brazovskii, S. A. and Dmitriev, S. G. (1975) *Zh. Eksp. Teor. Fis.*, **69**, 979.
Brazovskii, S. A. and Dmitriev, S. G. (1976) *Sov. Phys. J.É.T.P.*, **42**, 497.
Brochard, F. and de Gennes, P. G. (1970) *J. Physique*, **31**, 691.
Brown, W. F., Jr (1940) *Phys. Rev.*, **58**, 736.
Brown, W. F., Jr (1941) *Phys. Rev.*, **60**, 139.
Brown, W. F., Jr (1965) *J. Appl. Phys.*, **36**, 994.
Brown, W. F. Jr (1966) *Magnetoelastic Interactions* (Springer Tracts in Natural Philosophy, Berlin).
Brownsey, G. J. and Leadbetter, A. J. (1980) *Phys. Rev. Lett.*, **44**, 1608.
Brunet, M. (1975) *J. Physique Colloq.*, **36**, C 1-321.
Brunet, M. and Parodi, O. (1982) *J. Physique*, **43**, 515.
Brunet, M. and Williams, C. E. (1978) *Annal. de Phys.*, **3**, 237.
Callen, E. and Callen, H. B. (1965) *Phys. Rev.*, A **139**, 455.
Cano, R. (1967) *Bull. Soc. Franç. Minér. Cristall.*, **90**, 333.
Cano, R. (1968) *Bull. Soc. Franç. Minér. Cristall.*, **91**, 20.
Cano, R. and Châtelain, P. (1964) *C. R. Hebd. Séan. Acad. Sci.*, B **259**, 252.
Carey, R. and Isaac, E. D. (1966) *Brit. J. Appl. Phys.*, **17**, 279.
Caroli, C. and Dubois-Violette, E. (1969) *Solid State Commun.*, **7**, 779.
Chandrasekhar, S., Sadashiva, B. K. and Suresh, K. A. (1977) *Pramana*, **9**, 471.
Châtelain, P. (1943) *Bull. Soc. Franç. Minér.*, **66**, 105.
Chikazumi, S. (1966) *Physics of Magnetism* (John Wiley and Sons, New York).
Chikazumi, S. and Suzuki, K. (1955) *J. Phys. Japan*, **10**, 523.
Chou, T. W. (1971a) *J. Appl. Phys.*, **42**, 4092.
Chou, T. W. (1971b) *J. Appl. Phys.*, **42**, 4931.
Chou, T. W. and Lu, T. L. (1972) *Appl. Phys.*, **43**, 2562.
Chou, T. W. and Pan, Y. C. (1973) *J. Appl. Phys.*, **44**, 63.
Chu, K. C. and McMillan, W. L. (1975) *Phys. Rev.*, A11, 1059.

Chystiakov, I. G. (1967) *Sov. Phys. Usp.*, **9**, 551.
Cladis, P. E. (1973) *Phys. Rev. Lett.*, **31**, 1200.
Cladis, P. E. (1975) *Phys. Rev. Lett.*, **35**, 48.
Cladis, P. E. (1976) *J. Physique Colloq.*, **37**, C3-137.
Cladis, P. E. and Kléman, M. (1972a) *J. Physique*, **33**, 591.
Cladis, P. E. and Kléman, M. (1972b) *Mol. Cryst. Liq. Cryst.*, **16**, 1.
Cladis, P. E. and White, A. E. (1976) *J. Appl. Phys.*, **47**, 1256.
Cladis, P. E., Bogardus, R. K. and Aadsen, D. (1978) *Phys. Rev.*, **A 18**, 2292.
Cladis, P. E. and Torza, S. (1975) *J. Appl. Phys.*, **46**, 584.
Cladis, P. E., Kléman, M. and Piéranski, P. (1971) *C. R. Hebd. Séan. Acad. Sci.*, **273**, 275.
Clark, N. A. and Meyer, R. B. (1973) *Appl. Phys. Lett.*, **22**, 493.
Coates, D. and Gray, G. W. (1973) *Phys. Lett.*, **45A**, 115.
Colliex, C., Kléman, M. and Veyssié, M. (1974) Eighth. Inter. Congress on Electron Microscopy, Canberra, **1**, 718.
Corner, W. D. and Tanner, B. K. (1976) *J. Phys. C.*, **9**, 627.
Cosserat, E. and Cosserat, F. (1909) *Théorie des Corps Déformables* (Hermann, Paris).
Cottrell, A. H. and Bilby, B. A. (1951) *Phil. Mag.*, **42**, 573.
Craik, D. J. and Tebble, R. S. (1956) *Ferromagnetism* and *Ferromagnetic Domains* (North-Holland, Amsterdam).
Cross, M. C. (1975) *J. Low Temp. Phys.*, **21**, 525.
Dafermos, C. (1970) *Quat. J. Mech. & Appl. Math.*, **23**, S49.
Darboux, G. (1888) *Théorie Générale des Surfaces* (Gauthier-Villars, Paris), republished by (Chelsea Publishing Company, New York) 1954.
Das, E. S. P. and Marcinkowski, M. J. (1971) *J. Appl. Phys.*, **42**, 4107.
Das, E. S. P., Marcinkowski, M. J., Armstrong, R. W. and De Wit, R. (1973a) *Phil. Mag.*, **27**, 369.
Das, E. S. P., De Wit, R., Armstrong, R. W. and Marcinkowski, M. J. (1973b) *J. Appl. Phys.*, **44**, 4804.
Deloche, B. and Charvolin, J. (1975) Communication au Second Colloque Ampère de Spécialité (Budapest).
Delrieu, J. M. (1974) *J. Chem. Phys.*, **60**, 1081.
Delrieu, J. M. (1977) *J. Physique Lett.*, **38**, L-127.
Demus, D. (1975) *Krist. Tech.*, **10**, 933.
Demus, D. and Richter, L. (1978) *Textures of Liquid Crystals* (VEB Deutscher Verlag für Grundstoffindustrie, Leipzig).
Diele, S., Brand, P. and Sackmann, H. (1972a) *Mol. Cryst. Liq. Cryst.*, **16**, 105.
Diele, S., Brand, P. and Sackmann, H. (1972b) *Mol. Cryst. Liq. Cryst.*, **17**, 163.
Dirac, P. A. M. (1948) *Phys. Rev.*, **74**, 817.
Döring, W. (1968) *J. Appl. Phys.*, **39**, 1006.
Doucet, J., Keller, P., Levelut, A. M. and Porquet, P. (1978) *J. Phys.*, **39**, 458.
Doucet, J., Levulet, A. M. and Lambert, M. (1974) *Phys. Rev. Lett.*, **32**, 301.
Doucet, J., Levulet, A. M., Lambert, M., Liébert, M. and Strzelecki, S. (1975) *J. Physique Colloq.*, **36**, C 1-13.
Dreyfus, J. M. (1979) Third Cycle Thesis, Paris.
Dubois-Violette, E. and Parodi, O. (1969) *J. Physique Colloq.*, **30**, C4-57.
Duda, E. (1974) Rapport final Thomson-C.S.F., 91401, Orsay.
Duda, E., Désormière, B. and Volluet, G. (1974) *I.E.E.E. Trans. Mag.*, **MAG-10**, 3.
Dundurs, J. (1969) *J. Appl. Mech.*, **E36**, 650.
Durand, G. (1972) *C. R. Hebd. Séan. Acad. Sci.*, **275B**, 629.
Durand, G., Léger, L., Rondelez, F. and Veyssié, M. (1969) *Phys. Rev. Lett.*, **22**, 227.
Dzyaloshinskii, I. E. (1970) *Sov. Phys. J.E.T.P.*, **31**, 773.
Egami, T. (1973) *Phys. Status Solidi* (b), 211.
Ericksen, J. L. (1962) *Arch. Ratl. Mech. Anal.*, **10**, 189.
Ericksen, J. L. (1967) *Trans. Soc. Rheol.*, **11**, 5.
Ericksen, J. L. (1970) in *Liquid Crystals and Ordered Fluids* (Plenum Press, New York) 170.

312

Eshelby, J. D. (1956) *Solid State Phys.*, **3**, 79.

Fan, C. (1971) *Phys. Lett.*, **34A**, 335.

Feldtkeller, E. (1964) *Z. Angew. Phys.*, **17**, 121.

Feldtkeller, E. (1965) *Z. Angew. Phys.*, **19**, 530.

Feldtkeller, E. and Thomas, H. (1965) *Phys. Kondens. Mater.*, **4**, 8.

Fergason, J. (1964) *J. Scient. Amer.*, **211**, 77.

Feynman, R. P. (1955) *Prog. Low. Temp. Phys.*, **1**, 36.

Finkelstein, D. (1966) *J. Math. Phys.*, **7**, 1218.

Finkelstein, D. and Misner, C. (1959) *Ann. Phys.*, **6**, 230.

Finkelstein, F. and Weil, D. (1977) Magnetohydrodynamic Kinks in Astrophysics, preprint Yeshiva Universtiy.

Frank, F. C. (1949) *Research*, **2**, 542.

Frank, F. C. (1954) in *Report of the Conference on Defects in Crystalline Media* (Physical Society, London).

Frank, F. C. (1958) *Disc. Farad. Soc.*, **25**, 1.

Frank, F. C. (1977) *J. Physique Lett.*, **38**, L-207.

Freederiks, V. and Zwetkoff, V. (1934) *Sov. Phys.*, **6**, 490.

Friedel, E. (1925) *C. R. Hebd. Séan. Acad. Sci.*, **180**, 269.

Friedel, G. (1922) *Annis. de Phys.*, **18, 273.**

Freidel, G. (1929) *Lecons de Cristallographie* (Berger-Levrault, Paris).

Friedel, G. and Grandjean, F. (1910a) *Bull. Soc. Franç. Minér.*, **33**, 192, 409.

Friedel, G. and Grandjean, F. (1910b) *C. R. Hebd. Séan. Acad. Sci.*, **151**, 762.

Friedel, G. and Royer, L. (1922) *C. R. Hebd. Séan. Acad. Sci.*, **174**, 1607.

Friedel, J. (1964) *Dislocations* (Pergamon Press, London).

Friedel, J. (1980) in *Dislocations in Solids*, ed. F. R. N. Nabarro, vol. 1. (North-Holland, Amsterdam)

Friedel, J. and de Gennes, P. G. (1969) *C. R. Hebd. Séan. Acad. Sci.*, **268**, 257.

Friedel, J. and Kléman, M. (1970) in *Fundamental Aspects of dislocations Theory*, ed. J. A. Simmons, R. de Wit and R. Bullough, Nat. Stand. U.S. Spec. Pub., 317 **1**, 607.

Galligan, J. M. (1972a) *Scr. Metall.*, **6**, 161.

Galligan, J. M. (1972b) *Phys. Lett.*, A **39**, 407.

De Gennes, P. G. (1968a) *Solid State Commun.*, **6**, 163.

De Gennes, P. G. (1968b) *C. R. Hebd. Séan. Acad. Sci.*, **266**, 271.

De Gennes, P. G. (1969) *J. Physique Colloq.*, **30**, C4-65.

De Gennes, P. G. (1970) *Solid State Commun.*, **8**, 213.

De Gennes, P. G. (1971) *Symp. Farad. Soc.*, **5**, 16.

De Gennes, P. G. (1972a) *C. R. Hebd. Séan. Acad. Sci.*, **275B**, 319.

De Gennes, P. G. (1972b) *C. R. Hebd. Séan. Acad. Sci.*, **275B**, 549.

De Gennes, P. G. (1972c) *C. R. Hebd. Séan. Acad. Sci.*, **275B**, 939.

De Gennes, P. G. (1972d) *Solid State Commun.*, **10**, 753.

De Gennes, P. G. (1973a) *Mol. Cryst. Liq. Cryst.*, **21**, 49.

De Gennes, P. G. (1973b) Nobel 24, in *Collective Properties of Physical Systems*, 113.

De Gennes, P. G. (1974) *The Physics of Liquid Crystals* (Oxford University Press, Oxford).

De Gennes, P. G. (1975) *Phys. Fluids*, **17**, 1645.

De Gennes, P. G. and Dubois-Violette, E. (1975) *J. Phys. Lett.*, **36**, L-255.

De Gennes, P. G. and Pincus, P. A. (1969) *Solid State Commun.*, **7**, 339.

De Gennes, P. G. and Pincus, P. A. (1976) *J. de Physique*, **37**, 1359.

De Gennes, P. G. and Sarma, G. (1972) *Phys. Lett.*, **38A**, 219.

Gerritsma, G. and Van Zanten, P. (1971) *Mol. Cryst. Liq. Cryst.*, **15**, 267.

Ginsburg, V. L. (1975) *Nuovo Cimento*, **2**, 1234.

Ginsburg, V. L. and Pitayevskii, L. P. (1958) *Sov. Phys. J.E.T.P.*, **7**, 858.

Goodby, J. W. and Gray, G. W. (1976) *J. Physique Colloq.*, **37**, 3-17.

Goodby, J. W. and Gray, G. W. (1979) *J. de Physique Colloq.*, **40**, C3-363; C3-27.

Goscianski, M., Léger, L., Mircea-Roussel, A. and Steers, M. (1975) Oral Communication at the European Conference of Smectics, les Arcs, Savoie, France.

Grandjean, F. (1916) *Bull. Soc. Franc. Miner.*, **29**, 164.

Grandjean, F. (1921) *C. R. Hebd. Séan. Acad. Sci.*, **172**, 71.

Gray, G. W. (1962) *Molecular Structures and the Properties of Liquid Crystals* (Academic Press, London).

Gray, G. W. (1969) *Mol. Cryst. Liq. Cryst.*, **7**, 127.

Gray, G. W. (1973) *Mol. Cryst. Liq. Cryst.*, **21**, 161.

Groupe d'Orsay des cristaux liquides (1969a) *Phys. Lett.*, **28A**, 687.

Groupe d'Orsay des cristaux liquides (1969b) *J. Physique Colloq.*, **30**, C4-38.

Groupe d'Orsay des cristaux liquides (1971) *Solid State Commun.*, **9**, 653.

Grundy, P. J., Hothersall, D. C., Jones, G. A., Middleton, B. K. and Tebble, R. S. (1972) *Phys. Status Solidi*, **A 9**, 79.

Günther, H. (1958) *Abh. Braunschw. Wiss. Ges.*, **16**, 1.

Gürsey, F. (1964) *Ecole d'Eté des Houches: Relativity, Groups and Topology* (Gordon and Breach).

Guyon, E. and Urbach, W. (1976) in *Non-emissive Electrooptic Displays*, ed. A. R. Kmetz and F. K. von Willisen (Plenum, New York).

Hall, C. R. (1973) *Scr. Metall.*, **7**, 73.

Hall, M. E. (1976) preprint.

Haller, I. (1972) *J. Chem. Phys.*, **57**, 1400.

Halperin, B. I. and Nelson, D. R. (1978) *Phys. Rev. Lett.*, **41**, 121: 519(E).

Hardouin, F., Levelut, A. M., Sigaud, G., Achard, M. F., Nguyen Huu Tinh and Gasparoux, H. (September 1980) Colloque Pierre-Curie, Paris (to be published).

Hareng, M. and Le Berre, S. (1975) *Appl. Phys. Lett.*, **27**, 575.

Harris, W. F. (1970a) *Phil. Mag.*, **22**, 949.

Harris, W. F. (1970b) in *Fundamental Aspects of Dislocation Theory*, N.B.S. Spec. Publ., 317 **1**, 579.

Harris, W. F. (1970c) Thesis, University of Minnesota, Minneapolis.

Harris, W. F. (1974) in *Surface and Defect Properties of Solids* (Specialist Periodical Reports, Chemical Society, London) **3**, 57–92.

Harris, W. F. (1975) *Phil. Mag.*, **32**, 37.

Harris, W. F. and Scriven, L. E. (1970) *Nature*, **228**, 827.

Harris, W. F. and Scriven, L. E. (1971) *J. Appl. Phys.*, **42**, 3309.

Harris, W. F. and Thomas, S. L. (1975) *Phil. Mag.*, **32**, 929.

Harrison, C. G. and Leaver, K. D. (1973) *Phys. Status Solidi*, **15**, 415.

Heilmeier, G. H. and Goldmacher, J. E. (1968) *Appl. Phys. Lett.*, **13**, 132.

Heilmeier, G. H. and Goldmacher, J. E. (1969) *Proc. I.E.E.E.*, **57**, 34.

Helfrich, W. (1968) *Phys. Rev. Lett.*, **21**, 1518.

Helfrich, W. (1969) *Phys. Rev. Lett.*, **23**, 372.

Helfrich, W. (1970) *Appl. Phys. Lett.*, **17**, 531.

Helfrich, W. (1974) *Z. Naturforsch.*, **29C**, 692.

Helfrich, W. (1976) *Phys. Lett.*, **58A**, 457.

Helfrich, W. (1980) Defects in Lyotropic Phases, Les Houches, Summer School on Defects, to appear (NHPC).

Herpin, A. (1968) *Théorie du Magnétisme* (P.U.F. Paris).

Hilbert, D. and Cohn-Vossen, S. (1952) *Geometry and the Imagination* (Chelsea Publishing Company, New York).

Hirth, J. P. and Wells, R. G. (1970) *J. Appl. Phys.*, **41**, 5250.

Hornreich, R. M. and Shtrikman, S. (1980) *J. de Phys.*, **41**, 335.

Hu, S. (1959) *Homotopy Theory* (Academic Press, New York).

Huang, C. C., Pindack, R. S. and Ho, J. J. (1974) *Phys. Rev. Lett.*, **32**, 43.

Huang, W. and Mura, T. (1970) *J. Appl. Phys.*, **41**, 5175.

Huang, W. and Mura, T. (1972) *J. Appl. Phys.*, **43**, 239.

Huber, E. E., Smith, D. O. and Goodenough, J. B. (1958) *J. Appl. Phys.*, **29**, 294.

Huberman, B. A., Lublin, D. M. and Doniach, S. (1975) *Solid State Commun.*, **17**, 485.

Hubert, A. (1969) *Phys. Status Solidi*, **32**, 519.

314

Hubert, A. (1971) *Z. Angew Phys.*, **32**, 58.
Hurault, J. P. (1973) *J. Chem. Phys.*, **59**, 2068.
Jacobs, I. S. and Bean, C. P. (1958) *J. Appl. Phys.*, **29**, 537.
Jakubovicz, J. (1966) *Phil. Mag.*, **14**, 881.
Janak, J. J. (1966) *Appl. Phys. Lett.*, **9**, 225.
Jenkins, J. J. (1972) *Phys. Rev.*, **A 6**, 452.
Johnson, D. and Saupe, A. (1977) *Phys. Rev.*, **15A**, 2079.
Julia, G. (1954) *Cours de Géométrie de l'Ecole Polytechnique* (Gauthier-Villars, Paris).
Kaczer, J. and Zeleny, M. (1960) *Czech. J. Phys.*, **B 10**, 561.
Kassubeck, P. and Meier, G. (1969) *Mol. Cryst. Liq. Cryst.*, **8**, 305.
Keller, P., Liébert, L. and Strzelecki, L. (1976) *J. Physique Colloq.*, **37**, 3–27.
Kittel, C. and Galt, J. K. (1956) *Solid State Phys.*, **3**, 437.
Kléman, M. (1968) *J. Physique*, **29**, 329.
Kléman, M. (1969) *Phil. Mag.*, **19**, 285.
Kléman, M. (1970) *Phil. Mag.*, **22**, 178.
Kléman, M. (1973a) *Phil. Mag.*, **27**, 1057.
Kléman, M. (1973b) *J. Physique*, **34**, 931.
Kléman, M. (1974a) *J. Appl. Phys.*, **45**, 1377.
Kléman, M. (1974b) *J. Physique*, **35**, 595.
Kléman, M. (1976a) *Proc. R. Soc. London*, **A 347**, 387.
Kléman, M. (1976b) *J. Physique Lett.*, **37**, L-93.
Kléman, M. (1976c) *Phil. Mag.*, **34**, 79.
Kléman, M. (1976d) *Phys. Rev.*, **B 13**, 3091.
Kléman, M. (1977) *J. Physique*, **38**, 1511.
Kléman, M. and Bourret, A. (1968) *Bull. Soc. Franç. Minr. Cristal.*, **91**, 637.
Kléman, M. and Friedel, J. (1969) *J. Physique Colloq.*, **30**, C4-43.
Kléman, M. and Michel, L. (1978a) *J. Physique Lett.*, **39**, L-29.
Kléman, M. and Michel, L. (1978b) *Phys. Rev. Lett.*, **40**, 1387.
Kléman, M. and Parodi, O. (1975) *J. Physique*, **36**, 671.
Kléman, M. and Puchalska, I. B. (1980). *J. Mag. Mag. Mat.*, **15–18**, 1473.
Kléman, M. and Ryschenkow, G. (1976) *J. Chem. Phys.*, **64**, 413.
Kléman, M. and Schlenker, M. (1972) *J. Appl. Phys.*, **43**, 3184.
Kléman, M. and Williams, C. E. (1973) *Phil. Mag.*, **28**, 725.
Kléman, M. and Williams, C. E. (1974) *J. Physique Lett.*, **35**, L-49.
Kléman, M., Michel, L. and Toulouse, G. (1977) *J. Physique Lett.*, **38**, L-195.
Kléman, M., Williams, C. E., Costello, J. M. and Gulik-Krzywicki, T. (1977) *Phil. Mag.*, **35**, 33.
Kneller, E. (1956) *Beitrage zur Theorie der Ferromagnetismus* (Springer-Verlag, Berlin).
Kobayashi, K. K. (1970) *Phys. Lett.*, **31A**, 125.
Kondo, K. (1955) *R.A.A.G. Mem.*, **1**, 458.
Kondo, K. (1963) *Int. J. Eng. Sci.*, **1**, 71.
Kooy, C. and Enz, U. (1960) *Philips Res. Rep.*, **15**, 7.
Kosterlitz, J. M. and Thouless, D. J. (1972) *J. Phys.*, **C 5**, L-124.
Kosterlitz, J. M. and Thouless, D. B. (1973) *J. Phys.*, **C 6**, 1181.
Kröner, E. (1955) *Z. Phys.*, **142**, 463.
Kröner, E. (1958) *Kontinuumstheorie der Versetzunger und Eigenspannungen* (Springer-Verlag, Berlin).
Kröner, E. (1962) *Appl. Mech. Rev.*, **15**, 599.
Kröner, E. (1963) *Mat. Sci. Res.*, **1**, 281.
Kröner, E. (ed.) (1968) *Mechanics of Generalized Continua* (Springer-Verlag).
Kronmüller, H. (1959) *Zeit. Physik.*, **154**, 574.
Kronmüller, H. (1967) *Can. J. Phys.*, **45**, 631.
Kronmüller, H. and Seeger, A. (1961) *J. Phys. Chem. Sol.*, **18**, 93.
Kroupa, F. (1962) *Czech J. Phys.*, **B 12**, 191.
Kroupa, F. and Lejcek, L. (1972) *Phys. Status Solidi*, **51**, K121.

Kunin, I. A. (1965) *Sov. Phys. J. Appl. Mech. Tech. Phys.*, **65**, 6, 45. See also De Wit (1972) appendix II.B (delta functions on curves and surfaces).

La Bonte, A. E. (1969) *J. Appl. Phys.*, **40**, 2450.

Labrune, M. (1976) *J. Physique*, **37**, 1033.

Labrune, M. and Kléman, M. (1974) *J. Appl. Phys.*, **45**, 2716.

Lachampt, F. and Vila, R. M. (1969) *Parfum. Cosmét. Sav.*, **12**, 239.

Lagerwall, S. T., Meyer, R. B. and Stebler, B. (1978) *Annal. de Phys.*, **3**, 249.

Landau, L. and Lifshitz, I. (1967) *Physique Statistique* (Edition Mir, Moscow).

Landau, L. and Lifshitz, I. (1962) *Electrodynamics of Continuous Media* (Pergamon Press, London).

Lefèvre, M., Martinand, J., Durand, G. and Veyssié, M. (1971) *C.R. Hebd. Séan. Acad. Sci.*, **273B**, 403.

Léger, L. (1972) *Solid State Commun.*, **10**, 697.

Léger, L. (1973) *Phys. Lett.*, **44A**, 535.

Léger, L. (1976) Thesis, Orsay.

Legett, A. J. (1975) *Rev. Mod. Phys.*, **47**, 331.

Lehmann, O. (1904) *Flüssige Kristalle* (Verlag von Wilhelm Engelmann, Leipzig).

Lehmann, O. (1907) *Die Scheinbar lebenden Kristalle* (Esslingen).

Lehmann, O. (1909) *J. Physique*, **8**, 713.

Lehmann, O. (1910) *Bull. Soc. Franç. Minér.*, **33**, 300.

Levelut, A. M. (1976) *J. Physique Colloq.*, **37**, C3-51.

Levelut, A. M. (1979) *J. Physique Lett.*, **40**, 81.

Levelut, A. M. and Lambert, M. (1971) *C. R. Hebd. Séan. Acad. Sdi.*, **272B**, 1018.

Levelut, A. M., Doucet, J. and Lambert, M. (1974) *J. Physique*, **35**, 773.

Li, J. C. M. (1972) *Surface Sci.*, **31**, 12.

Li, J. C. M. and Gilman, J. J. (1970) *J. Appl. Phys.*, **41**, 4248.

Liébert, L. and Daniels, W. B. (1977) *J. Physique Lett.*, **38**, L.333.

Liébert, L. and Strzelecki, L. (1973) *Bull. Soc. Chimie Paris*, **2**, 603, 605.

Lifshitz, E. (1944) *J. Phys. (Moscow)*, **8**, 337.

Likhatchev, V. A. and Khairoff, P. Y. (1975) *Elements of the Theory of Disclinations* (in Russian) (University of Leningrad Press).

Lomont, J. S. (1959) *Applications of Finite Groups* (Academic Press, New York).

Love, A. E. H. (1944) *A Treatise on the Mathematical Theory of Elasticity* (Dover, New York).

Lubensky, T. (1972) *Phys. Rev.*, **A 6**, 452.

Luzzati, V. (1968) in *Biological Membranes*, ed. D. Chapman (Academic Press, New York) 71.

Luzzati, V. and Tardieu, A. (1974) *Ann. Rev. Phys. Chem.*, **25**, 79.

Luzzati, V., Ranck, J. L., Mateu, L., Salder, D. M., Tardieu, A. and Gulik-Krzywicki, T. (1974) *J. Mol. Biol.*, **85**, 249.

Ma, S. K. (1976) *Modern Theory of Critical Phenomena* (Benjamin, New York).

McClintock, F. A., André, P. A., Schwerdt, K. R. and Stoeckly, R. E. (1958) *Nature*, **182**, 652.

McMillan, W. L. (1971) *Phys. Rev.*, **A4**, 1238.

McMillan, W. L. (1972) *Phys. Rev.*, **A6**, 936.

McMillan, W. L. (1973a) *Phys. Rev.*, **A7**, 1419.

McMillan, W. L. (1973b) *Phys. Rev.*, **A8**, 1921.

Madariaga, R. (1980) *Dislocations and Earthquakes* (Les Houches Summer School on the Physics of Defects).

Malek, Z. (1957a) *Z. Angew. Phys.*, **9**, 279.

Malek, Z. (1957b) *Czech. J. Phys.*, **7**, 97.

Malek, Z. (1957c) *Czech. J. Phys.*, **7**, 152.

Malek, Z. (1957d) *Czech. J. Phys.*, **7**, 244.

Malek, Z. (1957e) *Czech. J. Phys.*, **7**, 335.

Malek, Z. (1959) *Czech. J. Phys.*, **9**, 613.

Malthète, J., Billard, J., Canceill, J., Gabard, J. and Jacques, J. (1976) *J. Physique Colloq.*, **37**, C3-1.

316

Malthète, J., Leclercq, M., Gabard, J., Billard, J. and Jacques, J. (1971) *C. R. Hebd. Séan. Acad. Sci.*, **C 273**, 265.
Mandelbrot, B. (1975) *Les Objets Fractals* (Flammarion, Paris).
Marcerou, J. P. and Prost, J. (1978) *Annal. de Phys.*, **3**, 269.
Marcinkowski, M. J., Das, E. S. P. and Sadananda, K. (1973) *Phys. Status Solidi*, **19**, 67.
Marcus, M. (1981) *J. Physique*, **42**, 61.
Maugin, C. (1911) *Bull. Soc. Franç. Minér. Cristal.*, **34**, 3.
Meiboom, S., Sethna, J. P., Anderson, P. W. and Brinkman, W. F. (1981) To be published.
Melmed, A. J. and Hayword, D. O. (1959) *J. Chem. Phys.*, **31**, 545.
Melzer, D. and Nabarro, F. R. N. (1977) *Phil. Mag.*, **35**, 901 and 907.
Mermin, N. D., Sussex Symposium on superfluid $^3$He (1976) preprint. Sanibel Symposium on quantum fluids and solids (1977) preprint. *Physica*, **90B + C** (1977), 1.
Mermin, N. D. (1979) *Rev. Mod. Phys.*, **51C**, 591.
Mermin, N. D. and Ho. T.-L. (1976) *Phys. Rev. Lett.*, **36**, 594.
Meunier, J. P. and Billard, J. (1969) *J. Mol. Cryst. Liq. Cryst.*, **7**, 421.
Meyer, R. B. (1968) *Appl. Phys. Lett.*, **14**, 208.
Meyer, R. B. (1969) *Phys. Rev. Lett.*, **22**, 918.
Meyer, R. B. (1972a) *Mol. Cryst. Liq. Cryst.*, **6**, 355.
Meyer, R. B. (1972b) *Solid State Commun.*, **12**, 585.
Meyer, R. B. (1973a) *Phil. Mag.*, **27**, 405.
Meyer, R. B. (1973b) in *Molecular Fluids les Houches* (Gordon and Breach, New York).
Meyer, R. B. (1976) VIth International conference on Liquid Crystals, Kent, U.S.
Meyer, R. B. and Lubensky, J. C. (1976) *Phys. Rev.*, **A14**, 2307.
Meyer, R. B., Liebert, L., Strzelecki, L. and Keller, P. (1975) *J. Physique Lett.*, **36**, L-69.
Michel, L. (1980) *Rev. Mod. Phys.*, **52**, 617.
Middelhoek, S. (1963) *J. Appl. Phys.*, **34**, 1054.
Migdal, A. A. (1975) preprint.
Miltat, J. (1976a) Thesis, Orsay AO 12760.
Miltat, J. (1976b) *Phil. Mag.*, **33**, 225.
Miltat, J. and Kléman, M. (1973) *Phil. Mag.*, **28**, 1015.
Mindlin, R. D. and Tiersten, H. F. (1962) *Arch. Rat. Mech. Anal.*, **11**, 415.
Mineyev, V. P. (1980) in Sov. Scient. Rev., Sect. A, vol. 2, ed. I. M. Khalatrikov (Harwood, London).
Mineyev, V. P. and Volovik, G. E. (1978) *Phys. Rev.*, **B18**, 3197.
Mizushima, S. (1960) *J. Phys. Soc. Jap.*, **15**, 70.
Moncton, D. E. and Pindak, R. (1979) *Phys. Rev. Lett.*, **43**, 701.
Mura, T. (1972a) The Relation Between Disclinations and Dislocations, Europhysics Conference on Disclinations, Aussois, France, 21–23 June.
Mura, T. (1972b) *Archiv. Mech.*, **24**, 449.
Nabarro, F. R. N. (1952) *Adv. Phys.*, **1**, 271.
Nabarro, F. R. N. (1967) *Theory of Crystal Dislocations* (Oxford University Press, Oxford), ch. X.
Nabarro, F. R. N. (1970) in *Fundamental Aspects of Dislocation Theory*, N.B.S. Spec. Publ., 317, **1**, 593.
Nabarro, F. R. N. (1969) in *Physics of Strength and Plasticity*, ed. A. S. Argon (MIT, Cambridge) 97.
Nabarro, F. R. N. (1972) *J. Physique*, **33**, 1089.
Nabarro, F. R. N. (ed.) (1982) *Dislocations in Solids*, vol. 5 (N.H.P.C.).
Néel, L. (1944) *J. Phys. & Radium*, **5**, 241 and 265.
Néel, L. (1945) *C. R. Hebd. Séan. Acad. Sci.*, **220**, 814.
Néel, L. (1948a) *J. Phys. & Radium*, **9**, 193.
Néel, L. (1948b) *J. Phys. & Radium*, **9**, 184.
Néel, L. (1955) *C. R. Hebd. Séan. Acad. Sci.*, **241**, 533.
Nelson, D. and Halperin, B. (1980) *Phys. Rev.*, **B19**, 2457.
Nohara, A. and Imura, T. (1969) *J. Phys. Soc. Japan*, **27**, 793.

Nourtier, C. (1973) *J. Physique*, **34**, 57.
Nourtier, C. and Kléman, M. (1973) *Int. J. Magn.*, **4**, 333.
Nye, J. F. (1953) *Acta Met.*, **1**, 153.
Ogburn, F., Paretzkin, B. and Peiser, H. S. (1964) *Acta Cryst.*, **17**, 774.
Onsager, L. (1949) *Nuovo Cimento*, Suppl. 2 to Vol. **6**, 249.
Oseen, C. W. (1933) *Trans. Far. Soc.*, **29**, 883.
Peach, M. and Koehler, J. S. (1959) *Phys. Rev.*, **80**, 436.
Pérez, A. (1980) Third Cycle Thesis, Montpellier.
Pérez, A., Brunet, M. and Parodi, O. (1978) *J. Phys. Lett.*, **39**, L-353.
Pershan, P. S. (1974) *J. Appl. Phys.*, **45**, 1590.
Pershan, P. S. and Prost, J. (1975) *J. Appl. Phys.*, **46**, 2343.
Pfeffer, H. (1967a) *Phys. Status Solidi*, **19**, 735.
Pfeffer, H. (1967b) *Phys. Status Solidi*, **20**, 395.
Pfeffer, H. (1967c) *Phys. Status Solidi*, **21**, 837.
Pfeffer, H. (1967d) *Phys. Status Solidi*, **21**, 857.
Poenaru, V. (1978) Ecole d'été de Physique Théorique, Les Houches, Session on Ill-condensed matter.
Poenaru, V. and Toulouse, G. (1977) *J. Physique*, **38**, 887.
Poincaré, H. (1885) *J. Mathém. Pures et Appl.*, **1**, 167.
Poincaré, H. (1886) *J. Mathém. Pures et Appl.*, **2**, 151.
Polcarova, M. (1969) *I.E.E.E. Trans. Magn.*, **5**, 536.
Polcarova, M. and Gemperlova, J. (1969) *J. Phys. Status Solidi*, **32**, 769.
Polcarova, M. and Kaczer, J. (1967) *J. Phys. Status Solidi*, **21**, 635.
Polyakov, A. M. (1976) *Sov. Phys. J.E.T.P.*, **41**, 988.
Porte, G. (1976) *J. de Phys.*, **37**, 1245.
Press, M. J. and Arrott, A. S. (1974) *Phys. Rev. Lett.*, **33**, 403.
Prost, J. (1980) 'Liquid crystals of one and two-dimensional order and their applications', Garmisch-Partenkirchen, Proceedings.
Prost, P. and Pershan, P. S. (1976) *J. Appl. Phys.*, **47**, 2298.
Proust, J. E. and Ter-Minassian-Saraga, L. (1977) *J. de Physique*, **36**, C1-77.
Proust, J. E., Perez, E. and Ter-Minassian-Saraga, L. (1978) *Colloid Surf. Sci.*, **256**, 666.
Proust, J. E., Ter-Minassian-Saraga, L. and Guyon, E. (1972) *Solid State Commun.*, **11**, 1227.
Prutton, J. (1964) *Thin Ferromagnetic Films* (Butterworths, London).
Puchalska, I. B. and Sadoc, J. F. (1976) *J. Appl. Phys.*, **47**, 333.
Rapini, A. (1972a) Third Cycle Thesis, Orsay.
Rapini, A. (1972b) *J. Physique*, **33**, 237.
Rapini, A. (1973) *J. Physique*, **34**, 629.
Rapini, A., Léger, L. and Martinet, A. (1975) *J. Physique Colloq.*, **36**, C 1-189.
Rault, J. (1971) *Solid State Commun.*, **9**, 1965.
Rault, J. (1972a) *Mol. Cryst. Liq. Cryst.*, **16**, 143.
Rault, J. (1972b) *J. Physique*, **33**, 383.
Rault, J. (1973) *Phil. Mag.*, **28**, 11.
Rault, J. (1974a) *Phil. Mag.*, **30**, 621.
Rault, J. (1974b) *Mol. Cryst. Liq. Cryst.*, **26**, 349.
Rault, J. (1974c) in *Liquid Crystals and Ordered Fluids*, vol. 2, ed. J. F. Johnson and R. S. Porter (Plenum Press, New York).
Rault, J. (1975) *Comptes Rendus Acad. Sci. (Paris)*, **B280**, 417.
Rault, J. and Cladis, P. E. (1971) *Mol. Cryst. Liq. Cryst.*, **15**, 1.
Rayleigh, J. W. S. (1894) *The Theory of Sound* (Macmillan, London).
Read, W. T. (1953) *Dislocations in Crystals* (McGraw-Hill).
Read, W. T. and Shockley, W. (1950) *Phys. Rev.*, **78**, 275.
Rey, C. and Saada, G. (1976) *Phil. Mag.*, **33**, 825.
Ribotta, R. (1974) *C. R. Hebd. Séan. Acad. Sci.*, **279B**, 295.
Ribotta, R. (1975) Thesis, Orsay.

318

Ribotta, R. (1976) *Phys. Lett.*, **56A**, 130.
Rieder, G. (1957) *Z. Angew. Phys.*, **9**, 187.
Rieder, G. (1959) *Abh. Braunschw. Wiss. Ges.*, **11**, 20.
Rieder, G. (1960) *Z. Angew. Math. Mech.*, **40 T**, 123.
Robinson, C. (1966) in *Liquid Crystals* (Gordon and Breach, London) 147.
Robinson, C. and Beevers, J. C. (1958) *Disc. Far. Soc.*, **25**, 29.
Rogers, J. and Winsor, P. A. (1969) *J. Colloid. Interf. Sci.*, **30**, 500.
Rogula, D. (1976) in *Trends in Applications of Pure Mathematics to Mechanics*, ed. by G. Fichera (Pitman Pub.).
Rondelez, F. (1973) Thesis, A. O. 1118, Orsay.
Rondelez, F. and Arnould, H. (1971) *C. R. Hebd. Séan. Acad. Sci.*, **B 273**, 549.
Rondelez, F. and Hulin, J. P. (1972) *Solid State Commun.*, **10**, 1009.
Rondelez, F., Arnould, H. and Gerritsma, C. J. (1972) *J. Phys. Rev. Lett.*, **28**, 735.
Rosenblatt, C. S., Pindak, R., Clark, N. A. and Meyer, R. B. (1977) *J. Physique*, **38**, 1105.
Rosevear, F. B. (1954) *J. Amer. Oil. Chem. Soc.*, **31**, 68.
Ruelle, D. and Takens, F. (1971) *Commun. Math. Phys.*, **20**, 167.
Ryschenkow, G. (1975) Third Cycle Thesis, Orsay.
Ryschenkow, G. and Kléman, M. (1976) *J. Chem. Phys.*, **64**, 404.
Saada, G. (1976) *Phil. Mag.*, **33**, 825.
Saada, G. (1979) *Acta Met.*, **27**, 921.
Sackmann, E. and Demus, D. (1966) *Mol. Cryst.*, **2**, 81.
Sackmann, E. and Demus, D. (1973) *Mol. Cryst. Liq. Cryst.*, **21**, 239.
Sackmann, E., Meiboom, S. and Snyder, L. C. (1967) *J. Amer. Chem. Soc.*, **89**, 5982.
Saupe, A. (1960) *Z. Naturforsch.*, **15A**, 815.
Saupe, A. (1964) *Z. Naturforsch.*, **A19**, 161.
Saupe, A. (1969) *Mol. Cryst. Liq. Cryst.*, **7**, 59.
Saupe, A. (1973) *Mol. Cryst. Liq. Cryst.*, **21**, 211.
Savage, J. C. (1980) in *Dislocations in Solids*, ed. F. R. N. Nabarro, vol. 3 (North-Holland, Amsterdam).
Schaeffer, H. (1967) *Z. Angew. Math. Mech.*, **47**, 319, 485.
Scheffer, T. J. (1972) *Phys. Rev.*, **A5**, 1327.
Scheffer, T. J., Grüler, H. and Meier, G. (1972) *Solid State Commun.*, **11**, 253.
Schlenker, M., Brissoneau, P. and Perrier, J. P. (1968) *Bull. Coc. Franç. Mineral Crist.*, **91**, 653.
Schmidt, D., Schadt, M. and Helfrich, W. (1972) *Z. Naturforsch.*, **A27**, 277.
Schön, L. and Buchenau, V. (1972) *Int. J. Magn.*, **3**, 145.
Schultz, J. M. and Kinloch, D. R. (1969) *Polymer*, **10**, 271.
Scott, A. C., Chu, F. Y. and McLaughlin, D. (1973) *Proc. I.E.E.E.*, **61**. 1443.
Scriven, L. E. (1977) in *Micellization, solubilization and microemulsions*, ed. K. L. Mittal, vol. 11, (Plenum, New York) 877.
Shankar, R. (1977) *J. Physique*, **38**, 1405.
Shtrikman, S. and Treves, D. (1960a) *J. Appl. Phys.*, **31**, 147S.
Shtrikman, S. and Treves, D. (1960b) *J. Appl. Phys.*, **31**, 1304.
Sigaud, G., Hardouin, F., Achard, M. F. and Gasparoux, H. (1979) *J. Phys. Colloq.*, **40**, C3-356.
Skoulios, A. (1967) *Adv. in Colloid Inter. Sci.*, **1**, 79.
Sleeswyk, A. W. (1966) *J. Physique Colloq.*, **27**, C 3-78.
Slonczewski, J. C. and Malozemoff, A. P. (1973) *A.I.P. Conf. Proc.*, **10**, 458.
Somigliana, C. (1914) *R. C. Acad. Lincei.*, **23**, 463.
Somigliana, C. (1915) *R.C. Acad. Lincei.*, **24**, 655.
Sommerfeld, A. (1964) *Mechanics of Deformable Bodies* (Academic Press, New York).
Steenrod, N. (1951) *Topology of Fiber Bundles* (Princeton University Press, N.I.).
Steers, M., Kléman, M. and Williams, C. E. (1974) *J. Physique Lett.*, **35**, L-21.
Stegemeyer, H. and Bergmann, K. (1980) in *Liquid Crystals of One- and Two-Dimensional Order*, ed. W. Helfrich and G. Heppke, Springer Series in Chem. Phys. **11**, (Springer-Verlag, Berlin) 161.

Stein, D. L., Pisarski, R. D. and Anderson, P. W. (1978) *Phys. Rev. Lett.*, **40**, 1269.
Strzelecki, L. and Liébert, L. (1973) *Bull. Soc. Chimie, Paris*, **2**, 597.
Tanner, B. K. (1976) *X-Ray Diffraction Topography* (Pergamon Press, Oxford).
Tardieu, A. and Billard, J. (1976) *J. Physique Colloq.*, **37**, C3-79.
Taylor, T. R., Arora, S. L. and Fergason, J. L. (1970) *Phys. Rev. Lett.*, **25**, 772.
Thiele, A. A. (1974) *J. Appl. Phys.*, **45**, 377.
Thiéry, J. P., Baudoin, R. and Gérôme, D. (1968) *J. Microscopie*, **7**, 81.
T'Hooft, G. (1974) *Nucl. Phys.*, **B79**, 276.
Thom, R. (1975) *Structural Stability and Morphogenesis* (Benjamin, New York).
Tiddy, J. T. (1980) *Phys. Rep.*, **57**, 1.
Timoshenko, S. (1951) *Theory of Elasticity* (MacGraw-Hill, New York).
Toulouse, G. (1977) *J. Physique Lett.*, **38**, L-67.
Toulouse, G. and Kléman, M. (1976) *J. Physique Lett.*, **37**, L-149.
Toulouse, G. and Pfeuty, P. (1975) *Introduction au Groupe de Renormalisation* (P.U.G., Grenoble).
Traüble, H. (1962) *Z. Metallk.*, **53**, 211.
Traüble, H. (1969) *Magn. Metall.*, **2**, 622.
Traüble, H. and Essmann, U. (1968) *J. Appl. Phys.*, **39**, 4052.
Urbach, W., Boix, M. and Guyon, E. (1974) *Appl. Phys. Lett.*, **25**, 479.
Vesely, D. (1975) *Scr. Metall.*, **9**, 233.
Vicena, F. (1954) *Czech. J. Phys.*, **4**, 419.
Vicena, F. (1955) *Czech. J. Phys.*, **5**, 480.
Vitek, V. and Kléman, M. (1975) *J. Physique*, **36**, 59.
Vladimirov, V. I. and Zhukovskii, I. M. (1975) *Sov. Phys. Solid State*, **17**, 772.
Volovik, G. E. (1978) *Zh.E.T.F. Pis'ma*, **28**, 605.
Volovik, G. E. and Mineyev, V. P. (1977) *Zh. Eksp. Teor. Fiz.*, **72**, 2256.
Volterra, V. (1907) *Ann. Ecol. Norm. Supér.*, **24**, 401.
De Vries, H. (1951) *Acta Cryst.*, **4**, 219.
Weatherburn, C. E. (1974) *Differential Geometry in Three Dimensions*, vol. 1 (Cambridge University Press).
Weiss, P. and Forrer, R. (1929) *Ann. Phys. (Paris)*, **12**, 279.
Wentorf, R. H. Jr (1963) *The Art and Science of Growing Crystals*, ed. J. J. Gilman (Wiley, New York).
Whelan, M. J., Hirsch, P. B., Horne, R. W. and Bollmann, W. (1957) *Proc. Roy. Soc.*, **A 240**, 524.
Williams, C. E. (1975) *Phil. Mag.*, **32**, 313.
Williams, C. E. (1976) Thesis, A.O. 12720, Orsay.
Williams, C. E. and Kléman, M. (1975) *J. Physique Colloq.*, **36**, Cl-321.
Williams, C. E. and Kléman, M. (1976) *Phil. Mag.*, **33**, 213.
Williams, C. E., Piéranski, P. and Cladis, C. E. (1972) *Phys. Rev. Lett.*, **29**, 90.
Williams, H. J. and Goertz, M. (1952) *J. Appl. Phys.*, **23**, 316.
de Wit, R. (1960) *Solid State Phys.*, **10**, 249.
de Wit, R. (1970) *Fundamental Aspects of Dislocation Theory*, Natl. Bur. Std. (U.S.) Special publication 317 (U.S., G.P.O., Washington D.C.), **1**, 651.
de Wit, R. (1971) *J. Appl. Phys.*, **42**, 3304.
de Wit, R. (1972a) *Arch. Mech.*, **24**, 499.
de Wit, R. (1972b) *J. Phys.*, **C 5**, 529.
de Wit, R. (1973a) *J. Res. Natl. Bur. Stand.*, **77A**, 49.
de Wit, R. (1973b) *J. Res. Nat. Bur. Stand.*, **77A**, 359.
de Wit, R. (1973c) *J. Res. Nat. Bur. Stand.*, **77A**, 607.
Zhukovskii, I. M. (1975) *Sov. Phys. Solid State*, **17**, 623.
Zisman, M. (1972) *Topologie Algébrique Elémentaire* (Armand Colin, Paris).
Zocher, H. (1933) *Trans. Far. Soc.*, **19**, 945.
Zvetkov, V. (1937) *Acta Phys. Chim. URSS*, **6**, 866.

# Index

amorphous ferromagnets, 275, 279
anchoring, 4, 57, 69
anisotropic fluids
   *see* liquid crystals, superfluids, chapter 6
anisotropy (magnetocrystalline), 241
antiferromagnets, 259

bend deformation, 39
bifurcation, 8
biological materials, 3, 78, 220
Block line, 261
Block wall, 9, 240, 241, 261, 284
blue phase, 101
boojum, 306
Borromian knot, 300
bubble domain, 266
Burgers circuit, 1, 180, 275, 282, 297, 298
Burgers vector, 6, 7, 9, 87, 172, 286

catastrophic theory, 8
chevrons, 100
cholesterics, 13, chapter 4
climb, 93
Codazzi–Mainardi equations, 204
congruence of lines, 175
conjugated lines, 4, 168
contortion tensor, 5, 6, 174, 178, 203, 257, 259
core, 46, 138, 304
Cosserat media, 5, 6, 182
cross-ties wall, 246
curvature, 202
   geodesic, 116
   normal, 117
   mean, 115
   principal, 115
cut surface, 16, 302

Darboux theorem, 168
Darboux–Ribaucour trihedron, 117, 180
deformation gradient
   *see* distortion tensor

de Gennes–Friedel process, 31
densities
   of rotation dislocations (disclinations), 5, 6, 236, chapters 6 and 8
   of translation dislocations, 6, 184, 229, 247, 268
directional media, 11, 28, 174
diremption, 6, 235, 238, 256
disgyration, 189
disclination
   in cholesterics, chapter 4
   in Cosserat media, 187
   in directional media, 28, chapter 6
   coupling of, 26, 84, 220
   in magnets, chapter 9
   motion of, 91, 224, 225
   in smectics A, 166, 294
   *see* focal lines
   in smectics C, 208, 291
   in smectics C*, 214
   surface-, 4, 54, 59, 244
dislocation
   of translation, 7, 8, 17, 18, 139, 210, 215, 240, 286
   of notation
      *see* disclination
   Somigliana-, 231, 233
   twin-, 272
   wall, 21, 158
discotic mesophases, 14
dispiration, 2, 215
distortion tensor, 5, 226, 233
   *see* chapter 6
distranslation, 6, 235, 254
Dupin cyclides, 130, 136, 176, 186, 206, 207

earthquakes, 3
elasticity
   *see* magnetoelasticity, magnetostriction
   of nematics, 38, 79
   one-constant, of nematics, 47
   of smectics A, 113, 119, 124, 149

320